Effects of Light Quantity and Quality on Horticultural Crops

Effects of Light Quantity and Quality on Horticultural Crops

Guest Editors

László Balázs
Gergő Péter Kovács

MDPI • Basel • Beijing • Wuhan • Barcelona • Belgrade • Novi Sad • Cluj • Manchester

Guest Editors

László Balázs
Institute of Agronomy
Hungarian University of
Agriculture and Life Sciences
Gödöllő
Hungary

Gergő Péter Kovács
Institute of Agronomy
Hungarian University of
Agriculture and Life Sciences
Gödöllő
Hungary

Editorial Office
MDPI AG
Grosspeteranlage 5
4052 Basel, Switzerland

This is a reprint of the Special Issue, published open access by the journal *Horticulturae* (ISSN 2311-7524), freely accessible at: https://www.mdpi.com/journal/horticulturae/special_issues/Light_Quantity_Quality.

For citation purposes, cite each article independently as indicated on the article page online and as indicated below:

Lastname, A.A.; Lastname, B.B. Article Title. *Journal Name* **Year**, *Volume Number*, Page Range.

ISBN 978-3-7258-4659-7 (Hbk)
ISBN 978-3-7258-4660-3 (PDF)
https://doi.org/10.3390/books978-3-7258-4660-3

Cover image courtesy of László Balázs

Contents

About the Editors

László Balázs

László Balázs, PhD, has been an assistant professor at the Institute of Agronomy of the Hungarian University of Agriculture and Life Sciences since 2022. His research interests include horticultural lighting, lighting recipes, crop monitoring, and the economics of vertical farming. Before his academic career, he worked for GE Lighting between 1994 and 2017 in various engineering and leadership positions. He holds a PhD and MSc degree from Eötvös Loránd University, Budapest.

Gergő Péter Kovács

Gergő Péter Kovács, PhD, is the Deputy Director at the Institute of Agronomy at the Hungarian University of Agriculture and Life Sciences, where his responsibilities include research, management, preparing agricultural reports, and education. His professional interests focus on sustainable agriculture, crop production, soil sciences, agricultural policy, and regenerative agriculture. Mr. Kovács has held several positions at Szent István University (the predecessor of his current institution) since 2011, including assistant lecturer and assistant professor, and engaging in active teaching, research, conference organization, and proposal writing. From 2012 to 2016, he was the Director at GAK Nonprofit Public Utility Ltd., where he managed daily crop production and research activities. Later, he served as the Deputy Director of Research at the National Agricultural Research and Innovation Centre's Research Institute of Agricultural Economics from 2019 to 2021. He has good experience in project and team management, having worked on H2020, Interreg, and other Hungarian projects. He holds an MSc in Agricultural Engineering, a PhD with a focus on soil cultivation and biomass utilization, and postgraduate specialist degrees in alternative energy management and soil sciences from Szent István University.

horticulturae

Editorial

Effects of Light Quantity and Quality on Horticultural Crops

László Balázs * and Gergő Péter Kovács

Institute of Agronomy, Hungarian University of Agriculture and Life Sciences, 2100 Gödöllő, Hungary; kovacs.gergo.peter@uni-mate.hu
* Correspondence: balazs.laszlo@uni-mate.hu

1. Introduction

Light plays a fundamental role in the growth and development of plants. It is the primary energy source of photosynthesis, enabling the process of carbon assimilation in the chloroplast [1]. On the other hand, light is an environmental signal that stimulates physiological processes and affects the synthesis of secondary metabolites in horticultural crops [2]. The rate of photosynthesis determines biomass accumulation in plant tissues, and it is a major driver of plant yield. In contrast, secondary metabolites significantly impact the phytonutrient profile and nutritional quality of crops [3].

The lighting environment in which plants thrive is also characterized by quantitative and qualitative parameters. In horticulture, the quantitative measures of light are the photoperiod and the photosynthetic photon flux density (PPFD), which correspond to the number of incident photons of photosynthetically active radiation (PAR) per area and per time interval [4]. PAR is defined as a waveband ranging from 400 to 700 nm [5]. PPFD is expressed in μmol m^{-2} s^{-1}.

The term "light quality" is associated with the spectral distribution of photon irradiance. Absorption spectra of various photoreceptors in higher plants cover a much broader wavelength range than PAR, spanning from 280 nm to 800 nm [6]. Photoreceptors sense the spectral differences in irradiance, trigger growth, and developmental processes, allowing plants to adapt to various environmental conditions. Light spectra in horticulture are often categorized by photon irradiance ratios [7]. The PAR waveband is divided into 100 nm wide wavelength intervals: B (blue, 400–500 nm), G (green, 500–600 nm), and R (red, 600–700 nm). Additional wavebands used in horticulture are the FR (far-red 700–800 nm), UV-A (315–400 nm), and UV-B (280–315 nm) wavebands. The full-spectrum white light spanning the PAR waveband is often abbreviated as W (400–700 nm). The quotient of photon irradiances, measured in two wavebands, e.g., the red/blue ratio (R/B), is regarded as a light quality attribute. Another often-used light parameter is the red/far-red (R/FR) photon ratio.

The quantity and quality of light are strongly correlated in nature. Daylight intensity and light color change with solar elevation, the altitude of the location, and meteorological conditions. Daylight intensity and spectral features, such as the R/B ratio or R/FR ratio, are not independent parameters and vary within a relatively narrow domain [8]. Shading nets [9] can reduce daylight intensity with a minor change in the spectral distribution of incident light. Supplementary lighting in greenhouses [10] extends the photoperiod and provides light treatments that are beneficial for crop growth and development [11]. LEDs, as a sole source of light, enable the complete separation of the quantitative and qualitative light parameters, enabling the testing of spectra that do not occur in nature [12].

 https://doi.org/10.3390/horticulturae11050512

This Special Issue, "Effects of Light Quantity and Quality on Horticultural Crops", was launched to collate research results on the interactive response of plants to variations in light intensity and spectral distributions.

2. Overview of Published Articles

The publications in the Special Issue cover a broad range of lighting solutions for horticultural crops. The quantitative and qualitative light parameters, the investigated crop, and the measured effects are summarized in Table 1 in order of contribution number. Five papers (Contributions 1, 2, 3, 5 and 8) described experiments using LEDs as the sole source of light. In three contributions (4, 6, and 7), shading nets were employed to reduce the light intensity of natural daylight. In Contribution 9, LEDs were employed as supplementary lighting in a greenhouse, and the final publication (Contribution 10) is a review of the phytochemical profile of peppers (*Capsicum* fruits) affected by a broad range of lighting conditions.

Table 1. Main findings of the Special Issue in the order of contribution number.

#	Light Quantity	Light Quality	Crop	Effect
1	PPFD: 90, 180 μmol m^{-2} s^{-1} Photoperiod: 12 h	R/B: 2.1, 3.1, 5.0	lettuce	growth traits nutrient profile
2	PPFD: 160 μmol m^{-2} s^{-1} Photoperiod: 12 h	B, R, W, B + R + W	Ethiopian kale	growth traits nutrient profile
3	PPFD: 200 μmol m^{-2} s^{-1} Photoperiod: 12, 14, 16, 18 h	W	Chinese cabbage	bolting and flowering time, gibberellin conc.
4	Relative solar irradiance: 75–100% Photoperiod: 10–14 h (daytime)	Control: no shade Gray, blue, black	piquin pepper	phytochemical profile
5	PPFD: 33–390 μmol m^{-2} s^{-1} Photoperiod: 16 h	R/B: 2.01–2.78 R/FR: 2.57–4.27	pea	growth traits
6	Relative solar irradiance: 50%, 75%, 80%, 100% Photoperiod: (daytime)	Control: no shade Black shading nets	tomato	biomass, photosynthesis rate, metabolism
7	Relative solar irradiance: 25%, 50%, 75%, 100% Photoperiod: (daytime)	Control: no shade Black shading nets	blueberry	hormone and enzyme activities
8	PPFD: 80, 160 μmol m^{-2} s^{-1} Photoperiod: 14 h	R/B = 4.1	lemon basil	growth traits, phenolic content
9	Supplementary LED light PPFD: 40, 170 μmol m^{-2} s^{-1} Photoperiod: 16 h, 0.5 h EOD	R/B = 3 FR R/B = 3 + FR	tomato	crop yield and quality
10	Review paper covering a broad range of light parameters		pepper	phytochemical profile

A hot chili pepper variant, piquin pepper (*Capsicum annuum* L. var. *glabriusculum*) was the focus of Contribution 4. Two publications (6 and 9) dealt with tomato (*Solanum lycopersicum* L.) cultivars. Other horticultural crops explored in this Special Issue were lettuce (*Lactuca sativa* L.) (1), Ethiopian kale (*Brassica carinata* A. Braun) (2), non-heading Chinese cabbage (*Brassica campestris* spp. *chinensis* Makino) (3), pea (*Pisum sativum* L., cv. Kleine Rheinländerin) (5), blueberry (Ericaceae, *Vaccinium*) (7) and lemon basil (*Ocimum citriodurum* Vis.) (8).

2.1. LEDs as a Sole Source of Light

In Contribution 1, Hernández-Adasme et al. tested three different light spectra (B + W, R + W and R + B) at two PPFD levels enabling the interaction of light quantity and quality to be studied. The high PPFD increased the fresh weight of lettuce and total phenolic and flavonoid content relative to low levels. On the other hand, antioxidant activity decreased with the increase in light intensity. These results highlight the importance of controlling light intensity to optimize the nutrient profile of the horticultural crop.

Substrate and light quality interactions were revealed in Contribution 2. The effect of four different spectra (B, R, W, and B + R + W) at 160 μmol m^{-2} s^{-1} on kale microgreens was studied using three types of substrates. The research is a good example for design space screening and identifying the cultivation parameters, leading to high and affordable yields.

Contribution 3 is the only paper in this Special Issue that uses the photoperiod as a quantitative light parameter. Liu et al. measured the bolting and flowering time as a function of the photoperiod and determined the lighting conditions required to achieve the optimum stem morphology. The duration of light/dark time intervals were set at four different levels (12/12, 14/10, 16/8, and 18/6 h), whereas the PPFD was kept constant at 200 μmol·m^{-2}·s^{-1}. Endogenous gibberellin concentrations measured in stem tips and young leaves indicated that the bolting and flowering of cabbage is regulated through the synthesis of gibberellin.

Contribution 5 by Balázs et al. is an outlier in a sense that the article presents a method of quantifying variations in lighting environments using pea microgreens as the sole test vehicle to demonstrate the effect of light intensity and light quality variations in a vertical farm. A broad PPFD range (33–390 μmol m^{-2} s^{-1}) was established on the cultivation trays by adjusting the power of the LEDs and the distance between the LED luminaires and the illuminated crop. The fresh weight of pea seedlings exhibited strong correlations with local PPFD levels measured in a high-spatial-resolution experiment. The study highlighted that the local light intensity accounted for 31% of the fresh weight variations, and the rest of the noise was attributed to the differences among the individual plants.

The growth, yield, and phenolic content of lemon basil was investigated by Daud et al. (Contribution 8) under controlled environmental conditions. Plants were grown under mixed red and blue LED light, with an R/B ratio of 4.1. Two PPFD levels, 80 and 160 μmol·m^{-2}·s^{-1}, were tested in the experiment. The electrical conductivity (EC) (concentration) of the nutrient solution was an additional factor set at four different levels. The experiment demonstrated that under low PPFD levels, plant development was limited by light availability, and there were minor differences in the fresh weights among the four different nutrient concentrations. The interaction between photon irradiance and EC became apparent at high PPFD. The maximum yield and the best phytochemical traits were measured in the same PPFD = 160 μmol·m^{-2}·s^{-1}, EC = 2.6 mS cm^{-1} treatment.

2.2. *Shading Net Experiments*

Colored shading nets affect several environmental parameters during crop growth: light intensity, spectral distribution of light, and microclimates, including air temperature and relative humidity. In the experiments of Jiménez-Viveros and Valiente-Banuet (Contribution 4), the phytochemical profile of piquin pepper cultivated under four shading conditions was investigated. The maximum reduction in light intensity was 25% in the case of black mesh relative to the control without shading. The air temperature reduction was 10% or less at the peak temperature during one day. This work highlighted the beneficial effect of shading on the fruit quality, but the relevance of the key conclusions is limited to growers in tropical/sub-tropical regions. In an additional review (Contribution 10), Jiménez-Viveros et al. summarized the effects of light on the nutritional properties of pepper.

In Contribution 7, An et al. studied the physiological response of blueberry in another shading net experiment. Anthocyanin content and key enzyme activities were measured in blueberry leaves cultivated under 25%, 50%, 75%, and 100% intensity of natural daylight. The endogenous hormone concentrations and enzyme activities positively correlated with light intensity. Anthocyanin concentration, however, exhibited a maximum at 75%. The molecular mechanism of anthocyanin synthesis through light intensity regulation was a key finding of this research.

The advantage of shaded tomato cultivation in warm tropical and sub-tropical climate was highlighted by Delgado-Vargas et al. in Contribution 6. Two tomato cultivars were grown under four solar irradiance conditions using three different types of shade mesh: 100% (without mesh crop cover), 80%, 75%, and 50%. The photosynthetic rate measurements revealed that open-field plants were exposed to the highest level of abiotic stress, limiting the growth and development of both cultivars. The photon irradiance exceeded the light saturation limit early in the morning and remained above the saturation level throughout the day. The shade nets were efficient in reducing the light intensity and decreasing the air temperature at the canopy, resulting in the alleviation of stress factors.

2.3. Supplementary LED Lighting

While designed shading was advantageous in tropical climates, the shade of the construction elements in a building integrated rooftop greenhouse created suboptimal lighting conditions for tomato cultivation in the Mediterranean area. Appolloni et al. tested three lighting strategies to supplement solar radiation and found 17% yield increase regardless of the type of treatment used (with or without far-red, during the whole day or at the end of the day). The paper investigated the economics of the supplementary lighting by estimating the specific production cost increase associated with the energy consumption of LED luminaires.

3. Conclusions

The ten publications in this Special Issue covered only a tiny proportion of the practical lighting challenges that horticulturalists are faced with. The lighting applications described ranged from shading nets to LEDs as a sole source of light. Several publications investigated the interactions between the lighting conditions and other environmental parameters, which are indispensable for optimizing closed indoor cultivation systems. The publications described methods to find the trade-off between the quantity and quality (nutritional profile) of the horticultural crop.

The use of qualitative attributes to describe the lighting environment significantly limits the transferability of experimental results. This limitation is particularly pronounced in shade net experiments. While relative solar irradiance allows for direct comparison of parallel tests differentiated by the shading net type only, the absence of absolute values for B, G, R, and FR wavebands hinders the direct comparison with results from other experiments conducted under different conditions. Future studies should incorporate absolute measurements of these wavebands as a function of time to enhance the comparability and robustness of findings across diverse experimental setups.

Conflicts of Interest: The authors declare no conflicts of interest.

List of Contributors

1. Hernández-Adasme, C.; Guevara, M.J.; Faicán-Benenaula, M.A.; Neira, R.; Delgadillo, D.; Muñoz, V.; Salazar-Parra, C.; Sun, B.; Yang, X.; Escalona, V.H. Effect of Light Conditions on Growth and Antioxidant Parameters of Two Hydroponically Grown Lettuce Cultivars (Green and Purple) in a Vertical Farm System. *Horticulturae* **2025**, *11*, 220. https://doi.org/10.3390/horticulturae11020220.
2. Maru, R.N.; Wesonga, J.; Okazawa, H.; Kavoo, A.; Neondo, J.O.; Mazibuko, D.M.; Maskey, S.; Orsini, F. Evaluation of Growth, Yield and Bioactive Compounds

of Ethiopian Kale (*Brassica carinata* A. Braun) Microgreens under Different LED Light Spectra and Substrates. *Horticulturae* **2024**, *10*, 436. https://doi.org/10.3390/horticulturae10050436.
3. Liu, S.; Lu, J.; Tian, J.; Cao, P.; Li, S.; Ge, H.; Han, M.; Zhong, F. Effect of Photoperiod and Gibberellin on the Bolting and Flowering of Non-Heading Chinese Cabbage. *Horticulturae* **2023**, *9*, 1349. https://doi.org/10.3390/horticulturae9121349.
4. Jiménez-Viveros, Y.; Valiente-Banuet, J.I. Colored Shading Nets Differentially Affect the Phytochemical Profile, Antioxidant Capacity, and Fruit Quality of Piquin Peppers (*Capsicum Annuum* L. Var. *glabriusculum*). *Horticulturae* **2023**, *9*, 1240. https://doi.org/10.3390/horticulturae9111240.
5. Balázs, L.; Kovács, G.P.; Gyuricza, C.; Piroska, P.; Tarnawa, Á.; Kende, Z. Quantifying the Effect of Light Intensity Uniformity on the Crop Yield by Pea Microgreens Growth Experiments. *Horticulturae* **2023**, *9*, 1187. https://doi.org/10.3390/horticulturae9111187.
6. Delgado-Vargas, V.A.; Hernández-Bolio, G.I.; Hernández-Núñez, E.; Gautier, H.; Ayala-Garay, O.J.; Garruña, R. Mesh Crop Cover Optimizes the Microenvironment in a Tropical Region and Modifies the Physiology and Metabolome in Tomato. *Horticulturae* **2023**, *9*, 636. https://doi.org/10.3390/horticulturae9060636.
7. An, X.; Tan, T.; Zhang, X.; Guo, X.; Zhu, Y.; Song, Z.; Wang, D. Effects of Light Intensity on Endogenous Hormones and Key Enzyme Activities of Anthocyanin Synthesis in Blueberry Leaves. *Horticulturae* **2023**, *9*, 618. https://doi.org/10.3390/horticulturae9060618.
8. Daud, Z.M.; Ismail, M.F.; Hakiman, M. Effects of LED Red and Blue Spectra Irradiance Levels and Nutrient Solution EC on the Growth, Yield, and Phenolic Content of Lemon Basil (*Ocimum citriodurum* Vis.). *Horticulturae* **2023**, *9*, 416. https://doi.org/10.3390/horticulturae9040416.
9. Appolloni, E.; Paucek, I.; Pennisi, G.; Stringari, G.; Gabarrell Durany, X.; Orsini, F.; Gianquinto, G. Supplemental LED Lighting Improves Fruit Growth and Yield of Tomato Grown under the Sub-Optimal Lighting Condition of a Building Integrated Rooftop Greenhouse (i-RTG). *Horticulturae* **2022**, *8*, 771. https://doi.org/10.3390/horticulturae8090771.
10. Jiménez-Viveros, Y.; Núñez-Palenius, H.G.; Fierros-Romero, G.; Valiente-Banuet, J.I. Modification of Light Characteristics Affect the Phytochemical Profile of Peppers. *Horticulturae* **2023**, *9*, 72. https://doi.org/10.3390/horticulturae9010072.

References

1. Stirbet, A.; Lazár, D.; Guo, Y.; Govindjee, G. Photosynthesis: Basics, History and Modelling. *Ann. Bot.* **2020**, *126*, 511–537. [CrossRef] [PubMed]
2. Thoma, F.; Somborn-Schulz, A.; Schlehuber, D.; Keuter, V.; Deerberg, G. Effects of Light on Secondary Metabolites in Selected Leafy Greens: A Review. *Front. Plant Sci.* **2020**, *11*, 497. [CrossRef] [PubMed]
3. Ashraf, M.A.; Iqbal, M.; Rasheed, R.; Hussain, I.; Riaz, M.; Arif, M.S. Chapter 8—Environmental Stress and Secondary Metabolites in Plants: An Overview. In *Plant Metabolites and Regulation Under Environmental Stress*; Ahmad, P., Ahanger, M.A., Singh, V.P., Tripathi, D.K., Alam, P., Alyemeni, M.N., Eds.; Academic Press: Cambridge, MA, USA, 2018; pp. 153–167. ISBN 978-0-12-812689-9.
4. Sipos, L.; Boros, I.F.; Csambalik, L.; Székely, G.; Jung, A.; Balázs, L. Horticultural Lighting System Optimalization: A Review. *Sci. Hortic.* **2020**, *273*, 109631. [CrossRef]
5. Pinho, P.; Jokinen, K.; Halonen, L. Horticultural Lighting—Present and Future Challenges. *Light. Res. Technol.* **2012**, *44*, 427–437. [CrossRef]
6. Kochetova, G.V.; Avercheva, O.V.; Bassarskaya, E.M.; Zhigalova, T.V. Light Quality as a Driver of Photosynthetic Apparatus Development. *Biophys. Rev.* **2022**, *14*, 779–803. [CrossRef] [PubMed]
7. Shelford, T.; Wallace, C.; Both, A.J. Calculating and Reporting Key Light Ratios for Plant Research. *Acta Hortic.* **2020**, *1296*, 559–566. [CrossRef]

8. Hernández-Andrés, J.; Romero, J.; Nieves, J.L.; Lee, R.L. Color and Spectral Analysis of Daylight in Southern Europe. *J. Opt. Soc. Am. A JOSAA* **2001**, *18*, 1325–1335. [CrossRef] [PubMed]
9. Castellano, S.; Mugnozza, G.S.; Russo, G.; Briassoulis, D.; Mistriotis, A.; Hemming, S.; Waaijenberg, D. Plastic Nets in Agriculture: A General Review of Types and Applications. *Appl. Eng. Agric.* **2008**, *24*, 799–808. [CrossRef]
10. Paradiso, R.; Proietti, S. Light-Quality Manipulation to Control Plant Growth and Photomorphogenesis in Greenhouse Horticulture: The State of the Art and the Opportunities of Modern LED Systems. *J. Plant Growth Regul.* **2022**, *41*, 742–780. [CrossRef]
11. Palmitessa, O.D.; Pantaleo, M.A.; Santamaria, P. Applications and Development of LEDs as Supplementary Lighting for Tomato at Different Latitudes. *Agronomy* **2021**, *11*, 835. [CrossRef]
12. Kozai, T.; Amagai, Y.; Hayashi, E. Towards Sustainable Plant Factories with Artificial Lighting (PFALs): From Greenhouses to Vertical Farms. In *Achieving Sustainable Greenhouse Cultivation*; Burleigh Dodds Science Publishing: Cambridge, UK, 2019; ISBN 978-0-429-26674-4.

horticulturae

MDPI

Article

Effect of Light Conditions on Growth and Antioxidant Parameters of Two Hydroponically Grown Lettuce Cultivars (Green and Purple) in a Vertical Farm System

Cristian Hernández-Adasme [1], María José Guevara [1], María Auxiliadora Faicán-Benenaula [2,3], Rodrigo Neira [1], Dakary Delgadillo [1], Violeta Muñoz [1], Carolina Salazar-Parra [4], Bo Sun [5], Xiao Yang [6] and Víctor Hugo Escalona [1,*]

1 Centro de Estudios de Postcosecha, Facultad de Ciencias Agronómicas, Universidad de Chile, La Pintana, Santiago 8820808, Chile; criherna@ug.uchile.cl (C.H.-A.); guevaramjm@gmail.com (M.J.G.); rodrigo.neira@uchile.cl (R.N.); dakary.delgadillo@uchile.cl (D.D.); violeta.munoz@uchile.cl (V.M.)
2 Instituto de Tecnología de Alimentos, Facultad de Ingeniería Química, Universidad Nacional del Litoral, Santa Fe 3000, Argentina; mfaican@gmail.com
3 Consejo Nacional de Investigaciones Científicas y Técnicas (CONICET), Santa Fe 3000, Argentina
4 Instituto de Investigaciones Agropecuarias (INIA), Centro Regional La Platina, Santiago 8831314, Chile; carolina.salazar@inia.cl
5 College of Horticulture, Sichuan Agricultural University, Chengdu 611130, China; bsun@sicau.edu.cn
6 Institute of Urban Agriculture, Chinese Academy of Agricultural Sciences, Chengdu National Agricultural Science and Technology Center, Chengdu 610000, China; yangxiao@caas.cn
* Correspondence: vescalona@uchile.cl

Abstract: The use of extended light spectra, including UV-A, green, and far-red, has been scarcely explored in vertical farming. This study evaluated the effects of full spectra under two intensities (90 and 180 μmol m^{-2} s^{-1}) on the growth and antioxidant properties of green and purple leaf lettuce. Three light spectra were tested: Blue-White (BW), Red-White (RW), and Red-Blue (RB). Fresh weight (FW), dry weight percentage (DWP), chlorophyll concentration (NDVI), and antioxidant parameters (total phenolic content (TPC), antioxidant capacity by DPPH and FRAP and total flavonoid content (TFC)) were assessed. Spectrum-intensity interactions significantly influenced FW, with RW-180 μmol m^{-2} s^{-1} yielding the highest FW (78.2 g plant^{-1} in green and 48.5 g plant^{-1} in purple lettuce). BW-90 μmol m^{-2} s^{-1} maximized DWP in green lettuce, while PAR intensity of 180 μmol m^{-2} s^{-1} favored DWP in purple lettuce. Chlorophyll concentration increased under PAR intensity of 180 μmol m^{-2} s^{-1}, and leaf color varied with spectrum, with RW producing lighter leaves. Antioxidant parameters declined over time, but a PAR intensity of 180 μmol m^{-2} s^{-1}, particularly under RW, boosted TPC and TFC contents in both lettuce cultivars during early stages (days 0 and 15). Conversely, a lower PAR intensity of 90 μmol m^{-2} s^{-1}, mainly under RW, enhanced antioxidant capacity by FRAP at 15 days and by the end of the cycle for both cultivars. Overall, RW-180 μmol m^{-2} s^{-1} interactions promoted the best characteristics in lettuce. Nonetheless, the findings emphasize the significance of fine-tuning both light spectrum and intensity to enhance lettuce growth and quality in vertical farming systems considering the cultivar, time and variable to be evaluated.

Keywords: LED light; spectrum; intensity; NDVI; phenolic; flavonoids; antioxidant capacity

1. Introduction

The demand for food is rising due to the continuous increase in the global population, which poses a significant challenge for agriculture, particularly amid decreasing arable land, resource scarcity, and the need for sustainable production [1]. According to Alexandratos and Bruinsma [2], the world population is expected to increase by more than one-third (2.3 billion people) between 2009 and 2050, driving a growing demand for food. Faced with this challenge, innovative alternatives have emerged that integrate agriculture, engineering, and architecture, creating vertical types of agriculture in cities [3], thereby optimizing the use of space, energy, and water.

Vertical farming has grown rapidly, combined with indoor farming production technologies such as hydroponics. This method produces a crop in water with nutrients, which offers numerous advantages, such as high yield, good quality, continuous production, and efficient use of resources, among others [4]. Among the most common crops in these systems are short-cycle, single-crop green leafy vegetables, such as lettuce [5]. This crop is notable for its rapid growth, short growing cycle, high planting density, and low energy demand [6–8].

Climatic conditions in vertical cultivation systems are highly dependent on energy consumption, with lighting being the main vector, responsible for 65–85% of total energy expenditure [9]. Increasing light intensity, for example, from 250 to 700 μmol m^{-2} s^{-1}, dramatically increases energy consumption [9–11], raising the need to optimize intensities to meet both plant photosynthetic demands and system energy efficiency.

Light provides energy for photosynthesis and regulates plant growth and development based on its intensity and spectral quality [12,13]. Photosynthetic pigments preferentially absorb light in the blue (430–453 nm) and red (642–663 nm) ranges, optimizing photon conversion and activating key metabolic pathways [14–18]. In particular, red and blue spectra are the most efficient for plant growth and development due to their impact on photosynthesis and regulation of physiological processes [19–22].

In addition to red and blue spectra, other light ranges, such as ultraviolet (UV), green, and far-red (FR), can induce specific responses in plants. For example, adding UV-A light can affect dry weight and leaf area, depending on the exact wavelength employed [23]. Low-intensity white light (55 μmol m^{-2} s^{-1}) has promoted an increase in fresh weight and leaf length in lettuce seedlings [24], while the combination of red and far-red light (R:FR in a 3:2 ratio) and intensities of 300 μmol m^{-2} s^{-1} significantly increased leaf area, fresh weight and gas exchange [25].

The light spectrum also influences the accumulation of phytochemicals. Blue and red light have increased compounds such as polyphenols and anthocyanins in lettuce [26,27]. For example, UV or green light supplementation over a basal spectrum (blue + red + FR) has increased the production of antioxidants and pigments such as α-carotene and anthocyanins [27]. Similarly, adding FR to red light has improved vitamin C and soluble sugar content [25]. However, lights with a high red fraction (150 μmol m^{-2} s^{-1}) have reduced chlorophyll and carotenoid levels in arugula and lamb's lettuce [28]. In comparison, spectra with high red:blue ratios (7.5:1) and intensities between 216 and 376 μmol m^{-2} s^{-1} have a reduced phenolic content in green lettuce [29].

Light intensity also regulates growth and nutritional quality. In microgreens, intensities of 120 and 160 μmol m^{-2} s^{-1} (23% blue + 75% red + 2% FR) showed enhanced yield compared to higher intensities (220 μmol m^{-2} s^{-1}) [30]. In lettuce, intensities of 250 μmol m^{-2} s^{-1} have promoted higher fresh biomass than low intensities of 60 μmol m^{-2} s^{-1} [31]. Furthermore, intensities between 350 and 450 μmol m^{-2} s^{-1} combined with red and blue spectra (R:B = 2:1) and concentrated nutrient solutions significantly improved polyphenol and anthocyanin levels [32]. Additionally, studies suggest that intensities between 150 and 300 μmol m^{-2} s^{-1} under blue and red light increase antioxidants, phenols, and sugars in species such as lettuce, cabbage, cucumber, and spinach [13].

The lower intensity of 60 μmol m^{-2} s^{-1} can also promote higher antioxidant capacity in lettuce when red + blue spectra with different R:B ratios were used [31]. Hernandez-Adasme et al. [30] showed that an intensity of 120 μmol m^{-2} s^{-1} under the spectrum of 23% blue + 75% red + 2% far-red and a photoperiod of 16 h of light promoted betalain accumulation in beet microgreens compared to 220 μmol m^{-2} s^{-1} by 35% and a photoperiod of 12 h of light by 96.8%.

Therefore, incorporating additional wavelengths to the blue and red spectrum, which amplifies the spectral quality of lighting, is a promising strategy to maximize vegetative growth, morphological development, and the antioxidant profile of vegetables. Although several studies have demonstrated the beneficial effects of blue and red light on crops like lettuce, the impact of more complete spectra that include ultraviolet (UV-A), green, and far-red (FR) light, in combination with varying light intensities, remains largely unexplored. Furthermore, previous studies have not sufficiently addressed the interaction between spectral quality and light intensity on agronomic parameters and functional quality in leafy vegetables grown in vertical hydroponic systems.

This study was set out with the aim of evaluating the interaction between complete light spectra and different light intensities on the morphological characteristics and antioxidant properties of green and purple leafy lettuce grown in vertical hydroponic systems. It is hypothesized that the combination of full light spectra, integrating ultraviolet (UV-A) or far-red (FR) light, at moderate intensities improves the fresh weight and antioxidant quality of lettuce grown in vertical hydroponic systems.

2. Materials and Methods

2.1. Plant Material and Growth Conditions

This study was conducted in a vertical farm system set up in adapted 3.5 × 4.0 × 6.0 m cold chambers at the Post-harvest Study Center (CEPOC) at the University of Chile (33°34′ S, 70°38′ W). Five 1.7 × 1.8 × 0.45 m metal shelves were arranged inside the chamber, with three levels per shelf. LED lamps for each light treatment were mounted on each level (Table 1). Dividers made of opaque, non-translucent material were placed between the experimental units to avoid overlapping between light treatments. This design ensured effective isolation, preventing light transmission between adjacent treatments and guaranteeing the independence of the evaluated light conditions.

Table 1. Treatments applied to green 'Bartimer' and purple 'Soltero' lettuces grown hydroponically in a vertical farm.

Light Treatments	Spectrum (%) UV:B:G:R:FR [2]	R:B Ratio	PAR [1] μmol m^{-2} s^{-1}	Photoperiod h
Blue-White (BW)	0:18:40:39:3	2.2:1.0	90 180	
Red-White (RW)	1:17:25:49:8	2.9:1.0	90 180	12
Red-Blue (RB)	1:17:4:76:2	4.5:1.0	90 180	

[1] Photosynthetically active radiation. [2] ultraviolet:blue:green:red:far-red.

Lettuce seeds (*Lactuca sativa* L.) of green and purple loose leaf type Lollo 'Bartimer' and 'Soltero', respectively, (Nuhmens, BASF) were used. The Bartimer variety is characterized by its bright green color, tender texture, vigorous growth, and high leaf quality [33]. 'Soltero', in contrast, is characterized by its reddish color and a high content of antioxidant compounds, in addition to showing positive results in using hydroponic crops [34].

Sowing was carried out in 105-cell plastic trays with a single seed allocated to each cell at the time of sowing. The substrate used was a mixture of DSM2 W R0632 peat (Kekkilä, Vantaa, Finland) and A6 perlite (Harborlite, Santiago, Chile) in a 1:2 (*v:v*) ratio. The sown trays were placed under each light treatment in the vertical cultivation system. Once the seedlings reached a 5 to 6 cm root length and three to four true leaves, they were transplanted to the floating root system and maintained until harvest (45 days after transplant). The floating root system consisted of plastic trays (0.40 × 0.30 × 0.06 m) on which a white acrylic sheet (0.45 × 0.35 × 0.07 m) with 14 perforations was placed. The nutrient solution (4 L per tray) was changed weekly in each tray, and its composition was mentioned in the studies by Hernández-Adasme et al. [24] and Lara et al. [35]. The nutrient solution was oxygenated by supplying air through 4 mm diameter silicone hoses connected to an air compressor (SOBO Electrical Appliance Co., Ltd., SB-748, Guangzhou, China), achieving a concentration that ranged between 8 and 10 mg L^{-1} in each tray. The pH of the nutrient solution was measured with a potentiometer (Hi99301, Hanna Instruments, Woonsocket, RI, USA), maintained between 5.8 and 6.0 and adjustments were made with an acid solution (1.2% phosphoric acid + 3.8% nitric acid + 95% water) when appropriate. The electrical conductivity was measured with a conductivity meter (Hi99301, Hanna Instruments, Woonsocket, RI, USA), maintained around 2.0 mS cm^{-1} by the addition of fertilizers to compensate for any variations in nutrient concentrations. Both parameters (pH and EC) were measured weekly at each solution change. The ambient temperature and relative humidity during crop growth were 22 ± 2 °C and 70–80%, respectively. Neither variable varied significantly during lettuce cultivation. The thermal energy emitted by the LED lamps was minimal (±2 °C), and the humidity variations were slight, not exceeding 5% between light and dark periods.

2.2. Light Treatments

The light treatments consisted of three different light spectra under two intensities, 90 and 180 μmol m^{-2} s^{-1}. A TG-14 plug-in analog timer (ManHua Electric Co., Ltd., Wenzhou, China) was used to program the 12-h photoperiod to save energy. The Blue-White (BW) treatment was given by a panel with a 32.5 × 19.5 cm dimmer (Samsung, LM301h Quantum LED, Suwon, Republic of Korea). The Red-White (RW) treatment consisted of two LED tubes 1.2 m long (Sonneteck Technology Co., Ltd., GL-TL040P12BF-01, Xiamen, China). Finally, Red-Blue (RB) was achieved with 36 × 30 cm LED lamps (ASYCAR, Santiago, Chile). Each treatment was initiated on the day of sowing.

2.3. Agronomic Characteristics

Agronomic characteristics were evaluated at harvest, i.e., 45 days after transplanting.

2.3.1. Fresh Weight (FW)

The fresh weight of the aerial part of three lettuce plants obtained from each replicate was measured at harvest. The result was expressed in grams per plant (g $plant^{-1}$).

2.3.2. Dry Weight Percentage (DWP)

The dry weight percentage was measured by drying the aerial part of the same three plants per replicate obtained for fresh weight. Drying was performed in an LFO-250F oven (LabTech, Gyeonggi-do, Republic of Korea) at 60 °C until the sample maintained a constant weight. The weight was obtained from a CMN3000-1 semi-analytical balance (Kern & Sohn GmbH, Balingen, Germany), and the result was presented as a percentage using the equation proposed by Hernández-Adasme et al. [26]:

$$DWP = (DW/FW) \times 100 \quad (1)$$

where FW and DW correspond to fresh and dried weight, respectively.

2.3.3. Leaf Number

The total number of leaves of three independent plants for each repetition and treatment was counted at harvest time.

2.3.4. Color

Lightness (L*), chroma (C*), and hue (h*) were measured on the adaxial side of all extended leaves of three plants per replicate. Three measurements were taken for each leaf using a compact tristimulus colorimeter minolta chroma meter model CM-2500d (Konica Minolta INC., Osaka, Japan).

2.3.5. Normalized Difference Vegetation Index (NDVI) as a Relative Index of Chlorophyll Concentration

This measurement was performed on two leaves from each of the three plants chosen per replicate in each treatment. Measurements were performed using a reflectance-based device (PlantPen NDVI 300, Photon Systems Instruments (PSI), Drásov, Czech Republic).

2.4. Antioxidant Parameters

Antioxidant parameters were evaluated at three harvest stages, at transplanting (day 0), 15 and 45 days after transplanting; each phenological stage was analyzed independently.

2.4.1. Total Phenolic Content (TPC)

Total phenolic content was determined according to the method proposed by Singleton and Rossi [36] and the modifications indicated by Hernández-Adasme et al. [29]. A calibration curve performed with gallic acid was used to obtain the total phenol concentration. The results were expressed as mg gallic acid equivalent (GAE) 100 g^{-1} FW.

2.4.2. Antioxidant Capacity

The FRAP (ferric reducing/antioxidant power) protocol was carried out according to the methods proposed by Benzie and Strain [37] following the modifications proposed by Hernández-Adasme et al. [30]. The method monitors the reaction of a ferric-TPTZ (2,4,6-tripyridyl-s-triazine) solution, which changes from a ferric to a ferrous form through contact with the antioxidant compounds. This reduced the total antioxidant compounds in the reaction, changing the absorbance ratios at 593 nm. The antioxidant capacity by FRAP was calculated through a calibration curve performed with Trolox. The results were expressed as mg Trolox equivalent (TE) 100 g^{-1} FW.

The measurement of antioxidant capacity by DPPH was carried out according to the method proposed by Brand-Williams et al. [38] and following the modifications used by Hernández-Adasme et al. [30]. To 250 μL of plant extract, 1 mL of 0.1 mM DPPH solution was added and incubated for 20 min. Subsequently, 200 μL of the mixture (extract and DPPH reagent) was taken and transferred to a spectrophotometer multi-cell plate (ASYS UVM340 Biochrom, Cambridge, UK), and readings were taken at 517 nm. After 2 h of incubation, the absorbance of the reaction was measured again. The antioxidant capacity was calculated using a calibration curve based on a Trolox stock solution. The results were expressed as mg Trolox equivalent (TE) 100 g^{-1} FW.

2.4.3. Total Flavonoid Content (TFC)

Total flavonoid content was determined with aluminum chloride as described by Flores et al. [39]. 100 μL of 5% $NaNO_2$ were added to 100 μL of the extract, and after 5 min, 10% $AlCl_3$ was added. After standing for 6 min at room temperature, 670 μL of

1 M NaOH were added. Finally, the reaction absorbance was measured at 510 nm using a spectrophotometer multi-cell plate (ASYS UVM340, UK). The results were expressed as milligrams of Rutin equivalent (RE) 100 g^{-1} FW.

2.5. Experimental Design and Statistical Analysis

The experiment was designed as a completely randomized 3 × 2 factorial structure, with three repetitions per treatment, with each repetition consisting of three plants. The first factor was the light spectrum, with three levels (Blue-White (BW; R:B = 2.2:1), Red-White (RW; R:B = 3.1:1), and Red-Blue (RB; R:B = 5:1). The second factor was the intensity, with two levels (90 and 180 μmol m^{-2} s^{-1}). The data were analyzed using linear mixed models for each variable evaluated and each lettuce cultivar independently. Finally, the differences between means were compared using Fisher's LSD test for the interaction of factors or independent factors when they corresponded with a significance level of 5% ($\alpha = 0.05$). Statistical analyses were performed with the InfoStat software (version 2020e) and R programming language (i386 3.6.3) version 2020 [40].

3. Results

3.1. Agronomic Characteristics

3.1.1. Fresh Weight

A significant interaction between light intensity and spectrum on FW was observed in both cultivars (Figure 1a, b). In green lettuce 'Bartimer', the highest fresh weights were recorded with RW-180 (78.2 g $plant^{-1}$) and BW-180 μmol m^{-2} s^{-1} (64.5 g $plant^{-1}$). For both cases, the increase in PAR intensity determined a 37% rise in FW in BW and RW whereas no differences were observed in RB treatments (Figure 1a). In purple 'Soltero' lettuce, the highest values were achieved with the treatments RW-180 (48.5 g $plant^{-1}$), BW-180 (46.1 g $plant^{-1}$), and BW-90 μmol m^{-2} s^{-1} (45.7 g $plant^{-1}$). In particular, increasing PAR intensity resulted in 41.4% more FW in RW while no significant differences were observed under BW and RB. On the other hand, these fresh weights were lower than those obtained by the green 'Bartimer' lettuce (Figure 1b).

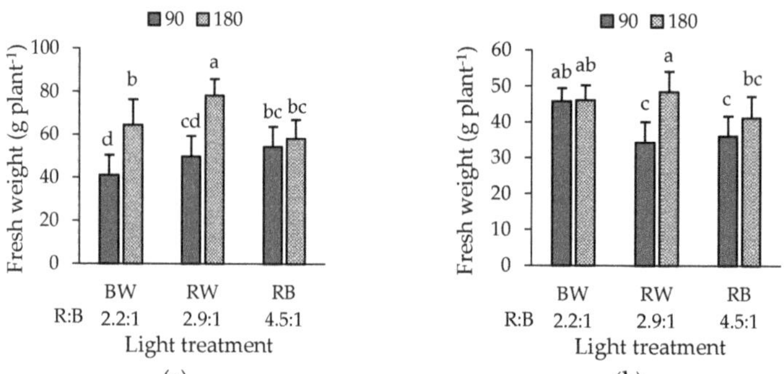

Figure 1. Fresh weight of (**a**) green 'Bartimer' and (**b**) purple 'Soltero' lettuces grown hydroponically in a vertical farm exposed to different light spectra and PAR intensities (μmol m^{-2} s^{-1}) at 45 days post-transplanting. Different letters indicate significant differences in the factor or the interaction between factors (Fisher's test, $p \leq 0.05$).

3.1.2. Dry Weight Percentage (DWP)

In green 'Bartimer' lettuce, a significant interaction between light intensity and spectrum on DWP was observed (Figure 2a). Treatments BW, RW, and RB-90 μmol m^{-2} s^{-1} showed the highest values; in particular, BW-90 μmol m^{-2} s^{-1} showed the highest value of

11.2%. Thus, the lower PAR intensity improved the DWP by 95.3, 93.8 and 69.1% under RW, BW and RB, respectively (Figure 2a). In purple 'Soltero' lettuce, significant differences were observed in the intensity factor only, with 180 μmol m^{-2} s^{-1} being the highest value (7.8%) compared to 90 μmol m^{-2} s^{-1} (6.9%) which meant a 13.0% increase in DWP (Figure 2b).

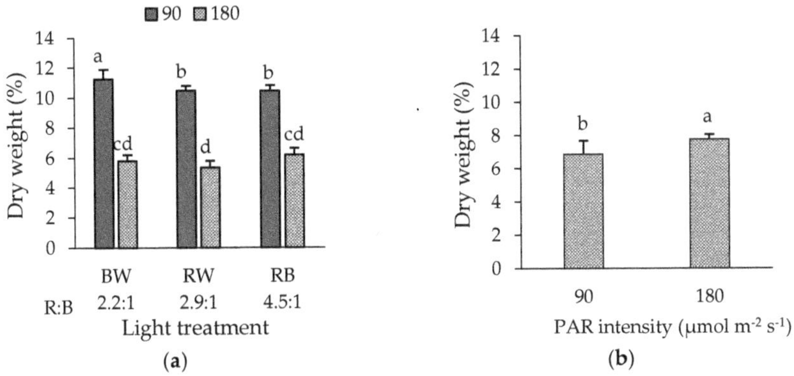

Figure 2. Dry weight percentage of (**a**) green 'Bartimer' and (**b**) purple 'Soltero' lettuces grown hydroponically in a vertical farm exposed to different light spectra and PAR intensities (μmol m^{-2} s^{-1}) at 45 days post-transplanting. Different letters indicate significant differences in the factor or the interaction between factors (Fisher's test, $p \leq 0.05$).

3.1.3. Number of Leaves per Plant

The number of leaves per plant of both cultivars showed significant differences in spectrum and intensity factors independently. In green 'Bartimer' lettuce, the highest values were observed under the RW treatment (25.0 leaves per plant). Specifically, the number of leaves increased significantly under RW versus RB by 6.4%, while no significant differences were found between BW and the other light spectra (RW and RB) (Table 2). On the other hand, PAR intensity of 180 μmol m^{-2} s^{-1} reached 25.6 leaves per plant, a significant increase of 12.8% compared to low PAR intensity (90 μmol m^{-2} s^{-1}) (Table 2). Similarly, in purple 'Soltero' lettuce, the maximum value was recorded under BW (22.0 leaves per plant). Thus, BW determined a 7.3% rise in the number of leaves compared to RB. In contrast, no differences were observed between BW and RW, and RW and RB (Table 2). Likewise, PAR intensity of 180 μmol m^{-2} s^{-1} significantly enhanced the number of leaves (22.6 leaves per plant) compared to low PAR intensity (90 μmol m^{-2} s^{-1}) by 13.0% (Table 2).

Table 2. Leaf number and NDVI (Normalized Difference Vegetation Index) of green 'Bartimer' and purple 'Soltero' lettuces grown hydroponically in a vertical farm exposed to different light spectra and PAR intensities (μmol m^{-2} s^{-1}) at 45 days post-transplanting.

Factor	Level	Leaf Number Plant^{-1}		NDVI	
		Bartimer cv.	**Soltero cv.**	**Bartimer cv.**	**Soltero cv.**
Spectrum (S)	BW	23.9 ± 0.7 ab [1]	22.0 ± 0.4 a	0.37 ± 0.010	0.44 ± 0.009
	RW	25.0 ± 0.7 a	21.3 ± 0.5 ab	0.37 ± 0.011	0.42 ± 0.011
	RB	23.5 ± 0.4 b	20.5 ± 0.6 b	0.38 ± 0.012	0.44 ± 0.007

Table 2. *Cont.*

Factor	Level	Leaf Number Plant^{-1}		NDVI	
		Bartimer cv.	Soltero cv.	Bartimer cv.	Soltero cv.
Significance		*	*	ns [2]	ns
Intensity (I)	90	22.7 ± 0.4 b	20.0 ± 0.4 b	0.40 ± 0.007 b	0.40 ± 0.007 b
	180	25.6 ± 0.4 a	22.6 ± 0.4 a	0.47 ± 0.004 a	0.47 ± 0.004 a
Significance		*	*	*	*
Interaction (S × I)	BW-90	22.5 ± 0.8	20.6 ± 0.4	0.34 ± 0.015	0.40 ± 0.009
	RW-90	23.5 ± 0.9	20.6 ± 0.6	0.35 ± 0.015	0.39 ± 0.014
	RB-90	22.3 ± 0.6	18.5 ± 0.6	0.33 ± 0.015	0.42 ± 0.008
	BW-180	25.4 ± 0.7	23.3 ± 0.4	0.40 ± 0.009	0.48 ± 0.007
	RW-180	26.6 ± 0.7	22.1 ± 0.8	0.39 ± 0.014	0.46 ± 0.007
	RB-180	24.8 ± 0.6	22.3 ± 0.8	0.42 ± 0.008	0.46 ± 0.008
Significance		ns	ns	ns	ns

[1] Different letters on the columns within each factor or interaction indicate significant differences (Fisher's test, * $p < 0.05$). [2] Indicates not significant.

3.1.4. NDVI (Normalized Difference Vegetation Index)

Significant differences were observed only for the intensity factor in both cultivars, with the highest values being found at 180 μmol m^{-2} s^{-1} (Table 2). Green 'Bartimer' and purple 'Soltero' lettuces reached an NDVI of 11.8% and 17.5% higher than the PAR intensity of 90 μmol m^{-2} s^{-1}, respectively (Table 2).

3.1.5. Color

The spectrum was the only factor significantly affecting lightness in green 'Bartimer' lettuce. In particular, the RW treatment significantly enhanced lightness, surpassing BW and RB by 3.1% and 8.6%, respectively (Table 3 and Figure 3). Conversely, no significant differences were detected in chroma or hue in this cultivar (Table 3). In purple 'Soltero' lettuce, significant differences were evident across all evaluated parameters—lightness, chroma, and hue—exclusively for the spectrum factor (Table 3). Specifically, RW (49) significantly increased lightness compared to RB (46) and BW (45) by 6.5% and 8.9%, respectively. Likewise, the chroma showed a notable rise under RW, exceeding RB and BW by 42.8% and 48.7%, respectively. Additionally, hue values were significantly higher in RW (117°) than RB (113°) and BW (112°). These findings suggest that purple 'Soltero' lettuce grown under RW conditions exhibited a greener and lighter appearance (Table 3 and Figure 4).

Table 3. Lightness (L*), Chroma (C*), and Hue (H°) of green 'Bartimer' and purple 'Soltero' lettuces grown hydroponically in a vertical farm exposed to different light spectra and PAR intensities (μmol m^{-2} s^{-1}) at 45 days post-transplanting.

Factor	Level	Lightness (L*)		Chroma (C*)		Hue (H°)	
		Bartimer cv.	Soltero cv.	Bartimer cv.	Soltero cv.	Bartimer cv.	Soltero cv.
Spectrum (S)	BW	65 ± 0.84 b [1]	45 ± 1.67 c	47 ± 2.48	26 ± 4.09 b	123 ± 1.99	112 ± 1.18 b
	RW	67 ± 0.85 a	49 ± 1.66 a	50 ± 2.41	39 ± 4.09 a	122 ± 1.96	117 ± 1.20 a
	RB	62 ± 0.86 c	46 ± 1.67 b	46 ± 2.79	27 ± 4.34 b	123 ± 2.10	113 ± 1.17 b
Significance		*	*	ns [2]	*	ns	*
Intensity (I)	90	64 ± 0.69	47 ± 1.69	49 ± 0.87	30 ± 3.04	123 ± 1.96	114 ± 1.45
	180	64 ± 0.76	47 ± 1.71	47 ± 0.85	31 ± 3.65	123 ± 1.97	114 ± 1.24

Table 3. *Cont.*

Factor	Level	Lightness (L*)		Chroma (C*)		Hue (H°)	
		Bartimer cv.	Soltero cv.	Bartimer cv.	Soltero cv.	Bartimer cv.	Soltero cv.
Significance		ns	ns	ns	ns	ns	ns
Interaction (S × I)	BW-90	65 ± 0.86	45 ± 0.55	47 ± 0.90	26 ± 0.56	123 ± 1.97	112 ± 0.86
	RW-90	67 ± 0.85	49 ± 0.51	54 ± 0.91	39 ± 7.92	122 ± 1.61	117 ± 0.40
	RB-90	62 ± 0.89	46 ± 0.69	46 ± 0.89	27 ± 0.70	123 ± 2.34	112 ± 0.83
	BW-180	65 ± 0.73	45 ± 0.56	47 ± 0.86	26 ± 0.65	123 ± 1.97	112 ± 0.85
	RW-180	67 ± 0.74	49 ± 0.54	47 ± 0.86	39 ± 7.91	122 ± 1.60	117 ± 0.41
	RB-180	62 ± 0.76	46 ± 0.65	46 ± 0.87	27 ± 0.68	123 ± 2.33	112 ± 0.79
Significance		ns	ns	ns	ns	ns	ns

[1] Different letters on the columns within each factor or interaction indicate significant differences (Fisher's test, * $p < 0.05$). [2] Indicates not significant.

Figure 3. Green 'Bartimer' lettuce grown hydroponically in a vertical farm exposed to (**a**) BW-90 μmol m^{-2} s^{-1}; (**b**) RW-90 μmol m^{-2} s^{-1}; (**c**) RB-90 μmol m^{-2} s^{-1}; (**d**) BW-180 μmol m^{-2} s^{-1}; (**e**) RW-180 μmol m^{-2} s^{-1}; (**f**) RB-180 μmol m^{-2} s^{-1} at 45 days post-transplanting.

Figure 4. Purple 'Soltero' lettuce grown hydroponically in a vertical farm exposed to (**a**) BW-90 μmol m^{-2} s^{-1}; (**b**) RW-90 μmol m^{-2} s^{-1}; (**c**) RB-90 μmol m^{-2} s^{-1}; (**d**) BW-180 μmol m^{-2} s^{-1}; (**e**) RW-180 μmol m^{-2} s^{-1}; (**f**) RB-180 μmol m^{-2} s^{-1} at 45 days post-transplanting.

3.2. Antioxidant Parameters

3.2.1. Total Phenolic Content (TPC)

In the initial evaluation (day 0), a significant interaction between the evaluated factors was observed in both cultivars. In green 'Bartimer' lettuce, the highest values were recorded in the treatments RW-180 μmol m^{-2} s^{-1} (351 mg GAE 100 g^{-1} FW), BW-180 μmol m^{-2} s^{-1} (321 mg GAE 100 g^{-1} FW), and RB-180 μmol m^{-2} s^{-1} (317 mg GAE 100 g^{-1} FW). Thus, increasing the PAR intensity promoted a 268.6, 191.8 and 112.7% increase in TPC in RB, BW and RW, respectively (Table 4). In purple 'Soltero' lettuce, the highest value was recorded under the RW-180 μmol m^{-2} s^{-1}, reaching 315 mg GAE 100 g^{-1} FW. Specifically, the enhancement of PAR intensity resulted in 164.7 and 50.0% rise in TPC under RW and RB, respectively, while no differences were observed in BW treatment (Table 5).

Table 4. Antioxidant parameters of green 'Bartimer' lettuce grown hydroponically in a vertical farm exposed to different light spectra and PAR intensities (μmol m^{-2} s^{-1}) at transplanting (day 0), 15, and 45 days post-transplanting.

Factor	Level	TPC [1] mg GAE100 g^{-1} FW			TFC [2] mg RE 100 g^{-1} FW			FRAP [3] mg TE 100 g^{-1} FW			DPPH [4] mg TE 100 g^{-1} FW		
		Days			Days			Days			Days		
		0	15	45	0	15	45	0	15	45	0	15	45
Spectrum (S)	BW	215 ± 17 ab [5]	223 ± 30 b	72 ± 4 b	380 ± 148	339 ± 57	30 ± 6 b	213 ± 80 b	78 ± 10 c	17 ± 2 b	1236 ± 4 b	1215 ± 21 b	1024 ± 7 a
	RW	258 ± 17 a	281 ± 33 a	81 ± 3 a	533 ± 149	437 ± 73	49 ± 3 a	337 ± 80 a	169 ± 26 a	38 ± 6 a	1355 ± 37 a	1276 ± 15 a	984 ± 10 b
	RB	202 ± 17 b	252 ± 31 ab	71 ± 2 b	440 ± 174	398 ± 72	36 ± 3 b	230 ± 84 b	117 ± 16 b	14 ± 2 c	1251 ± 25 b	1221 ± 22 b	1022 ± 15 a
Significance		*	*	*	ns [6]	ns	*	*	*	*	*	*	*
Intensity (I)	90	121 ± 12 b	96 ± 4 b	67 ± 3 b	113 ± 29 b	67 ± 2 b	39 ± 3	82 ± 21 b	213 ± 12 a	40 ± 3 a	1296 ± 35	1174 ± 11 b	1034 ± 10 a
	180	330 ± 15 a	408 ± 14 a	82 ± 2 a	788 ± 51 a	715 ± 36 a	37 ± 4	439 ± 24 a	29 ± 1 b	6 ± 0 b	1265 ± 15	1300 ± 15 a	986 ± 7 b
Significance		*	*	*	*	*	ns	*	*	*	ns	*	*
Interaction (S × I)	BW-90	110 ± 6 c	68 ± 4 d	51 ± 2 d	50 ± 5 c	69 ± 4 c	16 ± 2 c	37 ± 2 d	129 ± 6 c	28 ± 1 b	1239 ± 2 b	1151 ± 16 c	1026 ± 16 b
	RW-90	165 ± 8 b	132 ± 4 c	87 ± 3 a	227 ± 13 b	76 ± 3 c	60 ± 3 a	163 ± 15 c	307 ± 7 a	70 ± 2 a	1420 ± 43 a	1235 ± 20 b	1014 ± 15 bc
	RB-90	86 ± 6 c	89 ± 3 cd	64 ± 2 c	63 ± 5 bc	56 ± 4 c	41 ± 2 b	44 ± 1 d	204 ± 5 b	22 ± 1 c	1228 ± 40 b	1137 ± 8 c	1063 ± 23 a
	BW-180	321 ± 19 a	378 ± 17 b	93 ± 3 a	710 ± 21 a	608 ± 54 b	43 ± 11 b	389 ± 31 b	27 ± 1 d	7 ± 0 d	1231 ± 7 b	1278 ± 31 ab	1023 ± 11 b
	RW-180	351 ± 31 a	431 ± 34 a	74 ± 3 b	839 ± 131 a	798 ± 57 a	38 ± 5 b	511 ± 31 a	32 ± 2 d	5 ± 0 d	1289 ± 26 b	1317 ± 17 a	954 ± 8 d
	RB-180	317 ± 31 a	416 ± 15 ab	78 ± 2 b	816 ± 93 a	740 ± 70 a	30 ± 5 bc	416 ± 28 b	30 ± 1 d	6 ± 0 d	1274 ± 32 b	1306 ± 29 a	981 ± 10 cd
Significance		*	*	*	*	*	*	*	*	*	*	*	*

[1] Total phenolic content. [2] Total flavonoid content. [3] Antioxidant activity measured by FRAP assay. [4] Antioxidant activity measured by the DPPH assay. [5] Different letters on the columns within each factor or interaction indicate significant differences (Fisher's test, * $p < 0.05$). [6] Indicates not significant.

Table 5. Antioxidant parameters of purple 'Soltero' lettuce grown hydroponically in a vertical farm exposed to different light spectra and PAR intensities (µmol m^{-2} s^{-1}) at transplanting (day 0), 15, and 45 days post-transplanting.

Factor	Level	TPC [1] mg GAE 100 g^{-1} FW			TFC [2] mg RE 100 g^{-1} FW			FRAP [3] mg TE 100 g^{-1} FW			DPPH [4] mg TE 100 g^{-1} FW		
		Days			Days			Days			Days		
		0	15	45	0	15	45	0	15	45	0	15	45
Spectrum (S)	BW	191 ± 14	274 ± 36 b [5]	172 ± 5 a	439 ± 117 c	353 ± 37 b	144 ± 10	261 ± 58 c	72 ± 6 c	71 ± 11 b	1288 ± 83 b	1236 ± 20 c	1097 ± 19 b
	RW	217 ± 45	361 ± 27 a	169 ± 6 a	681 ± 187 a	395 ± 49 a	147 ± 7	491 ± 118 a	193 ± 27 a	97 ± 16 a	1587 ± 107 a	1364 ± 19 a	1159 ± 17 a
	RB	190 ± 18	261 ± 31 b	151 ± 4 b	556 ± 125 b	362 ± 57 ab	159 ± 13	371 ± 73 b	92 ± 10 b	67 ± 11 b	1425 ± 89 ab	1287 ± 23 b	1096 ± 24 b
Significance		ns [6]	*	*	*	*	ns	*	*	*	*	*	*
Intensity (I)	90	149 ± 11 b	143 ± 11 b	159 ± 3	242 ± 18 b	126 ± 8 b	157 ± 4	195 ± 17 b	198 ± 16 b	145 ± 5 a	1578 ± 79 a	1230 ± 16 b	1168 ± 16 a
	180	250 ± 19 a	459 ± 11 a	169 ± 5	875 ± 62 a	614 ± 18 a	143 ± 11	553 ± 58 a	41 ± 1 a	12 ± 0 b	1289 ± 77 b	1362 ± 15 a	1067 ± 15 b
Significance		*	*	ns	*	*	ns	*	*	*	*	*	*
Interaction (S × I)	BW-90	176 ± 19 cd	89 ± 3 d	168 ± 5	181 ± 4 e	168 ± 5 c	136 ± 6	137 ± 13 d	104 ± 4 c	129 ± 6 b	1425 ± 80	1142 ± 15	1123 ± 34
	RW-90	119 ± 12 e	234 ± 11 c	157 ± 6	264 ± 23 de	147 ± 9 c	156 ± 7	235 ± 17 d	345 ± 9 a	182 ± 8 a	1758 ± 155	1323 ± 27	1206 ± 23
	RB-90	152 ± 9 de	107 ± 5 d	153 ± 6	281 ± 23 d	62 ± 3 d	178 ± 6	216 ± 17 d	144 ± 6 b	124 ± 6 b	1551 ± 126	1226 ± 17	1175 ± 21
	BW-180	206 ± 18 bc	459 ± 18 ab	176 ± 9	696 ± 49 c	537 ± 27 b	152 ± 20	384 ± 34 c	40 ± 2 d	13 ± 0 c	1150 ± 98	1331 ± 15	1071 ± 16
	RW-180	315 ± 18 a	488 ± 22 a	182 ± 9	1098 ± 17 a	642 ± 32 a	138 ± 12	748 ± 59 a	41 ± 1 d	12 ± 1 c	1416 ± 60	1406 ± 23	1112 ± 19
	RB-180	228 ± 13 b	427 ± 18 b	150 ± 6	830 ± 41 b	662 ± 23 a	140 ± 25	527 ± 49 b	41 ± 2 d	10 ± 0 c	1300 ± 92	1348 ± 36	1017 ± 33
Significance		*	*	ns	*	*	ns	*	*	*	ns	ns	ns

[1] Total phenolic content. [2] Total flavonoid content. [3] Antioxidant activity measured by FRAP assay. [4] Antioxidant activity measured by the DPPH assay. [5] Different letters on the columns within each factor or interaction indicate significant differences (Fisher's test, * $p < 0.05$). [6] Indicates not significant.

At 15 days post-transplant, both cultivars continued to exhibit a significant interaction between factors. In green 'Bartimer' lettuce, the RW-180 µmol m^{-2} s^{-1} recorded the highest value at 431 mg GAE 100 g^{-1} FW. Likewise, the rise in PAR intensity determined a 455.9, 367.4 and 226.5% increase in TPC in BW, RB and RW, respectively (Table 4). In addition, TPC values under 180 µmol m^{-2} s^{-1} were higher than in the previous evaluation (Table 4). For purple 'Soltero' lettuce, the RW-180 µmol m^{-2} s^{-1} and BW-180 µmol m^{-2} s^{-1} treatments delivered the highest TPC levels, reaching 488 and 459 mg GAE 100 g^{-1} FW, respectively. Overall, the enhancement of PAR intensity improved in 415.7, 299.1 and 108.6% the TPC under BW, RB and RW, respectively. Notably, the application of 180 µmol m^{-2} s^{-1} yielded values up to five times greater than the PAR intensity of 90 µmol m^{-2} s^{-1} and higher values than the previous evaluation (Table 5).

By 45 days post-transplant, TPC levels had significantly decreased in all green 'Bartimer' lettuce treatments and those with a light intensity of 180 µmol m^{-2} s^{-1} in the purple 'Soltero' lettuce. In green 'Bartimer' lettuce, the BW-180 µmol m^{-2} s^{-1} and RW-90 µmol m^{-2} s^{-1} emerged as the top performers, reaching 93 and 87 mg GAE 100 g^{-1} FW, respectively (Table 4). Particularly, the increase in PAR intensity caused an 82.4% and 21.9% improvement in TPC under BW and RB, respectively. On the contrary, the decrease in PAR intensity determined a 17.6% rise in TPC under RW (Table 4). Meanwhile, differences in purple 'Soltero' lettuce were primarily linked to the light spectrum factor, with BW and RW achieving the highest levels at 172 and 169 mg GAE 100 g^{-1} FW, respectively. Specifically, the TPC under BW and RW increased compared to RB by 13.9% and 11.9%, respectively (Table 5).

3.2.2. Antioxidant Capacity

Antioxidant Capacity by FRAP Assay

The antioxidant capacity showed a significant interaction between factors in both cultivars at the different harvest times (Tables 4 and 5). On day 0, the RW-180 µmol m^{-2} s^{-1} significantly increased antioxidant capacity compared to the other treatments, reaching 511 mg TE 100 g^{-1} FW in green 'Bartimer' lettuce. On the other hand, the increase in PAR intensity determined a 951.4, 845.5 and 213.5% rise in antioxidant capacity in BW, RB and RW, respectively (Table 4). Similarly, in purple 'Soltero' lettuce, this same treatment (RW-180 µmol m^{-2} s^{-1}) resulted in the highest value, with 748 mg TE 100 g^{-1} FW. Furthermore, moderate PAR intensity (180 µmol m^{-2} s^{-1}) resulted in a 221.0, 180.3 and 144.0% enhancement of antioxidant capacity under RW, BW and RB, respectively, compared to lowest PAR intensity (90 µmol m^{-2} s^{-1}) (Table 5). Moreover, on day 0, the intensity of 180 µmol m^{-2} s^{-1} under the different spectra was associated with the greatest antioxidant capacity values by FRAP compared to 90 µmol m^{-2} s^{-1} in both cultivars (Tables 4 and 5).

At 15 and 45 days post-transplant, the highest antioxidant capacity was observed under the RW-90 µmol m^{-2} s^{-1} compared to the other treatments in both cultivars. In particular, in green 'Bartimer' lettuce, RW-90 µmol m^{-2} s^{-1} recorded 307 and 70 TE 100 g^{-1} FW on days 15 and 45, respectively. Thus, raising PAR intensity caused increases of 859.4, 580.0 and 377.8% of antioxidant capacity in RW, RB and BW, respectively, at day 15, whereas enhancing PAR intensity prompted the antioxidant capacity under RW, BW and RB by 1330.0, 300.0 and 266.7%, respectively at day 45 (Table 4). On the other hand, the highest antioxidant capacity of purple 'Soltero' lettuce was observed under RW-90 µmol m^{-2} s^{-1}, reaching 345 and 181 mg TE 100 g^{-1} FW at days 15 and 45, respectively. Overall, the lower PAR intensity (90 µmol m^{-2} s^{-1}) induced an increase in antioxidant capacity of 741.5, 251.2 and 169.2 at day 15 and 1408.2, 1140.0 and 892.3 at day 45 under RW, RB and BW, respectively (Table 5). Finally, at 180 µmol m^{-2} s^{-1}, a progressive decrease in antioxidant capacity was noted in both cultivars over time (Tables 4 and 5).

Antioxidant Capacity by DPPH Assay

The antioxidant capacity exhibited distinct patterns in both lettuce cultivars across the different evaluation stages. In the initial analysis (day 0), a significant interaction between factors was observed in green 'Bartimer' lettuce. Thus, RW-90 μmol m^{-2} s^{-1} achieving the highest TFC values (1420 mg TE 100 g^{-1} FW), i.e., the lower PAR intensity under RW increased antioxidant capacity compared to RW-180 μmol m^{-2} s^{-1} by 10.2%, meanwhile no differences were found between BW and RB spectra under the different intensities (Table 4). In contrast, the antioxidant capacity showed significant differences in each factor independently in the purple 'Soltero' lettuce. Specifically, the intensity of 90 μmol m^{-2} s^{-1} increased antioxidant capacity compared to 180 μmol m^{-2} s^{-1} by 16.4%. Additionally, the RW significantly enhanced antioxidant activity relative to BW by 23.2% while no differences were observed between RW and RB (Table 5).

At 15 days post-transplant, a significant interaction between factors persisted in green 'Bartimer' lettuce. The highest values were observed under RW- and RB-180 μmol m^{-2} s^{-1}, reaching 1317 and 1306 mg TE 100 g^{-1} FW, respectively. Thus, the rise in PAR intensity improved the antioxidant capacity under RB, BW and RW by 14.9, 11.0 and 6.6%, respectively (Table 4). In purple 'Soltero' lettuce, significant differences continued to occur independently for each factor. The PAR intensity of 180 μmol m^{-2} s^{-1} significantly increased the antioxidant capacity by DPPH compared to 90 μmol m^{-2} s^{-1} by 7.8%. Conversely, RW provided higher values compared to RB and BW by 6.0% and 10.4%, respectively (Table 5).

After 45 days post-transplant, a significant interaction between factors was again observed in green 'Bartimer' lettuce. In particular, RB-90 μmol m^{-2} s^{-1} recorded the highest antioxidant capacity (1063 mg TE 100 g^{-1} FW). Thus, PAR intensity of 180 μmol m^{-2} s^{-1} resulted in an 8.4 and 6.3% increase in antioxidant capacity under RB and RW, respectively, compared to lowest PAR intensity (90 μmol m^{-2} s^{-1}), whereas no differences were found in BW treatment (Table 4). In purple 'Soltero' lettuce, significant differences were maintained for each factor independently. The intensity of 90 μmol m^{-2} s^{-1} significantly raised antioxidant capacity compared to 180 μmol m^{-2} s^{-1} by 7.1%. Likewise, RW increased antioxidant capacity compared to both RB and BW by 5.8% (Table 5).

3.2.3. Total Flavonoid Content (TFC)

The TFC was significantly influenced by the interaction between the evaluated factors at all harvest stages in both green 'Bartimer' and purple 'Soltero' lettuces (Tables 4 and 5). Overall, a decreasing trend in TFC values was observed over time (Tables 4 and 5). On day 0, the RW-180 μmol m^{-2} s^{-1}, RB-180 μmol m^{-2} s^{-1}, and BW-180 μmol m^{-2} s^{-1} treatments in 'Bartimer' lettuce exhibited the highest TFC levels, reaching 839, 816, and 710 mg RE 100 g^{-1} FW, respectively. Therefore, increasing PAR intensity resulted in 1320.0, 1195.2 and 269.6% improved TFC in BW, RB and RW, respectively (Table 4). In purple 'Soltero' lettuce, the RW-180 μmol m^{-2} s^{-1} treatment exhibited the highest value, reaching 1098 mg RE 100 g^{-1} FW. On the other hand, PAR intensity of 180 μmol m^{-2} s^{-1} determined an increase of 315.9, 284.5, and 195.4% of TFC under RW, BW and RB, respectively (Table 5). In both cultivars, treatments under 180 μmol m^{-2} s^{-1} significantly increased TFC levels, proving to be the most effective compared to 90 μmol m^{-2} s^{-1} (Tables 4 and 5).

On day 15, the RW-180 μmol m^{-2} s^{-1} and RB-180 μmol m^{-2} s^{-1} showed the highest values in the TFC content in green 'Bartimer' lettuce, reaching 798 and 740 mg RE 100 g^{-1} FW, respectively. In addition, the increase in PAR intensity caused an increase of 1221.4, 950.0 and 781.2% in TFC under RB, RW and BW, respectively (Table 4). On the other hand, a decrease in TFC was recorded in the same lettuce cultivar under the spectra with an intensity of 180 μmol m^{-2} s^{-1} compared to the evaluation on day 0. Similarly, in purple 'Soltero' lettuce, a decrease in TFC was identified across all treatments compared to the assessment on day 0. In addition, RB-180 μmol m^{-2} s^{-1} (662 mg RE 100 g^{-1} FW) and RW-180 μmol m^{-2} s^{-1} (642 mg RE 100 g^{-1} FW) showed the highest TFC levels. Likewise, enhancing PAR intensity improved TFC under BW, RW and RB by 219.6, 336.7 and 967.7%, respectively (Table 5). In both lettuce cultivars, the intensity of 180 μmol m^{-2} s^{-1} under the different spectra positively influenced TFC accumulation, resulting in significantly higher values compared to 90 μmol m^{-2} s^{-1} (Tables 4 and 5).

Finally, on day 45 post-transplant, TFC was significantly affected by the interaction between the factors in green 'Bartimer' lettuce. In detail, RW 90 μmol m^{-2} s^{-1} treatment recorded the highest TFC, with a value of 60 mg RE 100 g^{-1} FW. On the other hand, the lower PAR intensity increased TFC in RW and RB by 57.9 and 36.7%, respectively. In contrast, PAR intensity of 180 μmol m^{-2} s^{-1} in BW significantly elevated TCF compared to lower intensity (90 μmol m^{-2} s^{-1}) by 168.8% (Table 4). Moreover, TFC levels in this cultivar decreased further compared to day 15. In purple 'Soltero' lettuce, no significant differences were found among the treatments (Table 5).

4. Discussion

4.1. Agronomic Characteristics

The results showed that the interaction between spectrum and intensity affected fresh weight (FW) in both lettuce cultivars and dry weight (DWP) in green 'Bartimer' lettuce. The RW-180 μmol m^{-2} s^{-1} treatment promoted the highest FW in both cultivars (Figure 1a,b), characterized by a UVA:B:G:R:FR spectrum = 1:17:25:49:8 and an R:B ratio of 2.9:1.0. This treatment also presented the highest proportion of far-red (FR, 8%) and the lowest R:FR ratio (6.1:1). Previous studies have reported that FR supplementation can increase biomass in lettuce plants [25,41–43]. Additionally, Tan et al. [44] pointed out that FR can regulate the photosynthetic capacity, facilitating biomass accumulation [45]. Then, the increase in FW under RW-180 μmol m^{-2} s^{-1} could be associated with the elevated proportion of FR in the spectrum.

In this study, BW, RW and RB spectra at low PAR intensity (90 μmol m^{-2} s^{-1}) favored the increase in the percentage of dry weight in green 'Bartimer' lettuce (Figure 2a). According to Ghorbanzadeh et al. [46] a lower PAR intensity (75 μmol m^{-2} s^{-1}) tends to develop a larger specific leaf area (cm^{-2} g^{-1}), i.e., thinner and wider leaves, which may improve light penetration and utilization within the canopy, resulting in higher dry weight accumulation. In contrast, the lowest DWP was observed at RW-180 μmol m^{-2} s^{-1} in 'Bartimer' (Figure 2a), indicating that the higher FW could be attributed to a higher water content. Furthermore, the number of leaves per plant increased significantly under RW-180 μmol m^{-2} s^{-1} (Table 2), which could also explain the increase in FW due to higher leaf production under FR treatments [42]. In purple lettuce 'Soltero', light intensity significantly impacted DWP, with higher values under 180 μmol m^{-2} s^{-1} (Figure 2b). Jin et al. [47] indicated that light intensity influences dry weight and is a key factor in determining photosynthesis [48]. This process converts light energy into chemical energy [49], mainly carbohydrates contributing to DWP [50]. Therefore, the higher intensity applied in this study probably promoted a greater accumulation of carbohydrates in purple 'Soltero' lettuce, increasing DWP. Thus, each cultivar responds differentially to the imposed light conditions, indicating that the effect of the light factors and/or their interaction on DWP is cultivar dependent.

Normalized difference vegetation index (NDVI) sensors equipped with red and NIR light detectors can estimate chlorophyll content by measuring light transmitted through leaves [51]. Alsiņa et al. [52] found that NDVI showed the best correlations for estimating chlorophyll in different species, including loose-leaf lettuce. In this study, higher light PAR intensity (180 μmol m^{-2} s^{-1}) increased chlorophyll concentration in green and purple lettuces, reflected in higher NDVI values compared to lower PAR intensity (90 μmol m^{-2} s^{-1}) (Table 2). These results align with those of Zhou et al. [53] and Pennisi et al. [54], who reported that a PAR intensity between 150 and 200 μmol m^{-2} s^{-1} enhances chlorophyll content in lettuce plants compared to 100 μmol m^{-2} s^{-1}. However, higher intensities can further increase chlorophyll content, with 250 μmol m^{-2} s^{-1} being the threshold beyond which no significant differences are observed [54]. Furthermore, Chen et al. [51] pointed out that a higher chlorophyll concentration reduces red light transmission, increasing absorption and generating higher NDVI values. In this study, NDVI responded significantly only to the light intensity factor in both cultivars; however, other studies, like that regarding Batavia cv. Blackhawk, showed that NDVI did not vary under different intensities (130–389 μmol m^{-2} s^{-1}) in a blue-red spectrum [20]. This discrepancy suggests that the effect of light intensity on NDVI might depend on the specific analyzed lettuce cultivar.

Leaf color is a key phenotypic characteristic in horticultural crops [55], affecting consumers' perception and choice of vegetables [56]. In this study, the light spectrum influenced lettuce leaf color. The RW spectrum produced a lighter color in the green 'Bartimer' lettuce by significantly increasing luminosity (Table 3). Similarly, in the purple 'Soltero' lettuce, the RW spectrum promoted higher luminosity, chroma, and hue (Table 3), generating greener and lighter leaves. Meanwhile, the BW and RB spectra induced more yellow and less green colors (Table 3). These effects could be due to the higher proportion of far-red (FR) in RW (8%), which is consistent with Carotti et al. [57], who found that increasing FR in RB light increased lightness and hue in red lettuce var. Canasta. However, the addition of FR showed a reduction in red coloration in other varieties, as reported by Meng et al. [58] and Meng and Runkle [59], who noted lower anthocyanin levels in 'Cherokee' and 'Rouxai' under similar conditions. In this study, the spectra used promoted low anthocyanins accumulation in 'Soltero' lettuce (Figure 4), possibly due to the low light intensities used.

4.2. Antioxidant Parameters

Lettuce is an important vegetable due to its high content of phytochemicals, such as phenolic acids [60–62], flavonoids [60–63], and anthocyanins [60,63], which provide essential antioxidant properties in the human diet. The biosynthesis and accumulation of these compounds are closely regulated by environmental factors such as light quality, intensity, and duration [62]. In this study, the interaction between light spectrum and intensity significantly influenced the antioxidant parameters. An intensity of 180 μmol m^{-2} s^{-1} promoted an increase in total phenolic (TPC) and flavonoid (TFC) content on days 0 and 15 in both cultivars, while at the final evaluation (day 45), a lower PAR intensity (90 μmol m^{-2} s^{-1}) favored these compounds in the green lettuce 'Bartimer'.

This differential effect may be attributed to the activation of specific metabolic pathways induced by the characteristics of the RW spectrum (UVA:B:G:R:FR = 1:17:25:49:8; R:B = 2.9:1.0), characterized by a high proportion of UV-A and far-red (FR). UV-A light has been shown to induce the expression of the *phenylalanine ammonia-lyase* (PAL) gene, a key point in the phenylpropanoid pathway, facilitating the synthesis of phenolic compounds and antioxidants [64]. Additionally, far-red (FR) is associated with increased photosynthetic capacity and biomass accumulation [41,44], which indirectly favors the synthesis of secondary metabolites such as flavonoids and anthocyanins.

In particular, the increase in TPC and TFC during the first 15 days under 180 μmol m^{-2} s^{-1} could be linked to the ability of light to induce the expression of key genes such as *chalcone synthase* (CHS), *flavonoid 3-hydroxylase* (F3H), and *UDP-glucose:flavonoid 3-O-glucosyltransferase* (UFGT), involved in the biosynthesis of flavonoids and phenols under blue and red light [65]. Reducing the intensity to 90 μmol m^{-2} s^{-1} at the final stages (day 45) might also have made it possible to preserve and stabilize antioxidant production by reducing light stress, as suggested by previous studies in lettuce grown under different light conditions [39,61].

Antioxidant capacity, as measured by FRAP and DPPH, showed a less clear but more prominent pattern under RW, especially at 90 μmol m^{-2} s^{-1} at day 45, suggesting that moderate intensities initially favor the activation of biosynthetic pathways. In contrast, lower intensities at later stages promote sustained antioxidant accumulation. Flores et al. [39] supported this observation by showing that intensities of 100 μmol m^{-2} s^{-1} significantly increased TPC and antioxidant capacity in green lettuce 'Romana Long Blonde Galaica' compared to lower intensities. Similarly, Song et al. [32] found that FRAP and DPPH were enhanced at irradiances between 350 and 450 μmol m^{-2} s^{-1} in contrast to lower irradiances (150–250 μmol m^{-2} s^{-1}). Furthermore, Hernández-Adasme et al. [29] indicated that low R:B ratios (0.4–1.6:1.0) increased TPC in green lettuce, while Naznin et al. [66] reported the opposite, with increases in antioxidants under higher R:B ratios (4.9:1.0). These findings suggest that the R:B ratio of 2.9:1.0 in this study may have optimized a balance between red and blue light, favoring the accumulation of antioxidant compounds in the first days of cultivation, while the intensity reduction to 90 μmol m^{-2} s^{-1} preserved these levels in later stages.

Regarding flavonoid content, the intensity of 180 μmol m^{-2} s^{-1} was effective in the first days (0 and 15) for both cultivars, especially under the RW spectrum, rich in UV-A and FR. This result agrees with studies highlighting the role of blue light in regulating biosynthetic genes such as CHS and F3H in lettuce plants [67], which drives flavonoid accumulation. On the other hand, Van Brenk et al. [42] showed that flavonoid content increased linearly with increasing blue in the R:B ratio (1.5:1.0 to 7.0:1.0). However, some studies, such as Naznin et al. [66], also pointed out that higher intensities may inhibit certain antioxidants, thereby necessitating a reduction in intensity to 90 μmol m^{-2} s^{-1} at later stages.

This behavior could also be related to differential gene activation under light intensities. Kitazaki et al. [65] revealed that blue and red wavelengths enhance the expression of genes such as PAL, CHS, and DFR, which are involved in flavonoid and phenolic metabolic pathways. Specifically, Hernández-Adasme et al. [26] observed enrichment in C3H expression under low R:B ratios (0.5:1.0), while other authors, such as Karami et al. [68] and Ouzounis et al. [69], highlighted that blue light increases flavonoid production, but with a greater effect on red lettuces.

The results suggest that the combination of a RW spectrum, rich in FR and UV-A, together with an initial intensity of 180 μmol m^{-2} s^{-1} followed by 90 μmol m^{-2} s^{-1}, optimizes the biosynthesis of antioxidants and flavonoids in green 'Bartimer' and purple 'Soltero' lettuces. This effect can be attributed to light-induced enzymatic and gene regulation, such as the activation of PAL, CHS, and UFGT, supported by previous studies on metabolic pathways under different spectra and intensities [70–72]. Therefore, this lighting strategy represents an effective tool to improve the nutritional value of lettuce through the precise management of light spectra and intensities.

5. Conclusions

Spectrum and light intensity significantly influence various growth characteristics and chemical quality of lettuce plants independently or by the interaction of both factors. Overall, RW spectrum (UV:B:G:R:FR = 1:17:25:49:8; R:B = 2.9:1.0) in combination with 180 μmol m^{-2} s^{-1} (RW-180 μmol m^{-2} s^{-1}) improved FW in both lettuce cultivars, although the effect was greater in 'Bartimer' green lettuce, indicating that the effect was cultivar-dependent. Independently, RW and 180 μmol m^{-2} s^{-1} positively promoted leaf number. Whereas, only the intensity of 180 μmol m^{-2} s^{-1} improved chlorophyll content in both lettuce cultivars. Thus, RW and the intensity of 180 μmol m^{-2} s^{-1} would be the most favorable factors to achieve better growth in lettuce plants. Regarding antioxidant parameters, the intensity of 180 μmol m^{-2} s^{-1}, especially under the RW spectrum, favored the content of total phenols (TPC) and flavonoids (TFC) in early stages (days 0 and 15) of green 'Bartimer' and purple 'Soltero' lettuce. While lower PAR intensity (90 μmol m^{-2} s^{-1}), mainly under RW, optimized antioxidant capacity only by FRAP at 15 days and at the end of the cycle (day 45), both in green 'Bartimer' and purple 'Soltero' lettuce. Thus, the effect on antioxidant parameters varied according to variable, time and cultivar. Finally, these results highlight the importance of optimizing both light spectrum and light intensity to maximize lettuce production and quality in vertical growing systems.

Author Contributions: Conceptualization, C.H.-A., M.J.G. and M.A.F.-B.; methodology, C.H.-A., M.J.G., M.A.F.-B. and V.M.; formal analysis, C.H.-A., R.N., D.D., C.S.-P.; investigation, C.H.-A., M.J.G. and M.A.F.-B.; data curation, C.H.-A. and M.A.F.-B.; writing—original draft preparation, C.H.-A., R.N., D.D. and V.M.; writing—review and editing, C.H.-A., R.N., D.D., C.S.-P., B.S., X.Y. and V.H.E.; supervision, C.H.-A. and V.H.E.; project administration, V.H.E.; funding acquisition, V.H.E. All authors have read and agreed to the published version of the manuscript.

Funding: This research was funded by FONDECYT REGULAR (ANID, Chile), number 1230703.

Data Availability Statement: No new data were created or analyzed in this study. Data sharing does not apply to this article.

Conflicts of Interest: The authors declare no conflicts of interest.

References

1. Wang, M.; Wan, D.; Xie, X.; Bai, Z.; Wang, R.; Zhang, X.; Yi-Zhou, G.; Zhiliang, T.; Yin, Y. Crop-livestock integration: Implications for food security, resource efficiency and greenhouse gas mitigation. *Innov. Life* **2024**, *2*, 100103-1. [CrossRef]
2. Alexandratos, N.; Bruinsma, J. *World Agriculture Towards 2030/2050: The 2012 Revision*; Food and Agriculture Organization of the United Nations: Rome, Italy, 2024; p. 147.
3. Soares, L.L.; Priore, R. Fazenda vertical como modelo sustentável de agricultura urbana. *Rev. Gest. Sustentabilidade Ambient.* **2023**, *12*, 1–15.
4. Li, J.; Wu, T.; Huang, K.; Liu, Y.; Liu, M.; Wang, J. Effect of LED spectrum on the quality and nitrogen metabolism of lettuce under recycled hydroponics. *Front. Plant Sci.* **2021**, *12*, 678197. [CrossRef]
5. Boros, I.F.; Székely, G.; Balázs, L.; Csambalik, L.; Sipos, L. Effects of LED lighting environments on lettuce (*Lactuca sativa* L.) in PFAL systems—A review. *Sci. Hortic.* **2023**, *321*, 112351. [CrossRef]
6. Bantis, F.; Smirnakou, S.; Ouzounis, T.; Koukounaras, A.; Ntagkas, N.; Radoglou, K. Current status and recent achievements in the field of horticulture with the use of light-emitting diodes (LEDs). *Sci. Hortic.* **2018**, *235*, 437–451. [CrossRef]
7. Zhang, X.; He, D.; Niu, G.; Yan, Z.; Song, J. Effects of environment lighting on the growth, photosynthesis, and quality of hydroponic lettuce in a plant factory. *Int. J. Agric. Biol. Eng.* **2018**, *11*, 33–40. [CrossRef]
8. Ahmed, H.A.; Yu-Xin, T.; Qi-Chang, Y. Optimal control of environmental conditions affecting lettuce plant growth in a controlled environment with artificial lighting: A review. *S. Afr. J. Bot.* **2020**, *130*, 75–89. [CrossRef]
9. Arcasi, A.; Mauro, A.W.; Napoli, G.; Tariello, F.; Vanoli, G.P. Energy and cost analysis for a crop production in a vertical farm. *Appl. Therm. Eng.* **2024**, *239*, 122129. [CrossRef]
10. Cui, J.; Song, S.; Yu, J.; Liu, H. Effect of daily light integral on cucumber plug seedlings in artificial light plant factory. *Horticulturae* **2021**, *7*, 139. [CrossRef]
11. Stanghellini, C.; Katzin, D. The dark side of lighting: A critical analysis of vertical farms' environmental impact. *J. Clean. Prod.* **2024**, *458*, 142359. [CrossRef]
12. Nájera, C.; Urrestarazu, M. Effect of the intensity and spectral quality of LED light on yield and nitrate accumulation in vegetables. *HortScience* **2019**, *54*, 1745–1750. [CrossRef]
13. Nájera, C.; Gallegos-Cedillo, V.M.; Ros, M.; Pascual, J.A. LED lighting in vertical farming systems enhances bioactive compounds and productivity of vegetables crops. *Biol. Life Sci. Forum* **2022**, *16*, 24. [CrossRef]
14. Hopkins, W.G.; Huner, N.P.A. *Introduction to Plant Physiology*, 3rd ed.; Wiley: Hoboken, NJ, USA, 2004; pp. 93–100.

15. Nelson, J.A.; Bugbee, B. Economic analysis of greenhouse lighting: Light emitting diodes vs. high intensity discharge fixtures. *PLoS ONE* **2014**, *9*, e99010. [CrossRef] [PubMed]
16. Wang, S.; Fang, H.; Xie, J.; Wu, Y.; Tang, Z.; Liu, Z.; Nivel, J.; Yu, J. Physiological responses of cucumber seedlings to different supplemental light duration of red and blue LED. *Front. Plant Sci.* **2021**, *12*, 709313. [CrossRef]
17. Son, K.H.; Oh, M.M. Leaf shape, growth, and antioxidant phenolic compounds of two lettuce cultivars grown under various combinations of blue and red light-emitting diodes. *HortScience* **2013**, *48*, 988–995. [CrossRef]
18. Stutte, G.W.; Edney, S.; Skerritt, T. Photoregulation of bioprotectant content of red leaf lettuce with light-emitting diodes. *HortScience* **2009**, *44*, 79–82. [CrossRef]
19. Lee, J.H.; Kwon, Y.B.; Choi, I.-L.; Yoon, H.S.; Kim, J.; Kim, Y.; Kang, H.-M. Changes in spectral reflectance, photosynthetic performance, chlorophyll fluorescence, and growth of mini green romaine lettuce according to various light qualities in indoor cultivation. *Horticulturae* **2024**, *10*, 860. [CrossRef]
20. Modarelli, G.C.; Paradiso, R.; Arena, C.; De Pascale, S.; Van Labeke, M.-C. High light intensity from blue-red LEDs enhance photosynthetic performance, plant growth, and optical properties of red lettuce in controlled environment. *Horticulturae* **2022**, *8*, 114. [CrossRef]
21. Lee, J.H.; Kwon, Y.B.; Roh, Y.H.; Choi, I.-L.; Kim, J.; Kim, Y.; Yoon, H.S.; Kang, H.-M. Effect of various LED light qualities, including wide red spectrum-LED, on the growth and quality of mini red romaine lettuce (cv. Breen). *Plants* **2023**, *12*, 2056. [CrossRef]
22. Brazaitytė, A.; Vaštakaitė-Kairienė, V.; Sutulienė, R.; Rasiukevičiūtė, N.; Viršilė, A.; Miliauskienė, J.; Laužikė, K.; Valiuškaitė, A.; Dene, L.; Chrapacienė, S.; et al. Phenolic compounds content evaluation of lettuce grown under short-term preharvest daytime or nighttime supplemental LEDs. *Plants* **2022**, *11*, 1123. [CrossRef]
23. Samuolienė, G.; Viršilė, A.; Miliauskienė, J.; Haimi, P.; Laužikė, K.; Jankauskienė, J.; Novičkovas, A.; Kupčinskienė, A.; Brazaitytė, A. The photosynthetic performance of red leaf lettuce under UV-A irradiation. *Agronomy* **2020**, *10*, 761. [CrossRef]
24. Hernández-Adasme, C.; Silva, H.; Escalona, V. In-door germination and seedling growth of green and red lettuce under LED-light spectrum and subsequent effect on baby leaf lettuce. *Ital. J. Agron.* **2022**, *17*, 1982. [CrossRef]
25. Bi, X.; Xu, H.; Yang, C.; Zhang, H.; Li, W.; Su, W.; Zheng, M.; Lei, B. Investigating the influence of varied ratios of red and far-red light on lettuce (*Lactuca sativa*): Effects on growth, photosynthetic characteristics and chlorophyll fluorescence. *Front. Plant Sci.* **2024**, *15*, 1430241. [CrossRef]
26. Hernández-Adasme, C.; Silva, H.; Peña, Á.; Vargas-Martínez, M.G.; Salazar-Parra, C.; Sun, B.; Escalona Contreras, V. Modifying the ambient light spectrum using LED lamps alters the phenolic profile of hydroponically grown greenhouse lettuce plants without affecting their agronomic characteristics. *Plants* **2024**, *13*, 2466. [CrossRef]
27. Samuolienė, G.; Brazaitytė, A.; Sirtautas, R.; Viršilė, A.; Sakalauskaitė, J.; Sakalauskienė, S.; Duchovskis, P. LED illumination affects bioactive compounds in romaine baby leaf lettuce. *J. Sci. Food Agric.* **2013**, *93*, 3286–3291. [CrossRef]
28. Frutos-Totosa, A.; Hernández-Adasme, C.; Martínez, V.; Mestre, T.; Díaz-Mula, H.M.; Botella, M.A.; Flores, P.; Martínez-Moreno, A. Light spectrum effects on rocket and lamb's lettuce cultivated in a vertical indoor farming system. *Sci. Hortic.* **2023**, *321*, 112221. [CrossRef]
29. Hernández-Adasme, C.; Silva, H.; Saavedra-Romero, J.; Martínez, V.; Escalona, V. Light supplementation and growing season affect the quality and antioxidant activity of lettuce. *Chil. J. Agric. Res.* **2023**, *83*, 320–333. [CrossRef]
30. Hernández-Adasme, C.; Palma-Dias, R.; Escalona, V.H. The effect of light intensity and photoperiod on the yield and antioxidant activity of beet microgreens produced in an indoor system. *Horticulturae* **2023**, *9*, 493. [CrossRef]
31. Mohamed, S.J.; Rihan, H.Z.; Aljafer, N.; Fuller, M.P. The impact of light spectrum and intensity on the growth, physiology, and antioxidant activity of lettuce (*Lactuca sativa* L.). *Plants* **2021**, *10*, 2162. [CrossRef]
32. Song, J.; Huang, H.; Hao, Y.; Song, S.; Zhang, Y.; Su, W.; Liu, H. Nutritional quality, mineral and antioxidant content in lettuce afected by interaction of light intensity and nutrient solution concentration. *Sci. Rep.* **2020**, *10*, 279.
33. Tobar, G.; Antúnez, A.; Corradini, F.; Vidal, M. Lettuce. In *Technical Aspects of Cultivation, Irrigation and Nutrition in Lettuce, Tomato and Melon for the Central Zone of Chile*; Blanco, C., Ed.; Instituto de Investigaciones Agropecuarias: Santiago, Chile, 2019; pp. 7–48. Available online: https://bibliotecadigital.ciren.cl/server/api/core/bitstreams/b9f82a07-65f6-47d8-a0c3-1b66d913d774/content (accessed on 20 November 2024).
34. Alucho, P.J.; Patin, Q.A. Agronomic and Productive Behavior of Three Varieties of Lettuce (*Lactuca sativa* L.), in a Hydroponic System (NFT), with the Application of Two Biostimulants, Under Two Types of Environments, in the Guanujo Parish, Bolívar Province. Bachelor's Thesis, Bolívar State University, Guaranda, Ecuador, 2023.
35. Lara, O.A.; Amoros, A.; Tapia, M.L.; Escalona, V.H. Effect of a photoselective filter on the yield and postharvest quality of 'Viroflay' baby spinach (*Spinacia oleracea* L.) leaves cultivated in a hydroponic system. *Sci. Hortic.* **2021**, *277*, 109804. [CrossRef]
36. Singleton, S.; Rossi, A. Colorimetry of total phenolics with phosphomolybdic-phosphotungstic and reagents. *Am. J. Enol. Vitic.* **1965**, *16*, 144–157. [CrossRef]

37. Benzie, I.F.; Strain, J.J. The ferric reducing ability of plasma (FRAP) as a canopy structure in wheat (*Triticum aestivum* L.) and wild oat (*Avena fatua* L.) exposed to enhanced ultraviolet-B radiation. *Funct. Ecol.* **1996**, *2*, 319–330.
38. Brand-Williams, W.; Cuvelier, M.E.; Berset, C.L. Use of a free radical method to evaluate antioxidant activity. *LWT-Food Sci. Technol.* **1995**, *28*, 25–30. [CrossRef]
39. Flores, M.; Urrestarazu, M.; Amorós, A.; Escalona, V. High intensity and red enriched LED lights increased growth of lettuce and endive. *Ital. J. Agron.* **2022**, *17*, 1915. [CrossRef]
40. Di Rienzo, J.A.; Casanoves, F.; Balzarini, M.G.; Gonzalez, L.; Tablada, M.; Robledo, C.W. InfoStat, version 2020. *InfoStat Group, FCA, National University of Córdoba: Córdoba, Argentina, 2020.* Available online: https://www.infostat.com.ar (accessed on 29 October 2024).
41. Lee, M.; Xu, J.W.; Wang, W.Q.; Rajashekar, C.B. The effect of supplemental blue, red and rar-red light on the growth and the nutritional quality of red and green leaf lettuce. *Am. J. Plant Sci.* **2019**, *10*, 2219–2235. [CrossRef]
42. Van Brenk, J.B.; Courbier, S.; Kleijweg, C.L.; Verdonk, J.C.; Marcelis, L.F.M. Paradise by the far-red light: Far-red and red:blue ratios independently affect yield, pigments, and carbohydrate production in lettuce, *Lactuca sativa*. *Front. Plant Sci.* **2024**, *15*, 1383100.
43. Zou, J.; Zhang, Y.; Zhang, Y.; Bian, Z.; Fanourakis, D.; Yang, Q.; Li, T. Morphological and physiological properties of indoor cultivated lettuce in response to additional far-red light. *Sci. Hortic.* **2019**, *257*, 108725. [CrossRef]
44. Tan, T.; Li, S.; Fan, Y.; Wang, Z.; Raza, M.A.; Shafiq, I.; Wang, B.; Wu, X.; Yong, T.; Wang, X.; et al. Far-red light: A regulator of plant morphology and photosynthetic capacity. *Crop J.* **2022**, *10*, 300–309. [CrossRef]
45. Orlando, M.; Trivellini, A.; Incrocci, L.; Ferrante, A.; Mensuali, A. The inclusion of green light in a red and blue light background impact the growth and functional quality of vegetable and flower microgreen species. *Horticulturae* **2022**, *8*, 217. [CrossRef]
46. Ghorbanzadeh, P.; Aliniaeifard, S.; Esmaeili, M.; Mashal, M.; Azadegan, B.; Seifet, M. Dependency of growth, water use efficiency, chlorophyll fluorescence, and stomatal characteristics of lettuce plants to light intensity. *J. Plant Growth. Regul.* **2021**, *40*, 2191–2207. [CrossRef]
47. Jin, W.; Ji, Y.; Larsen, D.H.; Huang, Y.; Heuvelink, E.; Marcelis, L.F.M. Gradually increasing light intensity during the growth period increases dry weight production compared to constant or gradually decreasing light intensity in lettuce. *Sci. Hortic.* **2023**, *311*, 111807. [CrossRef]
48. Wimalasekera, R. Effect of light intensity on photosynthesis. In *Photosynthesis, Productivity and Environmental Stres*; Ahmad, P., Ahanger, M.A., Alyemeni, M.N., Alam, P., Eds.; Wiley: Hoboken, NJ, USA, 2019; pp. 65–73.
49. Niu, G. Ligh. In *Plant Factory: An Indoor Vertical Farming System for Efficient Quality Food Production*, 2nd ed.; Kozai, T., Niu, N., Takagaki, M., Eds.; Academic Press Ltd.: Cambridge, MA, USA, 2020; pp. 115–128.
50. Brouwer, R. Nutritive influences on the distribution of dry matter in the plant. *Neth. J. Agric. Sei.* **1962**, *10*, 399–408. [CrossRef]
51. Chen, J.-J.; Zhen, S.; Sun, Y. Estimating leaf chlorophyll content of buffaloberry using normalized difference vegetation index sensors. *HortTechnology* **2021**, *31*, 297–303. [CrossRef]
52. Alsiņa, I.; Dūma, M.; Dubova, L.; Šenberga, A.; Daģis, S. Comparison of different chlorophylls determination methods for leafy vegetables. *Agron. Res.* **2016**, *14*, 309–316.
53. Zhou, J.; Li, P.; Wang, J. Effects of light intensity and temperature on the photosynthesis characteristics and yield of lettuce. *Horticulturae* **2022**, *8*, 178. [CrossRef]
54. Pennisi, G.; Pistillo, A.; Orsini, F.; Cellini, A.; Spinelli, F.; Nicola, S.; Fernandez, J.A.; Crepaldi, A.; Gianquinto, G.; Marcelis, L.F.M. Optimal light intensity for sustainable water and energy use in indoor cultivation of lettuce and basil under red and blue LEDs. *Sci. Hortic.* **2020**, *272*, 109508. [CrossRef]
55. Huo, J.; Zhang, N.; Gong, Y.; Bao, Y.; Li, Y.; Zhang, L.; Nie, S. Effects of different light intensity on leaf color changes in a Chinese cabbage yellow cotyledon mutant. *Front. Plant Sci.* **2024**, *15*, 1371451. [CrossRef]
56. Hoppu, U.; Puputti, S.; Sandell, M. Factors related to sensory properties and consumer acceptance of vegetables. *Crit. Rev. Food Sci. Nutr.* **2021**, *61*, 1751–1761. [CrossRef]
57. Carotti, L.; Pistillo, A.; Zauli, I.; Pennisi, G.; Martin, M.; Gianquinto, G.; Orsini, F. Far-red radiation management for lettuce growth: Physiological and morphological features leading to energy optimization in vertical farming. *Sci. Hortic.* **2024**, *33*, 113264. [CrossRef]
58. Meng, Q.; Kelly, N.; Runkle, E.S. Substituting green or far-red radiation for blue radiation induces shade avoidance and promotes growth in lettuce and kale. *Environ. Exp. Bot.* **2019**, *162*, 383–391. [CrossRef]
59. Meng, Q.; Runkle, E.S. Far-red radiation interacts with relative and absolute blue and red photon flux densities to regulate growth, morphology, and pigmentation of lettuce and basil seedlings. *Sci. Hortic.* **2019**, *255*, 269–280. [CrossRef]
60. Hameed, M.K.; Umar, W.; Razzaq, A.; Wei, S.; Niu, Q.; Huang, D.; Chang, L. Quantification of total polyphenols, antioxidants, anthocyanins and secondary metabolites by UPLC VION IMS QTOF MS/MS analysis in green and red lettuce cultivars. *Sci. Hortic.* **2023**, *315*, 111994. [CrossRef]

61. Materska, M.; Olszówka, K.; Chilczuk, B.; Stochmal, A.; Pecio, Ł.; Pacholczyk-Sienicka, B.; Piacente, S.; Pizza, C.; Masullo, M. Polyphenolic profiles in lettuce (*Lactuca sativa* L.) after $CaCl_2$ treatment and cold storage. *Eur. Food Res. Technol.* **2019**, *245*, 733–744. [CrossRef]
62. Yang, X.; Gil, M.I.; Yang, Q.; Tomás-Barberán, F.A. Bioactive compounds in lettuce: Highlighting the benefits to human health and impacts of preharvest and postharvest practices. *Compr. Rev. Food Sci. Food Saf.* **2021**, *21*, 4–45. [CrossRef]
63. Chen, R.; Wang, Z.; Liu, W.; Ding, Y.; Zhang, Q.; Wang, S. Side lighting of red, blue and green spectral combinations altered the growth, yield and quality of lettuce (*Lactuca sativa* L. cv. "Yidali") in plant factory. *Plants* **2023**, *12*, 4147. [CrossRef]
64. Lee, M.-J.; Son, J.E.; Oh, M.-M. Growth and phenolic compounds of *Lactuca sativa* L. grown in a closed-type plant production system with UV-A, -B, or -C lamp. *J. Sci. Food Agric.* **2014**, *94*, 197–204. [CrossRef]
65. Kitazaki, K.; Fukushima, A.; Nakabayashi, R.; Okazaki, Y.; Kobayashi, M.; Mori, T.; Nishizawa, T.; Reyes-Chin-Wo, S.; Michelmore, R.W.; Saito, K.; et al. Metabolic reprogramming in leaf lettuce grown under different light quality and intensity conditions using narrow-band LEDs. *Sci. Rep.* **2018**, *8*, 7914. [CrossRef]
66. Naznin, M.T.; Lefsrud, M.; Gravel, V.; Azad, M.O.K. Blue light added with red LEDs enhance growth characteristics, pigments content, and antioxidant capacity in lettuce, spinach, kale, basil, and sweet pepper in a controlled environment. *Plants* **2019**, *8*, 93. [CrossRef]
67. Soufi, H.R.; Roosta, H.R.; Stępień, P.; Malekzadeh, K.; Hamidpour, M. Manipulation of light spectrum is an efective tool to regulate biochemical traits and gene expression in lettuce under diferent replacement methods of nutrient solution. *Sci. Rep.* **2023**, *13*, 8600. [CrossRef]
68. Karami, A.; Ansari, N.A.; Hasibi, P. Evaluation of some chemical/biochemical compounds of leaf lettuce (*Lactuca sativa* L.) to the quality of radiant light in floating system. *Sci. Hortic.* **2022**, *304*, 111319. [CrossRef]
69. Ouzounis, T.; Parjikolaei, B.R.; Fretté, X.; Rosenqvist, E.; Ottosen, C.-O. Predawn and high intensity application of supplemental blue light decreases the quantum yield of PSII and enhances the amount of phenolic acids, flavonoids, and pigments in *Lactuca sativa*. *Front. Plant Sci.* **2015**, *6*, 1–14. [CrossRef] [PubMed]
70. Koh, M.X.; Singh, A. Effects of LED treatments on the growth and nutritional content of lettuce (*Lactuca sativa*) in a hydroponic vertical farming system. *Mal. J. Nutr.* **2024**, *30*, 257–270. [CrossRef]
71. Gerhardt, K.E.; Lampi, M.A.; Greenberg, B.M. The effects of far-red light on plant growth and flavonoid accumulation in *Brassica napus* in the presence of ultraviolet B radiation. *J. Photochem. Photobiol.* **2008**, *84*, 1445–1454. [CrossRef]
72. Vrábl, D.; Nezval, J.; Pech, R.; Volná, A.; Mašková, P.; Pleva, J.; Kuzniciusová, N.; Provazová, M.; Štroch, M.; Špunda, V. Light drives and temperature modulates: Variation of phenolic compounds profile in relation to photosynthesis in spring barley. *Int. J. Mol. Sci.* **2023**, *24*, 2427. [CrossRef]

horticulturae

MDPI

Article

Evaluation of Growth, Yield and Bioactive Compounds of Ethiopian Kale (*Brassica carinata* A. Braun) Microgreens under Different LED Light Spectra and Substrates

Ruth Nyambura Maru [1], John Wesonga [1], Hiromu Okazawa [2], Agnes Kavoo [1], Johnstone O. Neondo [3], Dickson Mgangathweni Mazibuko [4,5], Sarvesh Maskey [2] and Francesco Orsini [6,*]

[1] School of Agriculture and Environmental Sciences, Jomo Kenyatta University of Agriculture and Technology, Nairobi 00200, Kenya; nyamburaruthm70@gmail.com (R.N.M.); jwesonga@gmail.com (J.W.); agnes.kavoo@jkuat.ac.ke (A.K.)

[2] Faculty of Regional Environment Science, Tokyo University of Agriculture, Tokyo 156-8502, Japan; h1okazaw@nodai.ac.jp (H.O.); sm207888@nodai.ac.jp (S.M.)

[3] Institute of Biotechnology Research, Jomo Kenyatta University of Agriculture and Technology, Nairobi 00200, Kenya; jneondo@jkuat.ac.ke

[4] Graduate School of Agro-Environmental Science, Tokyo University of Agriculture, Tokyo 156-8502, Japan; dmazibuko@unima.ac.mw

[5] School of Natural and Applied Sciences, University of Malawi, Zomba P.O. Box 280, Malawi

[6] Department of Agricultural and Food Sciences (DISTAL), University of Bologna Alma Mater Studiorum, 40125 Bologna, Italy

* Correspondence: f.orsini@unibo.it

Abstract: Microgreens are innovative vegetable products whose production and consumption are gaining popularity globally thanks to their recognized nutraceutical properties. To date, the effects of lighting conditions and growing substrate on the performances of *Brassica carinata* microgreens (indigenous to Africa) remain underexplored. The present study aimed at providing insights into the influence of different lighting treatments provided by LEDs, namely monochromatic blue (B), red (R), cool white (W) and a combination of three color diodes (B + R + W), and substrates (cocopeat, sand and cocopeat–sand mix (v/v) (1:1)) on the growth, yield and bioactive compounds of *B. carinata* microgreens. Seeds were germinated in dark chambers and cultivated in growth chambers equipped with LED lighting systems for 14 days under a fixed light intensity of 160 ± 2.5 μmol m^{-2} s^{-1} and photoperiod of 12 h d^{-1}. The best performances were associated with the spectrum that combined B + R + W LEDs and with substrate resulting from the cocopeat–sand mix, including the highest yield (19.19 g $plant^{-1}$), plant height (9.94 cm), leaf area (68.11 mm^2) and canopy cover (55.9%). Enhanced carotenoid and flavonoid contents were obtained with B + R + W LEDs, while the B LED increased the total amount of chlorophyll (11,880 mg kg^{-1}). For plants grown under B + R + W LEDs in cocopeat, high nitrate levels were observed. Our results demonstrate that substrate and light environment interact to influence the growth, yield and concentration of bioactive compounds of *B. carinata* microgreens.

Keywords: African indigenous vegetables; healthy diets; light quality; functional foods; nutraceutical; phytochemical

1. Introduction

Microgreens are gaining attention and recognition as a new class of food due to their unique characteristics such as flavor, tenderness, color [1,2] and nutrient density [3]. Microgreens are young plants harvested shortly after the first true leaves emerge, usually between 7 and 21 days after sowing. They are harvested by cutting the stem just above the medium, or over the roots when soilless cultivation is adopted [4]. The harvested shoots are eaten raw, either alone or in mixed salads, or used as a garnish for dishes [2]. The superiority of microgreens over other plant stages of the same plant species is attributed to the germination process from dry seeds to growing plants which involves many metabolic activities and de novo synthesis of nutrients [5]. Microgreens are mainly grown in indoor hydroponic systems using different growing substrates and integrating supplemental lighting [2].

Ethiopian kale (*Brassica carinata* A. Braun) is one of the indigenous African leafy vegetables (ALVs) that are rich in nutrients and health-promoting secondary plant metabolites [6] with potential for use against non-communicable diseases (e.g., cancer). The leaves and seeds of *B. carinata* are rich in nutrients with high concentrations of glucosinolates, especially 2-propenyl glucosinolate (sinigrin), as well as phenolic compounds. *B. carinata* has been reported to reduce afb1-induced DNA damage [7]. *B. carinata* microgreens have been shown to contain flavonoids, phenols, tannins, saponins, alkaloids and terpenoids but not glycosides [8].

Growth substrate is critical in the production of microgreens as it is a major contributor to production costs [9]. Substrates will affect the growth, yield and environmental sustainability of microgreen production [10]. Locally available and inexpensive substrates that have good water-holding capacity and provide aeration are ideal for microgreen production. Those derived from renewable resources and/or those that can be recycled are to be preferred [11]. According to several authors, peat and peat-based mixes represent the most used growing substrates for the production of microgreens because of their good physicochemical properties, but coconut coir (also referred to as cocopeat) is common as well [10–13]. However, these substrates are quite expensive, and when they are not locally available, they require importation. The use of peat poses environmental concern due to its continuous extraction which contributes to the emission of carbon dioxide. On the other hand, cocopeat (derived from the coconut processing industry and its discarded fibers) is a renewable resource and could be used as an alternative to peat [11]. However, it can also be an expensive material and requires treatment for the removal of its concentrated salts before use, which increases costs. Accordingly, the exploration of alternative substrates or additives enabling a reduction in the amount of cocopeat needed may lead to the identification of sustainable, cheaper and renewable growing substrates for microgreens.

Light is another major factor in plant growth and influences the development and production of phytochemical and bioactive compounds [14]. Light quality (its composition in the spectral regions), quantity (intensity), direction and duration (photoperiod) are vital components in microgreen production. In plants such as lettuce, high light intensity results in the production of high amounts of phenolics, anthocyanins and carotenoids, among others, which could be beneficial to human health [15]. Regarding the effects of light on microgreen growth, research results vary across studies and for different vegetable species. For example, it has been found [16] that growth and phytochemical accumulation in *Brassica juncea* and *Brassica napus* using different R and B ratios differed depending on the species. The chlorophyll, carotenoid and soluble protein contents depended on photoperiod [17] in other *Brassica* species. Artificial light sources such as light-emitting diodes (LEDs) have been used as a source of supplemental lighting in controlled environments such as indoor spaces and greenhouses in the production of microgreens [17]. B, R and W LEDs used alone or in combination have been used to produce high-quality microgreens with various nutritional benefits [17]. However, the influence of LED grow lights on *B. carinata* microgreens is still unknown. In addition, it is unclear how plants respond to LEDs in combination with substrates since most of the previous studies assessed either LEDs or substrates alone. Therefore, this study aimed to investigate the influence of different LED lights and growing substrates on the growth, yield and phytochemical content of *B. carinata* microgreens. The results obtained from this study provide a baseline towards an understanding of the influence of the interactions between the substrate and LEDs on quality traits and bioactive accumulation of *B. carinata* microgreens.

2. Materials and Methods

2.1. Experimental Materials and Design

The experiment was conducted in a controlled environment in a locally fabricated walk-in growth chamber at Tokyo University of Agriculture between April and October 2023. The chamber was divided into four compartments using black opaque fabric to prevent light interference. Each compartment measured 100 cm by 100 cm. In each compartment, an LED fixture was placed 50 cm above the surface of the substrate. Ethiopian kale (*Brassica carinata*) seeds used in the study were sourced from a commercial vendor in Kenya. A phytosanitary certificate allowing entry of seeds to Japan was obtained from the Kenya Plant Health Inspectorate Service (KEPHIS). *B. carinata* was identified by a taxonomist at JKUAT GoK laboratories, and a voucher specimen (JMW/JKUAT/BOT/H001) is maintained at the JKUAT herbarium.

2.2. Growing Environment

Seeds of *B. carinata* were sown and grown using three substrates under four LED light spectra in a factorial experiment. The light spectra used were B (with a peak at 450 nm), R (with a peak at 650 nm), W, and B + R + W (managed by having one light with three diodes; B, R and W combined in the ratio of 1:1:1) LEDs in each compartment. The three substrate types (cocopeat, sand and a mix of cocopeat and sand) and one LED light were placed in each compartment to give a split plot design with light being the main plot factor and substrate the subplot factor. There were three replicates for light spectra and twelve for the substrate. The lights had a fixed light intensity of 160 ± 2.5 μmol m^{-2} s^{-1}, and a 12 h photoperiod was applied. The air temperature in the walk-in growth chamber was set and maintained at 26 ± 2 °C while relative humidity was maintained at approximately 60% during the experimental period. Temperature and relative humidity were monitored using a data logger (HOBO, OnSet Data Logging Solutions, Bourne, MA, USA). There were no nutrients supplied throughout the growing period. Irrigation was performed using capillary wick technology [18].

2.3. Growth Measurements

Growth was assessed at the end of the experiment (14 days after sowing) in terms of height, leaf area and canopy cover. Ten plants were randomly selected from each subplot and harvested for height and leaf area measurements. The plants were harvested by cutting above the substrate. The individual height of each plant was measured using a ruler. Leaf area values were estimated using ImageJ v.1.5 software [19]. Leaves from the ten selected plants were spread on a clean white sheet of paper, and photographs were taken against a ruler as a reference. Additionally, a square paper of known area (2×2 mm) was included for verification of the measurements obtained. Canopy cover was estimated using Canopeo software (version 1.1.7) [20]. This was done by taking aerial photographs of all the above-ground plant materials. To achieve uniformity in all the photographs, a 30 cm distance from the camera to the treatment was maintained. The photographs were processed with Canopeo software, and canopy cover was calculated as a percentage of the total surface area.

2.4. Yield and Biomass Analysis

Yield and dry biomass were obtained by weighing the whole harvested microgreen shoots 14 days after sowing (DAS). All above-ground parts including the leaves, stems and cotyledons were harvested by cutting them at the base, and fresh weight (yield) and dry biomass (after freeze drying at -41 °C for 24 h) were weighed using a weighing balance. The samples were further powdered and used for phytochemical analysis.

2.5. Phytochemical Analysis

2.5.1. Flavonoids

The estimation of total flavonoids in the sample was performed using the aluminum chloride method. Rutin was used as the standard [21]. The sample (0.1 mL) and standards were prepared in triplicates, vortexed and incubated for 5 min at room temperature. Afterward, 10% aluminum chloride was added, vortexed and incubated for 6 min at room temperature. The absorbance was measured against the blank at 510 nm using a spectrophotometer (Shimadzu model UV-1601 PC, Kyoto, Japan). The standard curve was plotted, and the total amount of flavonoids in the sample was expressed as mg of rutin equivalent (RE)/g of dry weight of the sample. Equation (1) was used to compute flavonoids (mg/100 g) from absorbance.

$$Flavonoids = 0.0001 * \frac{(A_s - A_b)}{0.0018 * W} * D \quad (1)$$

where A_b = absorbance of the blank, A_s = absorbance of the sample, D = dilution factor (30), W = weight of the sample (g), 0.0018 is the slope of the standard curve and 0.0001 is the factor for conversion to mg/100 g.

2.5.2. Carotenoids

Total carotenoids were extracted using acetone and analyzed using column chromatography (Rodriguez- Amaya and Kimura, 2004; AOAC, 1996) and a UV spectrophotometer (Shimadzu model UV-1601 PC, Kyoto, Japan) [22]. Approximately 0.08 g of dried sample was weighed and ground in a mortar containing 10 mL of acetone, and extraction was repeated until the residue turned colorless. Then, 25 mL of the extract was evaporated to dryness using a rotary evaporator; the residue was dissolved in 10 mL of petroleum ether, and the solution was introduced into a chromatographic column. Absorbance was read at 450 nm in a UV-Vis spectrophotometer. Equation (2) was used to calculate carotenoids (mg/100 g) from absorbance.

$$Carotenoids = 0.001 * \frac{A}{2592 * W} \quad (2)$$

where A = absorbance, W = weight of the sample (g) and 2592 is the absorption coefficient of β-carotene in petroleum ether.

2.5.3. Nitrates

The nitrate content in the test samples was determined by the calorimetric method using salicylic acid [22]. Samples of 0.3 g dry *B. carinata* were weighed and put in a test tube. Hot (90–95 °C) distilled water measuring 10 mL was added. The closed tubes were placed in a water bath at 80 °C and shaken for 30 min. The samples were then cooled and centrifuged at 4500 rpm. Chlorophyll in the sample was removed by adding 0.5 g $MgCO_3$ to the supernatant and centrifuging it again. The supernatant containing the nitrate extract was then treated with NaOH and a combination of salicylic acid and H_2SO_4. Nitrate standards were prepared using a sodium nitrate calibration curve. Absorbance was read at 410 nm in a UV-Vis spectrophotometer (Shimadzu model UV-1601 PC, Kyoto, Japan). The nitrate concentration was expressed on a dry weight basis (mg/100 g DW). Equation (3) was used to calculate nitrates (mg/100 g) from absorbance.

$$Nitrates = 0.1 * \frac{A_s - A_b}{0.0078 * W} * D \quad (3)$$

where A_b = absorbance of the blank, A_s = absorbance of the sample, D = dilution factor (30), W = weight of the sample (g), 0.0078 is the slope of the standard curve and 0.1 is the factor for conversion to mg/100 g.

2.5.4. Chlorophyll

Chlorophyll was extracted using acetone and analyzed using column chromatography (Rodriguez- Amaya and Kimura, 2004; AOAC, 1996) and a UV spectrophotometer (Shimadzu model UV-1601 PC, Kyoto, Japan) [23]. Approximately 0.08 g of a dry sample was weighed and ground in a mortar containing 10 mL acetone. The extraction was repeated until the residue turned colorless. An aliquot of 25 mL of the extract was evaporated to dryness using a rotary evaporator, and the residue was dissolved in 10 mL of petroleum ether. The solution was introduced into a chromatographic column, and absorbance was read at 645 nm and 663 nm in a UV-Vis spectrophotometer. Chlorophyll content was determined by computation from the absorbance using Equation (4).

$$\text{Total Chlorophyll (mg/100 g)} ChlA = 0.1 * (7.12 * A_{663} + 16.8 * A_{645}) * \frac{D}{W} \quad (4)$$

where A = absorbance at indicated wavelength (645 or 663), D = dilution factor (25), W = weight of the sample (g).

2.6. *Statistical Analysis*

Statistical analysis was performed using GenStat software, version 12.1. Growth measurements (leaf area and plant height) were analyzed based on the individual values of the 10 sampled plants from each subplot, while canopy cover, yield and dry weight were analyzed at the subplot level. All data were subjected to two-way ANOVA, and significant differences among means were determined by Tukey's multiple comparison test at $p < 0.05$.

3. Results

3.1. *Effect of LED Light and Substrate on Height, Leaf Area and Canopy Cover*

The results from the ANOVA indicated that the interaction between substrates and LED light treatments did not have a significant effect on plant morphological parameters. However, height differed significantly in response to both different substrates and LED light treatments (Table 1). The microgreens grown using monochromatic R were significantly shorter compared to those grown using other LEDs. More specifically, microgreens grown under monochromatic R were 8% shorter compared to those under monochromatic B. Microgreens grown under B, W and B + R + W did not differ significantly in height. Microgreens grown in either sand alone or cocopeat–sand mix were significantly taller ($F_{(3,108)} = 3.92$, $p < 0.001$) than those grown in cocopeat alone. Microgreens in cocopeat were shorter than those in sand and cocopeat–sand mix by 8%.

Table 1. Effect of LED light and substrate on height, leaf area and canopy cover.

Treatment	Height (cm)	Leaf Area (cm^2)	Canopy Cover (%)
LED Lights			
B	9.9 (0.16) ^{a}	57.62 (1.40) ^{c}	50.68 (4.51) ^{a}
R	9.2 (0.16) ^{b}	57.36 (1.46) ^{c}	44.45 (2.66) ^{b}
W	9.7 (0.18) ^{a}	63.43 (1.56) ^{b}	56.39 (2.85) ^{a}
B + R + W	9.8 (0.11) ^{a}	68.11 (1.96) ^{a}	55.15 (2.76) ^{a}
P	0.011	<0.001	<0.001
$LSD_{0.05}$	0.39	4.32	5.87
F Value	F (3,108) = 3.92	F (3,108) = 11.18	F (3,33) = 13.12
Substrates			
Sand	9.8 (0.13) ^{a}	60.0 (1.36) ^{b}	56.0 (3.26) ^{a}
Cocopeat	9.2 (0.12) ^{b}	59.1 (1.44) ^{c}	47.1 (2.07) ^{b}

Table 1. *Cont.*

Treatment	Height (cm)	Leaf Area (cm^2)	Canopy Cover (%)
Sand + Cocopeat	9.9 (0.14) ^{a}	65.6 (1.66) ^{a}	51.9 (3.394) ^{ab}
P	<0.001	0.001	0.005
$LSD_{0.05}$	0.34	3.74	5.08
F Value	F (3,108) = 11.86	F (3,108) = 7.28	F (3,33) = 12.02

Mean separation by the Tukey test at the 5% significance level. Values in brackets are standard errors of means. Values without a letter in common in a column within a factor are significantly different ($p < 0.05$).

Both substrate and LED treatment had a significant effect on leaf area (Table 1). Microgreens grown under B + R + W had significantly higher leaf area (68.11 mm^2) compared to microgreens grown under W (63.43 mm^2) and both under monochromatic B and R (57.62 mm^2 and 57.36 mm^2). Leaf area in the cocopeat–sand mix was significantly higher by 22% (65.75 mm^2) compared to microgreens produced using cocopeat alone (59.12 mm^2).

Both the growing media and LED treatments had a significant effect on canopy cover. Canopy cover values under B + R + W treatment were significantly higher (55.15%) than those produced in monochromatic R (44.45%). On the other hand, microgreens in sand had a significantly higher canopy cover (55.95%) compared to those in cocopeat (47.11%).

3.2. Effect of LED Light and Substrate on Yield and Dry Weight

The results from the ANOVA indicated that the interaction between substrates and LED light treatments was not significant. Similarly, no significant differences in yield were noted among LEDs. Regarding the effects of LEDs on dry weight, significant differences were noted between monochromatic R and all other LEDs. No differences were noted between B + R + W, W and monochromatic B LEDs (Table 2). Dry weight among the substrates ranged from about 1.0 g (cocopeat) to 1.3 g (sand). Regarding the yield, significant differences were found among the substrates but not the LED lights. The microgreen yield in sand and cocopeat–sand mix differed significantly from cocopeat alone ($p < 0.05$). Sand alone had a yield that was not significantly different from the cocopeat–sand mix.

Table 2. Effect of LED light and substrate on yield and dry weight of *Brassica carinata*.

Treatment	Yield (g)	Dry Weight (g)
LED Lights		
B	17.9 (1.94) ^{a}	1.2 (0.11) ^{ab}
R	16.0 (0.86) ^{a}	1.0 (0.06) ^{c}
W	18.8 (2.36) ^{a}	1.3 (0.17) ^{a}
B + R + W	19.5 (2.22) ^{a}	1.2 (0.10) ^{ab}
P	0.339	0.053
$LSD_{0.05}$	3.73	0.23
F (3,33)	1.28	5.38
Substrates		
Cocopeat	15.2 (1.75) ^{b}	1.0 (0.08) ^{a}
Sand	19.2 (1.54) ^{a}	1.3 (0.10) ^{b}
Cocopeat + sand	19.8 (1.76) ^{a}	1.2 (0.13) ^{ab}
P	0.013	0.016
$LSD_{0.05}$	3.23	0.20
F (2,33)	11.29	13.14

Mean separation by the Tukey test at the 5% significance level. Values in brackets are standard errors of means. Values without a letter in common in a column within a factor are significantly different ($p < 0.05$).

3.3. Effect of LED Light and Substrate on Phytochemical Content

Carotenoids: There were significant differences for carotenoids among LED lights (F (3,24) = 1270.56250, $p < 0.001$), substrates (F (2,24) = 50.24509, $p < 0.001$) and their interactions (F (6,24) = 1814.12864, $p < 0.001$). Microgreens under B + R + W light in cocopeat had the highest carotenoid content (644.4 mg kg^{-1} DW). Under monochromatic B and R, more carotenoids were found in sand compared to cocopeat and in the cocopeat–sand mix. Under W and B + R + W, cocopeat had higher carotenoids relative to those in sand alone and the cocopeat–sand mix (Figure 1A).

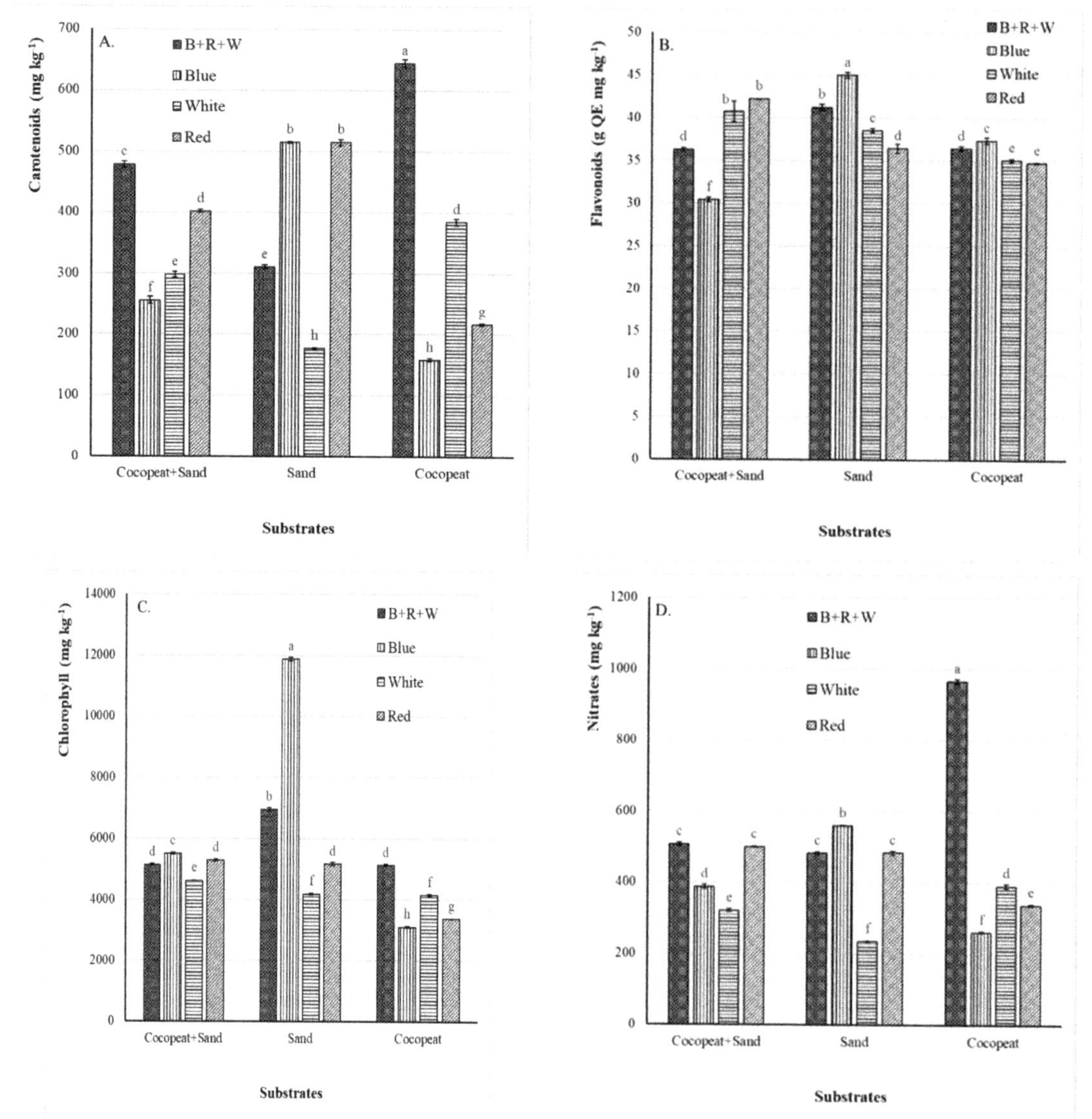

Figure 1. Effect of LED light on phytochemicals ((**A**) carotenoids, (**B**) flavonoids, (**C**) chlorophyll and (**D**) nitrates) under different substrates (cocopeat + sand, sand and cocopeat). Bars represent standard errors of means. Different letters indicate significant differences at $p < 0.05$.

Flavonoids: Flavonoids similarly showed significant differences among LED lights (F (3,24) = 100.7731207, $p < 0.001$), substrates (F (2,24) = 98.2264237, $p < 0.001$) and interactions (F (6,24) = 105.0911162, $p < 0.001$). Monochromatic B and B + R + W had higher flavonoid contents in sand than in cocopeat alone as well as in the cocopeat–sand mix. Under monochromatic B in sand, flavonoids were 16.8% higher than in cocopeat and 32.4% higher than in the cocopeat–sand mix. For B + R + W in sand, flavonoids were 11.5% higher than in cocopeat and 12.0% higher than in the cocopeat–sand mix. Monochromatic R had higher flavonoid contents in sand alone than in cocopeat alone by 4.6% but lower flavonoid contents in sand alone than in the cocopeat–sand mix by 15.7%. Similarly, under W, sand had 9.8% more flavonoids than cocopeat alone but less flavonoids by 6.3% than in the cocopeat–sand mix (Figure 1B).

Total Chlorophyll: Total chlorophyll content differed significantly among LED light (F (3,24) = 2690.467, $p < 0.001$) and substrates (F (2,24) = 6647.472, $p < 0.001$). In addition, the interaction between substrate and lights was significant (F (6,24) = 2957.422, $p < 0.001$). Except for W, total chlorophyll content under monochromatic B, R and B + R + W was higher in sand compared to cocopeat. The highest total chlorophyll content (11,880 mg kg^{-1}) was observed under monochromatic B in sand while the lowest (3100 mg kg^{-1}) was under monochromatic B in cocopeat, a reduction of 73.9%. The chlorophyll content under B + R + W was higher in sand by 26.1% compared to B + R + W in cocopeat substrate, while for monochromatic R it was 34.5% higher in sand than in cocopeat (Figure 1C).

Nitrates: There were significant differences for nitrates among LED lights (F (3,24) = 1696.0669, $p < 0.001$), substrates (F (2,24) = 110.4731, $p < 0.001$) and interactions (F (6,24) = 983.5374, $p < 0.001$). Microgreens under B + R + W in cocopeat had extremely higher nitrates (966.2 mg kg^{-1} DW) compared to other treatments. Except under W and B + R + W, nitrate contents were higher in sand than in cocopeat. Under monochromatic B, nitrate content in sand was higher by 53.4% compared to cocopeat, while for monochromatic R it was 30.3% higher compared to cocopeat (Figure 1D).

4. Discussion

4.1. Effect of LED Light and Substrate on Height, Leaf Area and Canopy Cover

In recent years, several scientific reports addressed the role of light in stimulating specific plant photoreceptors, allowing plants to be manipulated to produce desirable phytochemicals and nutrients. Lighting systems for indoor farming can therefore be designed to maximize growth, control morphology and optimize yield [24]. This study established that *B. carinata* grown under monochromatic B were significantly taller compared to those grown using a monochromatic R source. Such a result is surprising since it is commonly acknowledged that monochromatic B decreases hypocotyl elongation. For example, the stem length of baby lettuce decreased by 33% when a supplemental B treatment was provided [25]. Furthermore, lettuce grown using an increased ratio of red radiation had an increased shoot height and shoot/root ratio compared to that grown using a blue light source [26]. Inconsistencies in results on the effect of different spectral regions across plant species and phenological stages have been acknowledged as a gray area requiring further research [27]. Monochromatic B and B in combination with far-red light were found to increase mustard (*Brassica juncea*) and arugula (*Eruca sativa*) microgreen elongation (as defined as plant height) [28]. The results presented herein suggest that sand alone or the cocopeat–sand mix had better growth than cocopeat, indicating that these substrates provided a better growing environment. This could be due to the physiochemical properties such as low water retention capacity allowing good aeration as compared to cocopeat which could have retained excessive moisture potentially leading to anoxia conditions. Similarly, ref. [29] reported that using cocopeat-based mixes with other coarser materials such as burnt rice hull improved the growth of *Celosia cristata*.

The present research also found that B + R + W and white light resulted in better yield performances than monochromatic red or blue. This was previously associated with synergistic effects of the different spectral regions. Red light combined with varying ratios of blue has been reported to enhance the growth characteristics of lettuce, spinach, kale, basil and sweet pepper compared to red light alone [27]. Similarly, leaf area among other growth parameters of lettuce increased with an increase in the proportion of red light in combination with blue [30]. For leaf area and canopy cover, B + R + W LED in the ratio of 1:1:1 and cocopeat–sand mix enhanced the leaf growth of *B. carinata* microgreens. In this study, a cocopeat-based substrate (cocopeat–sand mix) showed increased leaf area of *B. carinata* microgreens. Similar results showed that cocopeat-based substrate increased plant growth, yield, nutritional, biochemical composition and antioxidant activity of various microgreen species [31]. These positive effects were attributed to enhanced nutrient acquisition, water retention and root development.

4.2. Effect of LED Light and Substrate on Yield and Biomass

Yield is an important parameter in microgreen production because microgreens are sold on a fresh weight basis [32]. One of the limiting factors in microgreen production continues to be low yield due to various elements [33]. Microgreen yield can be affected by seed quality [34], growing media [35], and light quality and intensity [36], among other factors. In our study, both substrate and light quality significantly affected the yield and dry matter accumulation for *B. carinata*. Notably, the yield of microgreens varied across the different light spectra used, being highest under W. The results obtained are similar to those reported in the literature where fresh weight, which was used as a measure of yield, responded differently in plants grown using different light spectra. On the other hand, in the experiments presented herein, the increase in yield also depended on the substrate used. For *B. carinata* microgreens, a higher yield was recorded in sand alone or in the cocopeat–sand mix. In previous research comparing different substrates, the yield of sunflower microgreens was significantly affected by the type of substrate used [12]. Dry mass yield is a good indicator of crop productivity and photosynthetic efficiency [37] in microgreens. In our study, the highest dry matter accumulation was in microgreens grown using W. Conversely, microgreens grown in cocopeat using R had the lowest dry matter accumulation. Therefore, a significant effect resulting from substrate was noticed in our trial indicating the importance of substrate and lighting on the yield of *B. carinata* microgreens. Other studies on dry matter assessment of microgreens seem to indicate interspecies variability. For example, ref. [38] found differences in dry mass accumulation within W and R for broccoli, cabbage and radish microgreens.

4.3. Effect of LED Light and Substrate on Phytochemical Content

4.3.1. Carotenoids

Microgreens grown using B + R + W and in cocopeat had higher amounts of carotenoids. This is consistent with previous observations on the effect of light treatments on carotenoid accumulation in plants, where the R + B combination increased carotenoid accumulation in lettuce, spinach and pepper [27], while in kale and basil, carotenoid accumulation was increased under monochromatic B. Earlier studies also demonstrated that R/B combinations positively influenced carotenoid accumulation in lettuce [26]. Conversely, however, enzymatic activities involved in the metabolic pathways of carotenoid pigments were largely increased under monochromatic B, resulting in higher carotenoid accumulation in Chinese cabbage [25]. For Brassica sprouts, carotenoid transcription of biosynthesis genes, namely PSY, βLCY and βOHASE1, was enhanced by a higher B percentage compared to R [39], therefore increasing the carotenoid accumulation in the sprouts. Similar results were associated with a combined spectrum (resulting from the integration of blue, red and amber diodes) that enhanced the transcription of a gene involved in carotenoid biosynthesis (PSY), leading to higher carotenoid accumulation in various *Brassica* plants [40]. In the present study, the results are consistent, as the treatment B + R + W often presented higher amounts of carotenoids. Such findings corroborate the concept that combined light spectra are superior to monochromatic B or R light supply. On the sand substrate, carotenoids were higher under monochromatic R and monochromatic B. We hypothesize that these two spectra may have boosted photosynthesis, and therefore leaf transpiration, a scenario that could have led to drought stress ultimately inducing carotenoid biosynthesis and accumulation. Further studies on water retention in sand (compared to other substrates) and how it influences carotenoid accumulation are needed to provide a conclusive explanation.

4.3.2. Flavonoids

Flavonoids are important plant compounds that are produced as a result of stress to prevent DNA damage [41]. Light quality triggers different transcriptional genes that are used for the biosynthesis of flavonoids and could cause differences in the levels of flavonoid accumulation in plants [42]. In the current study, both monochromatic B and B + R + W enhanced the accumulation of flavonoid content in *B. carinata* microgreens grown on sand and cocopeat substrates, just as monochromatic R and W did in those grown on the cocopeat–sand mix. An earlier study indicates that monochromatic B highly influenced the accumulation of flavonoids by modulating the phenylpropanoid pathway, a pathway in which most plant secondary metabolites are synthesized [43]. The adoption of R/B combinations at low intensities was formerly found to increase the accumulation of flavonoids in lettuce [44]. This could have resulted from the influence of different R/B ratios on the phenylalanine ammonia lyase (PAL), chalcone synthase (CHS) and other enzymes involved in the flavonoid biosynthesis, ultimately leading to the accumulation of flavonoids [45]. For *Scrophularia kakudensis*, ref. [46] reported that flavonoid accumulation was higher in monochromatic B and R than in W. Furthermore, these effects of light were also influenced by the substrate used (although different from those adopted in this study). While monochromatic B enhanced flavonoid accumulation in cocopeat and sand, R and W enhanced the same phytochemical in cocopeat–sand mix. These subtle differences point toward a substrate–light interaction, as also previously hypothesized [47].

4.3.3. Chlorophyll

Besides its role as photosynthetic pigment, total chlorophyll content is also one of the key indicators of quality in vegetables, as the green color indicates freshness, which leads to product acceptability or rejection by consumers. In microgreens, vivid and intense colors are particularly appreciated and tend to influence consumer preference [48]. Chlorophylls represent part of the light-harvesting complex and therefore play a significant role in photosynthesis. As reported in the literature, significant genotypic variations were observed for chlorophyll content in microgreens, with their level also being highly dependent on the lighting conditions [2,49]. In the present study, monochromatic B increased chlorophyll biosynthesis and accumulation in plant tissues. The role of B in boosting chlorophyll accumulation was evidenced in previous studies thanks to both increased photosynthetic efficiency and a concentration factor (e.g., as a consequence of lower leaf extension as compared with spectra with a higher R fraction) [49,50]. Blue light improves the expression of genes such as MgCH, GluTR and FeCH, involved in chlorophyll biosynthesis, while red light may lead to a reduction in 5-aminolevulinic acid, a tetrapyrrole precursor required for chlorophyll synthesis [51]. Furthermore, when a monochromatic R, a monochromatic B and a combination of R and B ratio (with R/B = 6) were alternatively applied to Chinese cabbage, a lower chlorophyll content was associated with monochromatic R, as a result of reductions in the synthesis of chlorophyll precursors including ALA, Proto IX, Mg-Proto IX and protochlorophyllide [51]. In an analysis of the effect of the tested substrates, higher chlorophyll content was observed in *B. carinata* grown using sand compared to those grown using cocopeat, which could have contributed to the higher yield observed for the same treatments. The use of sand for microgreen production is not common. Elsewhere, the use of sand as a substrate is reported as an additive to another substrate [35]. The effects of sand as a microgreen substrate may thus require some further investigation, e.g., by using different mixture combinations.

4.3.4. Nitrates

Nitrates are among the main compounds that may negatively affect food safety. Vegetables can accumulate nitrates which are associated with harmful effects on human health, with toxic effects of methemoglobinemia and the possibility of causing an endogenous formation of carcinogenic N-nitroso compounds. Accumulation of nitrates in vegetables may vary depending on the species, the substrate used for production or the stage of plant growth at harvest. Several studies reported that microgreens recorded lower levels of nitrates compared to their mature counterparts [52,53]; therefore, microgreens are commonly considered safe to consume within a healthy diet. As reported earlier, lighting conditions can influence the accumulation of nitrates in vegetables, thus affecting their quality [52]. Regarding the substrates, the result contrasts with what was reported by two studies that evaluated microgreens grown on different substrates and found significantly lower concentrations of nitrates in microgreens grown using cocopeat substrate [2,10]. In our case, cocopeat showed a higher nitrate content compared to the other substrates. This could possibly be because of the differences in the lighting sources during cultivation. Notably, no such results have been reported for microgreens, and this assumption could be further investigated.

4.4. Interactive Effects of Light and Substrate on Phytochemicals

The current study reports some significant interactions between lighting treatments and substrate composition. For example, the interaction between cocopeat and B + R + W and the interaction between sand and B enhanced the production of all phytochemicals investigated here. Further, the cocopeat–sand mix and R exhibit a strong interaction except in the accumulation of carotenoids. This suggests that the effect of light was dependent on the substrate. No such results have been previously reported for microgreens. Possibly, the cause of these interactive effects may be associated with either reflective or absorptive attributes of the substrates. This could be better studied, e.g., by measuring the light intensity in a sealed box with light turned on and only one substrate at a time. The incident radiation could be absorbed or reflected depending on the substrate, leading to differences in lighting conditions experienced by the microgreens. Sand for instance is known to have the capacity to cause light scattering [54], while cocopeat due to its color and texture would be expected to absorb light. The light absorption and reflection are further affected by moisture content, which varies across different substrates. It will be good to test this assumption to understand the mechanisms involved in the noted interactive effects.

5. Conclusions

This study aimed to investigate the influence of LED light quality and different substrates on the growth, yield and accumulation of selected bioactive compounds of *Brassica carinata* microgreens. Our results demonstrate that substrate and light environment interact to influence the growth, yield and concentration of bioactive compounds of *B. carinata* microgreens, enabling improved cultivation strategies. A combination of various light spectra (B + R + W) offers a better chance of obtaining higher yields and better-quality *B. carinata* microgreens. A combination of cocopeat with sand is a viable alternative to cocopeat considering the additional benefits of lower costs and ubiquitous availability of sand. Further studies are needed to elucidate media-related physical and biochemical dynamics that could potentially influence how different lighting systems lead to the varied accumulation of phytochemicals. Since *B. carinata* microgreens have not been extensively studied (compared to other species), such exploratory studies should first focus on the most commonly studied microgreen taxa. Such an understanding would help to describe the specific influence of the interactions between substrates and LED ratios on the quality traits (nutritional value, color, texture, taste, etc.) of microgreens.

Author Contributions: Conceptualization, R.N.M. and J.W.; methodology, R.N.M. and J.W.; validation, A.K. and J.O.N.; formal analysis, R.N.M., J.W. and D.M.M.; investigation, R.N.M.; D.M.M., S.M. and H.O.; resources, F.O., H.O. and J.W.; data curation, R.N.M.; writing—original draft preparation, R.N.M., J.W., D.M.M. and H.O.; writing—review and editing, R.N.M., J.W., H.O., A.K., J.O.N., D.M.M., S.M. and F.O.; supervision, J.W., H.O., A.K. and J.O.N.; project administration, J.W. and H.O.; funding acquisition, F.O., H.O. and J.W. All authors have read and agreed to the published version of the manuscript.

Funding: The research leading to this publication has received funding from the European Union's Horizon Europe research and innovation program under grant agreement No. 101083790 (project InCitis-Food). Part of the work was supported by the Africa ai Japan project under the JICA capacity-building program at JKUAT.

Data Availability Statement: The original data presented in the study are openly available in AMS Acta repository, at https://doi.org/10.6092/unibo/amsacta/7688.

Acknowledgments: The authors acknowledge the Inter-University Exchange Program of Tokyo University of Agriculture and Jomo Kenyatta University of Agriculture and Technology (JKUAT) and further mention the JASSO scholarship and Tokyo University of Agriculture for logistical support.

Conflicts of Interest: The authors declare that the research was conducted in the absence of any commercial or financial relationships that could be construed as a potential conflict of interest.

References

1. Appolloni, E.; Pennisi, G.; Zauli, I.; Carotti, L.; Paucek, I.; Quaini, S. Beyond vegetables: Effects of indoor LED light on specialized metabolite biosynthesis in medicinal and aromatic plants, edible flowers, and microgreens. *J. Sci. Food Agric.* **2022**, *102*, 472–487. [CrossRef]
2. Bulgari, R.; Negri, M.; Santoro, P.; Ferrante, A. Quality evaluation of indoor-grown microgreens cultivated on three different substrates. *Horticulturae* **2021**, *7*, 96. [CrossRef]
3. Zhang, Y.; Xiao, Z.; Ager, E.; Kong, L.; Tan, L. Nutritional quality and health benefits of microgreens, a crop of modern agriculture. *J. Future Foods* **2021**, *1*, 58–66. [CrossRef]
4. Verlinden, S. Microgreens: Definitions, product types, and production practices. *Hortic. Res.* **2020**, 85–124. [CrossRef]
5. Loedolff, B.; Brooks, J.; Stander, M.; Peters, S.; Kossmann, J. High light bio-fortification stimulates de novo synthesis of resveratrol in *Diplotaxis tenuifolia* (wild rocket) micro-greens. *J. Funct. Food Health Dis.* **2017**, *7*, 859–872. [CrossRef]
6. Neugart, S.; Baldermann, S.; Ngwene, B.; Wesonga, J.; Schreiner, M. Indigenous leafy vegetables of Eastern Africa—A source of extraordinary secondary plant metabolites. *Food Res. Int.* **2017**, *1*, 411–422. [CrossRef]
7. Odongo, G.A.; Schlotz, N.; Herz, C.; Hanschen, F.S.; Baldermann, S.; Neugart, S. The role of plant processing for the cancer preventive potential of *Ethiopian kale* (*Brassica carinata*). *Food Nutr. Res.* **2017**, *1*, 31–61. [CrossRef]
8. Nakakaawa, L.; Gbala, I.D.; Cheseto, X.; Bargul, J.L.; Wesonga, J.M. Oral acute, sub-acute toxicity and phytochemical profile of *Brassica carinata* A. Braun microgreens ethanolic extract in Wistar rats. *J. Ethnopharmacol.* **2023**, *6*, 305. [CrossRef]
9. Chen, H.; Tong, X.; Tan, L.; Kong, L. Consumers' Acceptability and Perceptions toward the Consumption of Hydroponically and Soil Grown Broccoli Microgreens. *J. Agric. Food Res.* **2020**, *2*, 100051. [CrossRef]
10. Poudel, P.; Duenas, A.E.K.; Di Gioia, F. Organic waste compost and spent mushroom compost as potential growing media components for the sustainable production of microgreens. *Front. Plant Sci.* **2023**, *14*, 1229157. [CrossRef]
11. Di Gioia, F.; De Bellis, P.; Mininni, C.; Santamaria, P.; Serio, F. Physicochemical, agronomical and microbiological evaluation of alternative growing media for the production of rapini (*Brassica rapa* L.) microgreens. *J. Sci. Food Agric.* **2017**, *97*, 1212–1219. [CrossRef] [PubMed]
12. Thepsilvisut, O.; Sukree, N.; Chutimanukul, P.; Athinuwat, D.; Chuaboon, W.; Poomipan, P. Efficacy of Agricultural and Food Wastes as the Growing Media for Sunflower and Water Spinach Microgreens Production. *Horticulturae* **2023**, *9*, 876. [CrossRef]
13. Kyriacou, M.C.; El-Nakhel, C.; Pannico, A.; Graziani, G.; Soteriou, G.A.; Giordano, M. Phenolic constitution, phytochemical and macronutrient content in three species of microgreens as modulated by natural fiber and synthetic substrates. *Antioxidants* **2020**, *9*, 252. [CrossRef] [PubMed]
14. Ying, Q. Exploration on Using Light-Emitting Diode Spectra to Improve the Quality and Yield of Microgreens in Controlled Environments. Ph.D. Thesis, The University of Guelph, Guelph, ON, Canada, 2020.
15. Craver, J.K.; Gerovac, J.R.; Lopez, R.G.; Kopsell, D.A. Light intensity and light quality from sole-source light-emitting diodes impact phytochemical concentrations within brassica microgreens. *J. Am. Soc. Hortic.* **2017**, *142*, 3–12. [CrossRef]
16. Brazaityte, A.; Viršile, A.; Jankauskiene, J.; Sakalauskiene, S.; Samuoliene, G.; Sirtautas, R. Effect of supplemental UV-A irradiation in solid-state lighting on the growth and phytochemical content of microgreens. *Int. Agrophys.* **2015**, *29*, 13–22. [CrossRef]
17. Liu, K.; Gao, M.; Jiang, H.; Ou, S.; Li, X.; He, R. Light Intensity and Photoperiod Affect Growth and Nutritional Quality of Brassica Microgreens. *Molecules* **2022**, *27*, 883. [CrossRef] [PubMed]
18. Semananda, N.P.K.; Ward, J.D.; Myers, B.R. A Semi-Systematic Review of Capillary Irrigation: The Benefits, Limitations, and Opportunities. *Horticulturae* **2018**, *4*, 23. [CrossRef]
19. Schneider, C.A.; Rasband, W.S.; Eliceiri, K.W. NIH Image to ImageJ: 25 years of image analysis. *Nat. Methods* **2012**, *7*, 671–675. [CrossRef] [PubMed]
20. Patrignani, A.; Ochsner, T.E. Canopeo: A powerful new tool for measuring fractional green canopy cover. *Agron. J.* **2015**, *107*, 2312–2320. [CrossRef]
21. Baba, S.A.; Malik, S.A. Determination of total phenolic and flavonoid content, antimicrobial and antioxidant activity of a root extract of *Arisaema jacquemontii* Blume. *J. Taibah Univ. Sci.* **2015**, *9*, 449–454. [CrossRef]

22. Nyonje, W.A.; Makokha, A.O.; Abukutsa-Onyango, M.O. Anti-Nutrient, Phytochemical and Antiradical Evaluation of 10 Amaranth (*Amaranthus* spp.) Varieties Before and After Flowering. *J. Agric. Sci.* **2014**, *6*, 68. [CrossRef]
23. Ritchie, R.J. Universal chlorophyll equations for estimating chlorophylls a, b, c, and d and total chlorophylls in natural assemblages of photosynthetic organisms using acetone, methanol, or ethanol solvents. *Photosynthetica* **2008**, *46*, 115–126. [CrossRef]
24. Davis, P.A.; Burns, C. Photobiology in protected horticulture. *Food Energy Secur.* **2016**, *5*, 223–238. [CrossRef]
25. Qian, H.; Liu, T.; Deng, M.; Miao, H.; Cai, C.; Shen, W. Effects of light quality on main health-promoting compounds and antioxidant capacity of Chinese kale sprouts. *Food Chem.* **2016**, *196*, 1232–1238. [CrossRef] [PubMed]
26. Son, K.H.; Oh, M.M. Growth, photosynthetic and antioxidant parameters of two lettuce cultivars as affected by red, green, and blue light-emitting diodes. *Hortic. Environ. Biotechnol.* **2015**, *56*, 639–653. [CrossRef]
27. Naznin, M.T.; Lefsrud, M.; Gravel, V.; Azad, M.O.K. Blue light added with red LEDs enhance growth characteristics, pigments content, and antioxidant capacity in lettuce, Spinach, Kale, Basil, and sweet pepper in a controlled environment. *Plants* **2019**, *8*, 93. [CrossRef] [PubMed]
28. Ying, Q.; Kong, Y.; Zheng, Y. Applying blue light alone, or in combination with far-red light, during nighttime increases elongation without compromising yield and quality of indoor-grown microgreens. *HortScience* **2020**, *55*, 876–881. [CrossRef]
29. Awang, Y.; Shazmi Shaharom, A.; Mohamad, R.B.; Selamat, A. Chemical and Physical Characteristics of Cocopeat-Based Media Mixtures and Their Effects on the Growth and Development of Celosia cristata. *Am. J. Agric. Biol. Sci.* **2009**, *4*, 63–71. [CrossRef]
30. Son, K.H.; Oh, M.M. Leaf Shape, Growth, and Antioxidant Phenolic Compounds of Two Lettuce Cultivars Grown under Various Combinations of Blue and Red Light-emitting Diodes. *HortScience* **2013**, *48*, 988–995. [CrossRef]
31. Gunjal, M.; Singh, J.; Kaur, J.; Kaur, S.; Nanda, V.; Mehta, C.M. Comparative analysis of morphological, nutritional, and bioactive properties of selected microgreens in alternative growing medium. *S. Afr. J. Bot.* **2024**, *165*, 188–201. [CrossRef]
32. Lanoue, J.; St. Louis, S.; Little, C.; Hao, X. Continuous lighting can improve yield and reduce energy costs while increasing or maintaining nutritional contents of microgreens. *Front. Plant Sci.* **2022**, *13*, 983222. [CrossRef] [PubMed]
33. Bulgari, R.; Baldi, A.; Ferrante, A.; Lenzi, A. Yield and quality of basil, Swiss chard, and rocket microgreens grown in a hydroponic system. *N. Z. J. Crop Hortic. Sci.* **2017**, *45*, 119–129. [CrossRef]
34. Nolan, D.A. *Effects of Seed Density and Other Factors on the Yield of Microgreens Grown Hydroponically on Burlap*; Virginia Tech: Blacksburg, VA, USA, 2018; pp. 1–44.
35. Thuong, V.T.; Minh, H.G. Effects of growing substrates and seed density on yield and quality of radish (*Raphanus sativus*) microgreens. *Res. Crops* **2020**, *21*, 579–586.
36. Jones-Baumgardt, C.S. *The Use of Light-Emitting Diodes for Microgreen Production in Controlled Environments*; The University of Guelph: Guelph, ON, USA, 2019; pp. 1–111.
37. Liu, J.; Pattey, E.; Miller, J.R.; McNairn, H.; Smith, A.; Hu, B. Estimating crop stresses, aboveground dry biomass and yield of corn using multi-temporal optical data combined with a radiation use efficiency model. *Remote Sens. Environ.* **2010**, *114*, 1167–1177. [CrossRef]
38. Demir, K.; Sarıkamış, G.; Çakırer Seyrek, G. Effect of LED lights on the growth, nutritional quality and glucosinolate content of broccoli, cabbage and radish microgreens. *Food Chem.* **2023**, *401*, 134088. [CrossRef]
39. Frede, K.; Winkelmann, S.; Busse, L.; Baldermann, S. The effect of LED light quality on the carotenoid metabolism and related gene expression in the genus Brassica. *BMC Plant Biol.* **2023**, *23*, 328. [CrossRef]
40. Alrifai, O.; Hao, X.; Liu, R.; Lu, Z.; Marcone, M.F.; Tsao, R. LED-Induced Carotenoid Synthesis and Related Gene Expression in Brassica Microgreens. *J. Agric. Food Chem.* **2021**, *69*, 4674–4685. [CrossRef]
41. Samuoliene, G.; Sirtautas, R.; Brazaityte, A.; Duchovskis, P. LED lighting and seasonality effects antioxidant properties of baby leaf lettuce. *Food Chem.* **2012**, *134*, 1494–1499. [CrossRef]
42. Harbart, V.; Frede, K.; Fitzner, M.; Baldermann, S. Regulation of carotenoid and flavonoid biosynthetic pathways in *Lactuca sativa* var capitate L. in protected cultivation. *Front. Plant Sci.* **2023**, *14*, 1124750. [CrossRef]
43. Landi, M.; Zivcak, M.; Sytar, O.; Brestic, M.; Allakhverdiev, S.I. Plasticity of photosynthetic processes and the accumulation of secondary metabolites in plants in response to monochromatic light environments. *Biochim. Biophys. Acta Bioenerg.* **2020**, *1861*, 148131. [CrossRef]
44. Jiang, H.; Li, Y.; He, R.; Tan, J.; Liu, K.; Chen, Y. Effect of Supplemental UV-A Intensity on Growth and Quality of Kale under Red and Blue Light. *Int. J. Mol. Sci.* **2022**, *23*, 6819. [CrossRef]
45. Wu, X.; Zhang, S.; Liu, X.; Shang, J.; Zhang, A.; Zhu, Z. Chalcone synthase (CHS) family members analysis from eggplant (*Solanum melongena* L.) in the flavonoid biosynthetic pathway and expression patterns in response to heat stress. *PLoS ONE* **2020**, *15*, e0226537. [CrossRef]
46. Manivannan, A.; Soundararajan, P.; Park, Y.G.; Jeong, B.R. Physiological and Proteomic Insights Into Red and Blue Light-Mediated Enhancement of in vitro Growth in *Scrophularia kakudensis*—A Potential Medicinal Plant. *Front. Plant Sci.* **2021**, *11*, 607007. [CrossRef]
47. Saleh, R. Growing Media Amendments and LED Light Interaction Effect on Microgreens Plant Growth and Biochemical Composition. Dalhouse University: Halifax, NS, USA, 2023.
48. Barrett, D.M.; Beaulieu, J.C.; Shewfelt, R. Color, flavor, texture, and nutritional quality of fresh-cut fruits and vegetables: Desirable levels, instrumental and sensory measurement, and the effects of processing. *Crit. Rev. Food Sci. Nutr.* **2010**, *50*, 369–389. [CrossRef]

49. Lobiuc, A.; Vasilache, V.; Pintilie, O.; Stoleru, T.; Burducea, M.; Oroian, M. Blue and red LED illumination improves growth and bioactive compounds contents in acyanic and cyanic *Ocimum basilicum* L. Microgreens. *Molecules* **2017**, *22*, 2111. [CrossRef] [PubMed]
50. Pennisi, G.; Blasioli, S.; Cellini, A.; Maia, L.; Crepaldi, A.; Braschi, I. Unraveling the role of Red: Blue LED lights on resource use efficiency and nutritional properties of indoor grown sweet basil. *Front. Plant Sci.* **2019**, *10*, 00305. [CrossRef] [PubMed]
51. Fan, X.X.; Zang, J.; Xu, Z.G.; Guo, S.R.; Jiao, X.L.; Liu, X.Y. Effects of different light quality on growth, chlorophyll concentration and chlorophyll biosynthesis precursors of non-heading Chinese cabbage (*Brassica campestris* L.). *Acta Physiol. Plant.* **2013**, *35*, 2721–2726. [CrossRef]
52. Ferrón-Carrillo, F.; Luis Guil-Guerrero, J.; María González-Fernández, J.; Lyashenko, S.; Battafarano, F. LED Enhances Plant Performance and Both Carotenoids and Nitrates Profiles in Lettuce. *Plant Foods Hum. Nutr.* **2021**, *76*, 210–218. [CrossRef] [PubMed]
53. Pinto, E.; Almeida, A.A.; Aguiar, A.A.; Ferreira, I.M.P.L.V.O. Comparison between the mineral profile and nitrate content of microgreens and mature lettuces. *J. Food Compos. Anal.* **2015**, *37*, 38–43. [CrossRef]
54. Hanrahan, P.; Krueger, W. Reflection from Layered Surfaces due to Subsurface Scattering. *Semin. Graph. Pap.* **2023**, *2*, 279–288.

horticulturae

MDPI

Article

Effect of Photoperiod and Gibberellin on the Bolting and Flowering of Non-Heading Chinese Cabbage

Shuping Liu [1,†], Junyang Lu [1,†], Jun Tian [1], Ping Cao [2], Shuhao Li [1], Haicui Ge [1], Mingxuan Han [1] and Fenglin Zhong [1,*]

[1] College of Horticulture, Fujian Agriculture and Forestry University, Fuzhou 350002, China
[2] Fujian Jinpin Agricultural Technology Co., Ltd., Fuzhou 350100, China
[*] Correspondence: zhong591@fafu.edu.cn; Tel.: +86-13774563218
[†] These authors contributed equally to this work.

Abstract: Non-heading Chinese cabbage (cabbage) is an essential green leafy vegetable, and bolting and flowering are necessary for reproduction. However, further research is needed to study the effect of photoperiod on the bolting and flowering of cabbage, particularly on the development of the stem. In this study, we performed phenotypic analysis and measured endogenous gibberellin levels in the cabbage. We carried out these experiments under four different photoperiodic treatments, 12 h (light)/12 h (dark), 14 h (light)/10 h (dark), 16 h (light)/8 h (dark), and 18 h (light)/6 h (dark). The results showed that the time of bolting and flowering gradually decreased with increasing light duration. The development of stems was optimal under the 16 h (light)/8 h (dark) photoperiod treatment, and the same result was obtained via cytological observation. In addition, the changes in the endogenous gibberellin3 (GA_3) content under different photoperiodic treatments were consistent with the development of stems and peaked at 16 h (light)/8 h (dark). At the same time, qRT-PCR analysis showed that the relative expression of the key gibberellin synthase genes, *BcGA3ox2* and *BcGA20ox2*, exhibited upregulation. When treated with exogenous GA_3 and its synthesis inhibitor, paclobutrazol (PAC), exogenous gibberellins significantly promoted bolting; conversely, gibberellin inhibitors suppressed the bolting, flowering, and stem elongation of cabbage. Therefore, the photoperiod may regulate cabbage bolting by regulating endogenous GA_3.

Keywords: cabbage; photoperiodic; gibberellin; bolting; flowering; stem development

1. Introduction

Non-heading Chinese cabbage (cabbage) (*Brassica campestris* spp. *chinensis* Makino), a vernalization-responsive, long-day (LD) plant of the Brassica genus in the Cruciferae family [1], is an important leafy vegetable cultivated worldwide [2]. The life cycle of higher plants can be divided into two stages, namely, vegetative growth and reproductive growth [3]. Completing the transition from vegetative growth to reproductive growth at the right time is essential for plant reproduction [4], which occurs in the meristem within the rosette [5]. Bolting and flowering are landmark features of cabbage entering the reproductive growth stage and critical agronomic traits in production [6]. Bolting refers to the phenomenon of the flower moss gradually elongating and growing from the rosette of leaves after the completion of floral bud differentiation [7], which is a characteristic of cabbage stem development [8]. In cruciferous vegetables, bolting is an important process in the transition to flowering, with flowering being an evolutionary component [9]. Endogenous and environmental signals jointly manipulate the timing of bolting and flowering [10–12], and these signals form a complex regulatory network to determine the transition of reproductive growth. Environmental signals, such as light and temperature, particularly through the photoperiodic pathway and the vernalization pathway, and endogenous developmental signals, including phytohormones like salicylic acid (SA), jasmonic acid (JA), gibberellin (GA), and auxin (IAA) [13–15], play critical roles in this process.

Light is an essential environmental condition for normal plant development, being an important factor in crop growth and development and quality formation [16]. Plants can sense changes in the photoperiod with their biology and regulate the time of bolting and flowering according to the duration of light exposure, as well as provide energy for plant development [17]. Plants regulate growth in response to environmental variations, which rely on the interaction between endogenous factors and environmental signals, such as hormone and light signaling pathways [18]. On the one hand, plants absorb different light levels through a series of photoreceptors, inducing light signaling to regulate plant flowering via sensing the presence or absence, direction, and intensity of light [19]. On the other hand, related studies have demonstrated that light signaling synergistically with gibberellin signaling mediates plant flowering [20]. Wang found that GA induced the expression of the *CO* (*CONSTANS*) gene and thus facilitated the expression of *FT* (*FLOWERING LOCUS T*) under LD conditions [21]. However, other studies have also shown that flowering is associated with the apical bioactivity of gibberellins under short-day light (SD) conditions [22]. In addition, bolting can affect the structure of the plant, and a longer light duration can favorably induce the initial elongation of flowering stems [23]. The stem is not only one of the important production organs but also a storage organ for supplying nutrients to the plant [24]. The photoperiodic response to flowering is a hot research topic today. At the same time, the study of light and stem development has also attracted increasing attention. Current research has focused on onion species [25], and there is still a demand to deepen the study in *Brassica*.

Active gibberellins (GAs) are involved in various processes of plant development, including seed germination, stem and leaf development, and flowering time; they also play a pivotal role in plant cell expansion and elongation, and they can respond to endogenous and environmental signals in plants [26–29]. Secondly, promoting branch tip elongation and plant height is also a more prominent feature of GAs [30]. Wang found in cabbage that the knockout of the *BraRGL1* (DELLA protein) gene using the CRISPR/Cas9 gene editing system resulted in a delay in the process of bolting initiation and flowering in the mutant material [31]. It has been shown that in the gibberellin synthesis pathway, three classes of oxidases, *GA3ox*, *GA20ox*, and *GA2ox*, positively regulate the synthesis and inactivation of active GAs, and the overexpression of the *GA20ox* gene in Arabidopsis was found to be effective in promoting branch elongation [32]. Although many studies have shown that GAs are involved in the flowering process [11], the changes and functions of endogenous GAs in the photoperiodic regulation of cabbage bolting need to be further investigated, especially under extended light duration conditions.

In this study, we investigated the effects of photoperiod on the time of bolting and flowering, stem development, changes in endogenous GA_3 content, and the expression of related genes using morphology, cytology, and molecular biology to explore the regulation of photoperiod on the bolting and flowering of cabbage and the mechanism of the endogenous gibberellin response.

2. Materials and Methods

2.1. Plant Material and Growing Conditions

The cabbage cultivar 'Qingtai No. 4' was obtained from Fujian Jinpin Agricultural Technology Co. (Fujian, China). The cabbage seeds were sown into the substrate in a 25 °C growth chamber until they grew to 4–5 true leaves. Afterward, the plants were placed in a climate chamber (Jiangnan Instrument Factory, Ningbo, China) for ten days of low-temperature vernalization at 8 °C/6 °C (day/night), 14 h/10 h photoperiod, photosynthetic photon flux density (PPFD) of 200 $\mu mol \cdot m^{-2} \cdot s^{-1}$, and relative humidity of 75%.

2.2. Photoperiod Processing

Uniform and healthy plants were selected for photoperiod treatment after vernalization. Four different photoperiod treatment groups were set up as 12 h (light)/12 h (dark) (recorded as Ph12), 14 h (light)/10 h (dark) (Ph14), 16 h (light)/8 h (dark) (Ph16), and

18 h (light)/6 h (dark) (Ph18) in a special growth chamber. The plants were grown at a temperature of 25 °C and white light (LED with a wavelength of 400–700 nm) (Fujian Jiupu Biotechnology Co. Fuzhou, China) with PPFD of 200 μmol·m^{-2}·s^{-1}.

2.3. Treatment with Exogenous GA_3 and Inhibitors

The cabbage plants with the same growth after vernalization were selected for treatment. The exogenous GA_3 (300 mg/L) and PAC (20 mg/L) treatments were sprayed once every two days individually under the 18 h (light)/6 h (dark) photoperiod. Each treatment application was sprayed only twice during the entire growth period, with water spray as the control. The experiment was over when the plants bloomed.

2.4. Plant Phenotyping and Morphological Characterization

The beginning of sowing was recorded as the first day, and we recorded the time taken for the emergence of green flower buds as the time of the squaring stage. The bolting stage was the time of the elongation of the flower moss and rosette leaf flush. The time that elapsed until the first flower fully opened was recorded as the flowering stage. The growth indices of the different treatments were determined at different stages. Plant height and stem height were measured using a ruler. We took the length from the cotyledon to the top as the plant height. Stem height was defined as the distance from above the rosette to the stem tip. A Vernier caliper was used to determine the stem diameter, and the diameter of the stem with the first internodal distance greater than 1cm was the thickness of the stem. Fresh weight and dry weight were determined using an electronic balance. Chlorophyll content was determined using the ethanol extraction colorimetric method [3] from the young leaf to the outer third mature functional leaf.

2.5. Measurement of Endogenous GA Content

The endogenous GA_3 content was determined at the apical of the cabbage stems, including the stem tips and young leaves, at the bolting and flowering stages under different photoperiod treatments. The endogenous gibberellin content was determined using an enzyme-linked immunosorbent assay with a gibberellin (GA_3) ELISA kit (Enzyme-linked Biotechnology, Shanghai, China).

2.6. Cytologic Observations

Stems from the same nodulation locus (where the internode distance on the rosette first appeared to be greater than 1 cm nodulation) of each treatment at the flowering stage were immersed in FAA fixative (70% ethanol: acetic acid: formaldehyde = 90:5:5) for 24 h. Afterward, paraffin section preparations were embedded in paraffin wax and stained with Senna red and solid green. Finally, the cells were observed and photographed using a fluorescence inverted microscope (Leica, Wetzlar, Germany). Data measurements were performed using ImageJ 1.8.0 software.

2.7. Gene Expression Analysis

Total RNA was extracted from plants of different treatments at the bolting and flowering stages using the Plant Total RNA Isolation Kit (Vazyme, Nanjing, China). First-strand cDNA was synthesized with a FastKing gDNA Dispelling RT SuperMix Kit (Tiangen, Beijing, China). The real-time PCR analysis was performed using the 2× RealStar Green Fast Mixture reagent (Genstar, Beijing, China) in a LightCycler® 96 Real-Time PCR instrument (Roche, Basel, Switzerland). The β-actin gene was used as an internal control, and relative expression was calculated using the $2^{-\Delta\Delta Ct}$ calculation method. Gene-specific primers are listed in Table S1.

2.8. Statistical Analysis

Three replicates were set up for each treatment, and each replicate consisted of twenty plants. Statistical analysis was performed using SPSS 26.0 software, and the significance of differences between treatments was compared using a one-way ANOVA ($p = 0.05$). Plotting was performed using Origin 2022 software.

3. Results

3.1. Bolting and Flowering Time of Cabbage under Different Photoperiod Treatments

The development of buds, bolting, and flowering of cabbage are closely related to light duration. The experimental results show (Figure 1) that photoperiodic treatments caused positive effects on the bolting and flowering stages of the cabbage. And the bud appearance, bolting, and flowering stages were gradually shortened with increasing light duration. Among them, the Ph18 treatment took the shortest time in the flowering process, followed by the Ph16 treatment, but the difference between the two treatments was slight. In contrast, the Ph12 treatment delayed bolting and flowering time the longest compared with the Ph18 treatment. It could be seen that prolonging the light duration had a promoting effect on bolting and flowering, but the promotion effect was weakened beyond the Ph16 treatment. Flowering was advanced by 8.6%, 12.1%, and 13.8% (median as a reference) under Ph14, Ph16, and Ph18 treatments, respectively, compared with the Ph12 treatment.

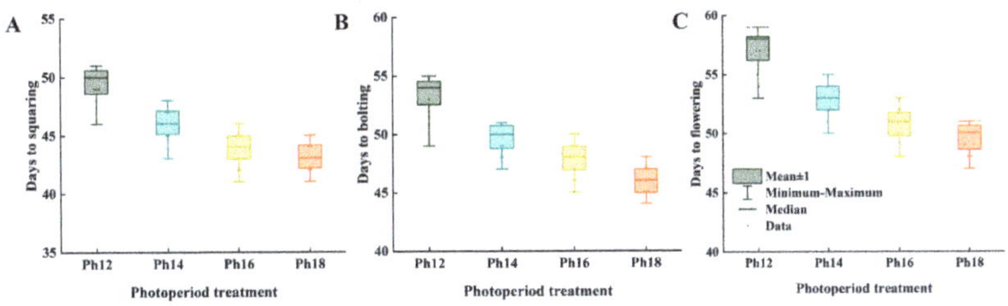

Figure 1. Effect of photoperiod on flowering of non-heading Chinese cabbage (cabbage). (**A**) Analysis of the squaring time under different photoperiod treatments. (**B**) Analysis of the bolting time. (**C**) Analysis of the flowering time. The horizontal line in the figure indicates the median ($n > 20$).

Photoperiod affected the accumulation of photosynthetic pigments in cabbage. This study determined changes in the photosynthetic pigments content of the third mature functional leaf from the young leaf outward. As shown in Figure 2A,B, the total chlorophyll and carotenoid content showed the same trend of change under the same light duration treatment as the growth process advanced. The photosynthetic pigments all showed a gradual decrease under the Ph12 treatment. In contrast, they showed a gradual increase in the Ph14 and Ph16 treatments. Beyond that, they all increased before bolting and decreased after bolting under the Ph18 treatment. In terms of the developmental period, there was no marked difference in the pigment content between the different light treatment groups at the squaring stage. At the bolting stage, the Ph18 treatment showed a distinct advantage, and the carotenoid content and total chlorophyll content were 1.5 and 1.3 times higher than those under the Ph12 treatment, respectively. During the flowering stage, the pigment content significantly increased under the Ph14 and Ph16 treatments, whereas it was downregulated under the Ph12 treatment, and its content decreased by 29.3% and 12.6%, respectively, compared with that at the squaring stage.

Plant dry and fresh weights are also indicators for evaluating plant quality and yield. The growth morphology and weight changed after the cabbage entered the carex stem development stage. In this study, the dry and fresh weights of the aboveground parts of the cabbage with different light duration treatments were measured at the bolting and flowering stages. There were significant differences in fresh weight but not in dry weight among the treatments, while both dry and fresh weights peaked under the Ph16 treatment, as shown in Figure 3.

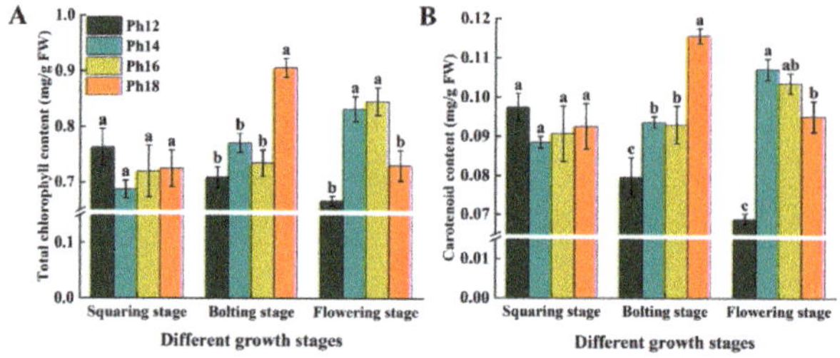

Figure 2. Effect of photoperiod on the photosynthetic pigments of cabbage. (**A**,**B**) Total chlorophyll content and carotenoid content of the third mature functional leaf of cabbage at squaring, bolting, and flowering stages. Data shown are means and error lines indicate SE ($n = 3$). All data were evaluated statistically using one-way ANOVA ($p \leq 0.05$). Different lowercase letters indicate a significant difference.

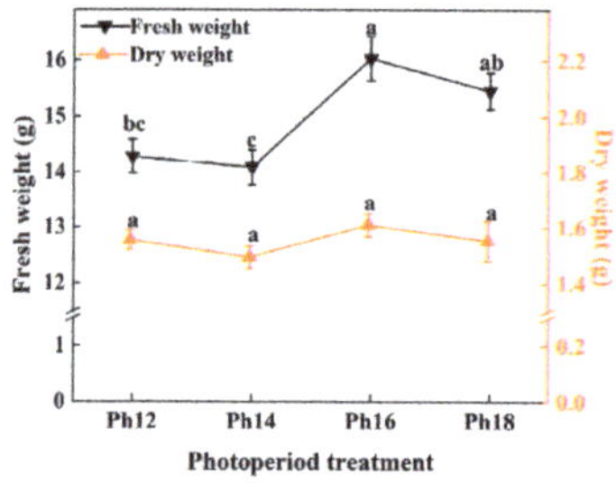

Figure 3. Effect of photoperiod on the aboveground fresh and dry weights of cabbage at flowering. The data shown are means and error lines indicate SE ($n > 15$). All data were evaluated statistically using one-way ANOVA ($p \leq 0.05$). Different lowercase letters indicate a significant difference.

3.2. Analysis of Stem Phenotypes and Endogenous GA$_3$

In the experiment, it was found that photoperiod not only affected the time of bolting and flowering but also had a significant effect on plant height. According to the measured plant height data (Figure 4A), at the squaring stage, the plant height showed a decreasing and then increasing trend with increasing light duration. However, the aboveground height at flowering showed a trend of increasing and then decreasing, peaking under the Ph16 treatment, which was 16.5%, 11.2%, and 4.3% higher than that under the Ph12, Ph14, and Ph18 treatments, respectively. The flowering phenotype is shown in Figure 4B.

Photoperiod affects the development of stems in cabbage. In the present study, we observed the phenotypes of carex stems in the upper part of the rosette at the bolting (Figure 5A) and flowering stages (Figure 5B). As shown in Figure 5C, the results of stem length indicated that the maximum was achieved under the Ph16 treatment, which was 5.4% higher than that of the Ph12 treatment at the bolting stage. At the flowering stage, it showed a unimodal trend, peaking under the Ph16 treatment and being 18.5% higher than that under the Ph12 treatment. Stem diameter is also one of the indicators of carex stem development. Moreover, the stem diameter in the upper part of the rosette (Figure 5D) showed a tendency to increase and then decrease at both the bolting and flowering stages and peaked under the Ph16 treatment. This indicates that carex stems develop both horizontally and vertically under the influence of the photoperiod.

Figure 4. Effect of photoperiod on the phenotype of cabbage. (**A**) Aboveground plant heights at squaring stage and flowering stage, data shown are means and error lines indicate SE ($n > 20$). (**B**) Phenotypes at the time of 1–3 flowers in each photoperiod treatment, scale bar = 5 cm. All data were evaluated statistically using one-way ANOVA ($p \leq 0.05$). Different lowercase letters indicate a significant difference.

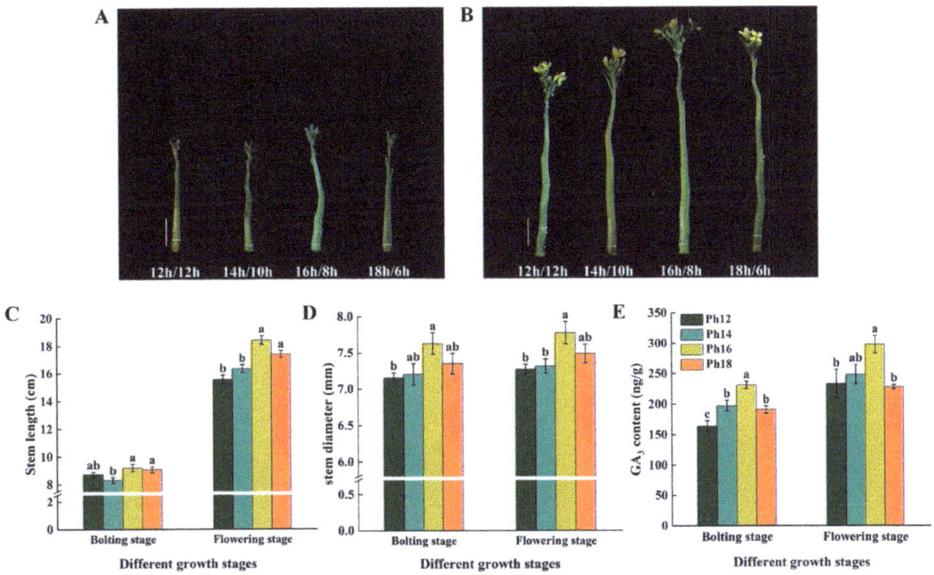

Figure 5. Effect of photoperiod on the development of stems during the bolting and flowering stages of cabbage. (**A**,**B**) Stems on the rosette at the bolting and flowering stages, scale bar = 2 cm. (**C**) Stem length on the rosette at bolting stage and flowering stage, data shown are means, and error lines indicate SE ($n > 20$). (**D**) Stem diameter at the bolting stage and flowering stage, data shown are means and error lines indicate SE ($n > 20$). (**E**) Changes in endogenous GA_3 content in cabbage at the bolting and flowering stages under different photoperiod treatments, the data shown are means, and error lines indicate SE ($n = 3$). All data were evaluated statistically using one-way ANOVA ($p \leq 0.05$). Different lowercase letters indicate a significant difference.

To investigate the role of gibberellin in development under different photoperiodic treatments, we measured the levels of endogenous GA_3 content at both the bolting and flowering stages of the cabbage. The data showed that the GA_3 content was at its maximum under the Ph16 treatment, both at the bolting and flowering stages. However, further extension of light duration caused the gibberellin content to decrease under the Ph18 treatment (Figure 5E). At the bolting stage, compared to the Ph16 treatment, endogenous GA_3 was

29.2%, 14.7%, and 17.4% lower in the Ph12, Ph14, and Ph18 treatments, respectively. At the flowering stage, endogenous GA_3 under the Ph16 treatment was 27.7%, 19.8%, and 30.5% higher than under the Ph14, Ph16, and Ph18 treatments.

3.3. Cellular Observation

Morphological indicators showed that different photoperiod treatments affected the development of the cabbage stems. For this reason, we further observed the rosette stem nodes at the flowering stage from a cytological point of view in paraffin sections. The morphology of the transverse cut cells is shown in Figure 6A, and the longitudinal cut cells are shown in Figure 6B. By observing the cells in the pith of the stems between different treatments, we found that both the area of the cells in the transverse section (Figure 6C) and the length of the cells in the longitudinal section (Figure 6D) were higher under the Ph16 treatment than under the other three treatments. In addition, the cell area and length under the Ph16 treatment reached 1.2 times that of the minimum value under the Ph12 treatment. It is evident that the photoperiod induces the development of stems by affecting the elongation and division of stem cells.

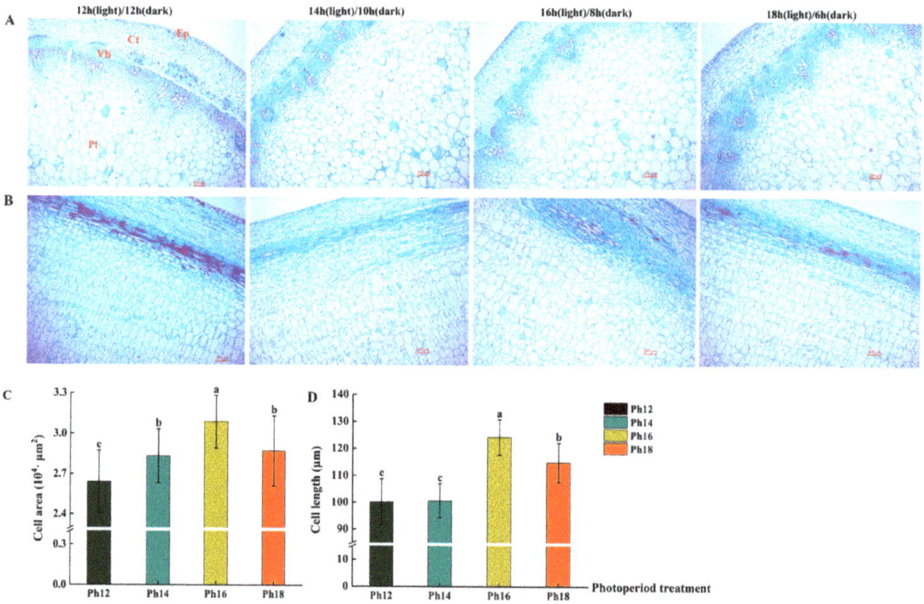

Figure 6. Photoperiodic effects on carex stem cells at the flowering stage of cabbage. (**A**) Anatomical drawings of the cross-section of the stem of cabbage at the flowering stage, EP, epidermis; Ct, cortex; Vb, vascular bundle; Pi: pith. Scale bar = 200 μm. (**B**) Anatomy of the longitudinal interface of carex stems at the flowering stage. Scale bar = 200 μm. (**C**,**D**) Cell area and cell length of cabbage stems at the flowering stage. Three medullary regions were photographed in each tissue section, no less than 50 cells were randomly selected from each region for counting, and the data shown are means and error lines indicate SE ($n > 150$). All data were evaluated statistically using one-way ANOVA ($p \leq 0.05$). Different lowercase letters indicate a significant difference.

3.4. Molecular Characterization of Related Genes in the Process of Bolting and Flowering in Cabbage

Photoperiod affected the expression levels of related genes. To further investigate the influence of photoperiod on stem development, according to previous research [33] and prior laboratory study [34], we analyzed the expression of the key enzyme-encoding genes for GA synthesis (*BcGA20ox2* and *BcGA3ox2*), the flowering-associated genes (*BcCO*,

BcSOC1, and *BcFT*), and the cell elongation genes (expansion protein (*BcEXPA10*) and xyloglucan endotransferase (*BcXTH4*)) at both the bolting (Figure 7) and flowering (Figure 8) stages under different photoperiodic treatments. The results showed that the gibberellin synthase genes *BcGA3ox2* and *BcGA20ox2* exhibited distinct expression patterns in photoperiodic regulation. In contrast, the flowering-related genes were gradually upregulated with the prolongation of light duration at both the bolting and flowering stages. The experimental results show that the photoperiod also affected the relative expression of the cell expansion-related genes *BcEXPA10* and *BcXTH4*.

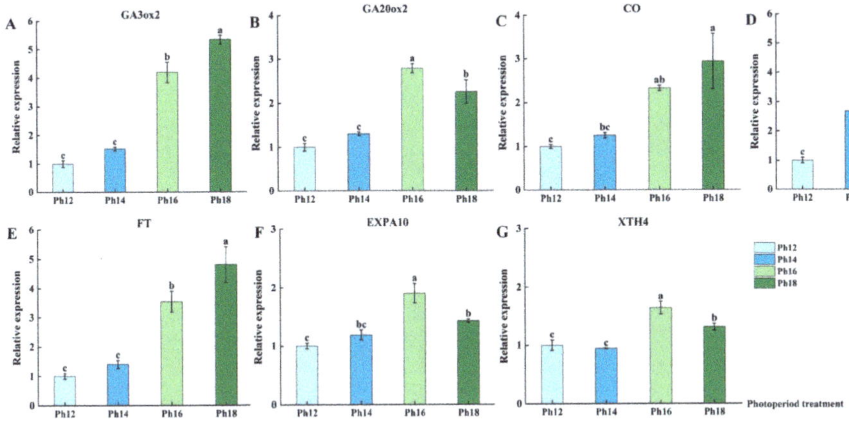

Figure 7. During the bolting stage, (**A**,**B**) relative expression levels of photoperiod-regulated gibberellin synthesis genes, (**C**–**E**) flowering-related genes, (**F**,**G**) cell elongation-related genes in the cabbage. The data shown are means and error lines indicate SE (n = 3). All data were evaluated statistically using one-way ANOVA ($p \leq 0.05$). Different lowercase letters indicate a significant difference.

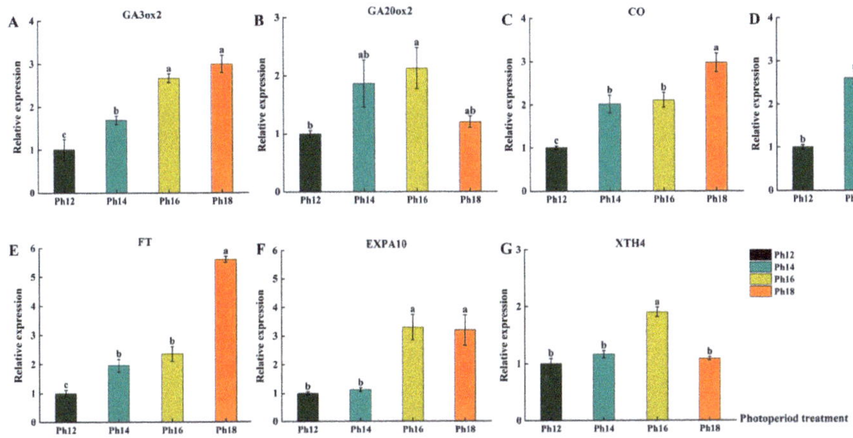

Figure 8. During the flowering stage, (**A**,**B**) relative expression levels of photoperiod-regulated gibberellin synthesis genes, (**C**–**E**) flowering-related genes, (**F**,**G**) cell elongation-related genes in the cabbage. The data shown are means and error lines indicate SE (n = 3). All data were evaluated statistically using one-way ANOVA ($p \leq 0.05$). Different lowercase letters indicate a significant difference.

3.5. Effect of Exogenous GA_3 on Bolting, and Flowering, and Stem Development in Cabbage

To further evaluate the role of GA in bolting and flowering, we exogenously sprayed GA_3 and PAC on the vernalized cabbage. As shown in Figure 9A, the spraying of GA_3 promoted bolting initiation and flowering under the 18 h (light)/6 h (dark) light condition. However, the PAC treatment significantly delayed carex stem development and flowering time. The externally applied GA_3 treatment significantly affected the morphology of the carex stems in the cabbage (Figure 9B). In addition, the differences in carex stem development were also significant. Compared with the control, the GA_3 treatment significantly increased the stem diameter by 9.6% (Figure 9C) and the elongation by 24.5% (Figure 9D). In contrast, the sample under the PAC treatment until flowering did not reach the flush with the rosette of the cabbage and became significantly thinner and weaker. Moreover, internode was not obvious, but the number of leaves increased. The data showed a 20.3% reduction in stem diameter and a 61.4% reduction in stem length compared to the control. The above results indicate that GA_3 is more prominent in the role of carex stem development.

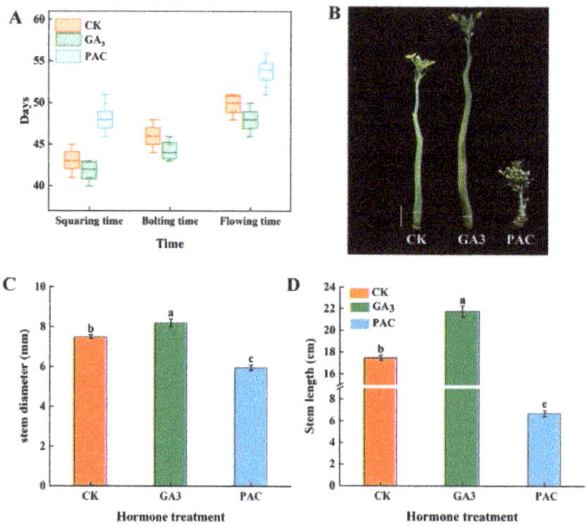

Figure 9. Effects of externally applied GA_3 and PAC on the time of bolting and flowering and the development of carex stem in cabbage. (**A**) Quantitative analysis of the time of squaring, bolting, and flowering in GA_3 treatment, PAC treatment and control, and the horizontal line in the figure indicates the median ($n > 20$). (**B**) Morphological phenotypes of stems at flowering stage after external application of treatments, scale bar = 2 cm. (**C**,**D**) Stem diameter and length of stems at flowering stage, the data shown are means and error lines indicate SE ($n > 20$). All data were evaluated statistically using one-way ANOVA ($p \leq 0.05$). Different lowercase letters indicate a significant difference.

We further observed the alterations in the paraffin sections of stem cells. The transverse cell morphology of the stems under different treatments is shown in Figure 10A, and the longitudinal cell morphology is shown in Figure 10B. By comparing the area (Figure 10C) and length (Figure 10D) of pith cells, the results show that the area and length of the pith cells in the GA_3 treatment increased by 31.2% and 33.3%, respectively, compared with those in the CK. However, the area and length of the pith cells in the PAC treatment decreased by 69% and 53%, respectively, compared with those in the CK group, and the cells were more tightly arranged. Therefore, it follows that exogenous GA_3 can influence the bolting and stem development of cabbage.

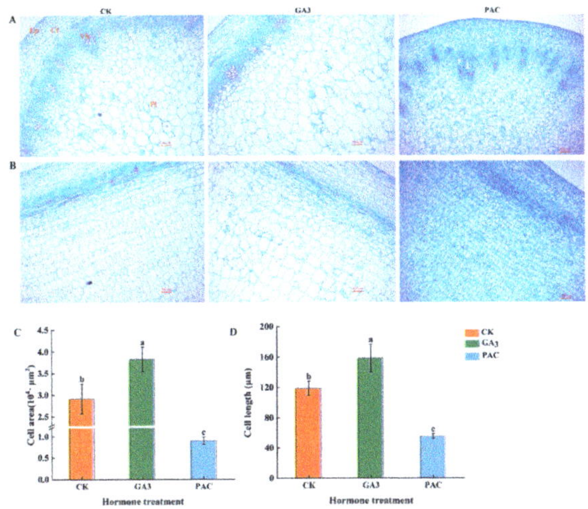

Figure 10. Effects of externally applied GA_3 and PAC on the cells of stems during the flowering period of cabbage. (**A**) Anatomical drawings of stems in cross-section at the flowering stage, EP: epidermis, Ct: cortex, Vb: vascular bundles, Pi: pith. Scale bar = 200 μm. (**B**) Anatomical drawings of longitudinal sections of stems at flowering stage. (**C**,**D**) Stem cell area and cell length of cabbage under different exogenous treatments, data shown are means and error lines indicate SE ($n > 150$). All data were evaluated statistically using one-way ANOVA ($p \leq 0.05$). Different lowercase letters indicate a significant difference.

4. Discussion

In this study, we demonstrated that photoperiod affects the growth of bolting, flowering, and stem development in cabbage and that gibberellin plays an important role in this process. Non-heading Chinese cabbage is an LD plant. However, the length of the bolting and flowering time varied under different LD conditions. Moreover, the time of bolting and flowering not only directly affects the yield and quality of cabbage [35] but also features critical impacts on the breeding of floral regulation among different varieties. Therefore, proper bolting and flowering under suitable light conditions can ensure the quality of cabbage and reduce energy consumption. In this study, we found that prolonging the light duration could accelerate the bolting and flowering of cabbage under LD. Santos [36] also found that the flowering time of cassava was significantly advanced when the light duration was extended via artificial supplementation. The photoperiodic change also affected the photosynthesis of leaves. Relatively high chlorophyll could be more favorable for photosynthesis, providing a material basis for the next stage of the flowering process [37]. It was also found in rapeseed that prolonging photoperiod could promote plant growth [38]. The findings of this study revealed that the pigment content under the Ph18 treatment, which caused a short flowering time, peaked at the bolting stage, possibly indicating a reserve of nutrition for the early flowering stage, thus facilitating early flowering. However, the pigment content under the Ph14 and Ph16 treatments increased gradually with the advancement of the bolting and flowering time, and thus, under these two treatments, blossoming successively occurred following the Ph18 treatment. The pigment content under the Ph12 treatment gradually decreased with the advancement of the floral transition. Reduced pigmentation may not be favorable for photosynthesis, so the time of bolting and flowering was also longer.

The process of the bolting and flowering of cabbage is accompanied by the development of carex stems. Stems are also considered as nutrient storage organs of plants [24]. Therefore, the quality of carex stems can be used as an index to evaluate the quality of bolting. In this study, we found from the morphological observation that the stem development was optimum under the Ph16 treatment at the bolting and flowering stages. This indicates that photoperiod affects stem development in the process of flower formation. When the light duration was sufficient, incomplete stem growth may have been due to premature flowering and nutrient bias in preparing for the flowering of cabbage. In contrast, the time of bolting and flowering was delayed under shorter photoperiod conditions, while the development of the carex stems was also hindered. The proliferation and expansion of the internodal meristematic tissue cells significantly affected stem elongation and development [39]. The results of paraffin sections of the stems showed consistency with the phenotype. We can hypothesize that photoperiod affects the elongation and growth of the cells inside carex stems inside the cells, which causes the differences in stem length and stem diameter phenotypes.

In previous studies, GAs have been found to respond positively to the photoperiodic regulation of flowering, especially under SD [40,41]. However, significant effects of GAs have also been demonstrated in Viola philippica [42], garlic [25], spinach [43], and sugar beet [44]. They also proved the actions of GAs to be remarkable under LD, while the GAs requirement was reduced by CO and FT relative to under SD [45]. Temperature and light have also jointly affected plant elongation and growth in Brassica juncea [46]. However, using shade and PAC to treat Arabidopsis, Alabadi found that endogenous GAs repress photomorphogenesis in the dark [47]. Therefore, the functional role of GA varies with light conditions. The present study, however, focused on LD to investigate the function of GAs in the photoperiodic regulation of bolting and flowering. GAs are essential hormones that harmonize plant development, with the most critical functions including flowering, stem development, and cell elongation [48]. GAs have been reported to be involved in the photoperiodic induction of the flowering pathway [21,49]. GAs and SLs have been shown to cooperate in the regulation of stem development in cucumber [50]. Measurements of endogenous GA_3 showed that within a specific range of light duration, a higher GA_3 content promoted bolting and flowering. Beyond this range, the GA_3 content declined, while flowering still advanced. This may be attributable to the synergistic action of GA_3 with other species of active gibberellins in inducing flowering under more prolonged photoperiodic conditions [51,52]. It is also assumed that feedback regulation by endogenous GAs resulted in a lower endogenous GA_3 content in the cabbage under longer light durations. In addition, we found that changes in the endogenous GA_3 content coincided with the development of carex stems, and we hypothesized that endogenous GA_3 is involved in photoperiod induction of stem growth in cabbage.

The feedback regulation of *GA20ox*, *GA3ox*, and *GA2ox* can mediate GAs level homeostasis [53]. In this experiment, data from the corresponding relative quantitative analysis of genes showed that the expression tendency of *BcGA3ox2* was the same as that of *BcFT*, which may be involved in the flowering process of cabbage. Osnato showed that, in Arabidopsis, *GA3ox2* affected flowering by regulating *FT* with the TEM1 transcription factor [41]. We hypothesize that a conserved function exists since cabbage belongs to the same cruciferous family as Arabidopsis to some extent. *BcGA20ox2* prefers to regulate carex stem development, while the different gibberellin synthase oxidase genes are expressed in separate patterns during the bolting and flowering process. It is hypothesized that *BcGA20ox2* may promote active GAs synthesis to accelerate the rapid growth of the stems of cabbage and bolting. *EXPAs* and *XTHs* performed cell wall expansion by decreasing the viscosity of the polysaccharides between cell walls and cleaving xyloglucan chains to regulate cell wall relaxation, respectively [54]. Alabadi [55] proposed that GAs regulate cell expansion by integrating light signals. Related studies have also shown that GAs can promote cell wall relaxation and elongation by stimulating the expression of *EXPAs* and *XTHs* [56]. The data from this experiment showed that the relative expression of *BcEXPA10* and *BcXTH4* responded prominently to photoperiod. Therefore, the photoperiod may further affect the expression of cell expansion genes by regulating endogenous GA_3 content, leading to stem elongation and stem thickness.

To verify the above inference, we subjected cabbage to exogenous GA_3 and PAC treatments. The results confirmed that GA_3 significantly affected the elongation and development of inflorescence stems. Exogenous GA_3 promoted stem elongation and thickness, while PAC significantly inhibited stem growth. It was also confirmed in lettuce, showing a pronounced stem elongation after 12 days of exogenous spray treatment with 25 mg/L GA_3 [57]. Wang also demonstrated that exogenous GA_3 could increase the height of dwarfed plants [58]. Therefore, we speculate that the photoperiod affects the bolting and flowering of cabbage with the regulation of endogenous GA_3, which may be a key target for intervention in stem development. This was also shown in a recent study of gibberellin involvement in the photoperiodic regulation of chrysanthemum flowering [59].

5. Conclusions

In summary, the photoperiodic regulation of the bolting and flowering of cabbage showed that the longer the light duration, the shorter the bolting and flowering time. Cabbage stems developed preferably under 16 h (light)/8 h (dark) photoperiodic conditions. Meanwhile, endogenous GA_3 responded positively to the bolting process. Exogenous GA_3 induced a more prominent development of stems in the cabbage under LD conditions. Our results provide a scientific basis for the rational use of light facilities to regulate the breeding process and the mechanism of stem development in cabbage.

Supplementary Materials: The following supporting information can be downloaded at: https://www.mdpi.com/article/10.3390/horticulturae9121349/s1. Table S1: Primer sequences used for qRT-PCR amplification of relevant genes.

Author Contributions: Conceptualization, F.Z.; investigation, S.L. (Shuping Liu) and P.C.; data curation, J.T. and S.L. (Shuhao Li); visualization, H.G. and M.H.; writing—original draft preparation, S.L. (Shuping Liu); writing—review and editing, J.L.; project administration, F.Z. All authors have read and agreed to the published version of the manuscript.

Funding: This research was funded by the Seed Industry Innovation and Industrialization Project in Fujian Province (zycxny2021009) and the Suitable for Mechanized Harvesting Facilities in Green Leafy Vegetables Introduction, Selection and Demonstration Promotion (2022-P-001).

Data Availability Statement: The data are not publicly available as they can be obtained by contacting the corresponding author.

Conflicts of Interest: Author Ping Cao was employed by the company Jinpin Agricultural Technology. The remaining authors declare that the research was conducted in the absence of any commercial or financial relationships that could be construed as a potential conflict of interest.

References

1. Dai, Y.; Sun, X.; Wang, C.; Li, F.; Zhang, S.; Zhang, H.; Li, G.; Yuan, L.; Chen, G.; Sun, R.; et al. Gene co-expression network analysis reveals key pathways and hub genes in Chinese cabbage (*Brassica rapa* L.) during vernalization. *BMC Genom.* **2021**, *22*, 236. [CrossRef] [PubMed]
2. Li, Y.; Liu, G.-F.; Ma, L.-M.; Liu, T.-K.; Zhang, C.-W.; Xiao, D.; Zheng, H.-K.; Chen, F.; Hou, X.-L. A chromosome-level reference genome of non-heading Chinese cabbage [*Brassica campestris* (syn. *Brassica rapa*) ssp. chinensis]. *Hortic. Res.* **2020**, *7*, 212. [CrossRef] [PubMed]
3. Han, Y.; Chen, Z.; Lv, S.; Ning, K.; Ji, X.; Liu, X.; Wang, Q.; Liu, R.; Fan, S.; Zhang, X. *MADS-Box* genes and gibberellins regulate bolting in lettuce (*Lactuca sativa* L.). *Front. Plant Sci.* **2016**, *7*, 1889. [CrossRef] [PubMed]
4. Yan, J.; Li, X.; Zeng, B.; Zhong, M.; Yang, J.; Yang, P.; Li, X.; He, C.; Lin, J.; Liu, X.; et al. FKF1 F-box protein promotes flowering in part by negatively regulating DELLA protein stability under long-day photoperiod in *Arabidopsis*. *J. Integr. Plant Biol.* **2020**, *62*, 1717–1740. [CrossRef]
5. Chen, Z.; Han, Y.; Ning, K.; Ding, Y.; Zhao, W.; Yan, S.; Luo, C.; Jiang, X.; Ge, D.; Liu, R.; et al. Inflorescence development and the role of *LsFT* in regulating bolting in lettuce (*Lactuca sativa* L.). *Front. Plant Sci.* **2017**, *8*, 2248. [CrossRef] [PubMed]
6. Liu, S.; Wang, R.; Zhang, Z.; Li, Q.; Wang, L.; Wang, Y.; Zhao, Z. High-resolution mapping of quantitative trait loci controlling main floral stalk length in Chinese cabbage (*Brassica rapa* L. ssp. *pekinensis*). *BMC Genom.* **2019**, *20*, 437. [CrossRef] [PubMed]
7. Liang, N.; Cheng, D.; Liu, Q.; Cui, J.; Luo, C. Difference of proteomics vernalization-induced in bolting and flowering transitions of *Beta vulgaris*. *Plant Physiol. Biochem.* **2018**, *123*, 222–232. [CrossRef]

8. Wang, Y.; Huang, X.; Huang, X.; Su, W.; Hao, Y.; Liu, H.; Chen, R.; Song, S. *BcSOC1* promotes bolting and stem elongation in flowering Chinese cabbage. *Int. J. Mol. Sci.* **2022**, *23*, 3459. [CrossRef]
9. Pouteau, S.; Albertini, C. The significance of bolting and floral transitions as indicators of reproductive phase change in *Arabidopsis*. *J. Exp. Bot.* **2009**, *60*, 3367–3377. [CrossRef]
10. Andrés, F.; Coupland, G. The genetic basis of flowering responses to seasonal cues. *Nat. Rev. Genet.* **2012**, *13*, 627–639. [CrossRef]
11. Mutasa-Göttgens, E.; Hedden, P. Gibberellin as a factor in floral regulatory networks. *J. Exp. Bot.* **2009**, *60*, 1979–1989. [CrossRef] [PubMed]
12. Wang, J.-W.; Czech, B.; Weigel, D. miR156-regulated spl transcription factors define an endogenous flowering pathway in *Arabidopsis thaliana*. *Cell* **2009**, *138*, 738–749. [CrossRef] [PubMed]
13. Davis, S.J. Integrating hormones into the floral-transition pathway of *Arabidopsis thaliana*. *Plant Cell Environ.* **2009**, *32*, 1201–1210. [CrossRef]
14. Kazan, K.; Lyons, R. The link between flowering time and stress tolerance. *J. Exp. Bot.* **2015**, *67*, 47–60. [CrossRef] [PubMed]
15. Arikan, B.; Yildiztugay, E.; Ozfidan-Konakci, C. Responses of salicylic acid encapsulation on growth, photosynthetic attributes and ROS scavenging system in *Lactuca sativa* exposed to polycyclic aromatic hydrocarbon pollution. *Plant Physiol. Biochem.* **2023**, *203*, 108026. [CrossRef] [PubMed]
16. De Wit, M.; Galvão, V.C.; Fankhauser, C. Light-mediated hormonal regulation of plant growth and development. *Annu. Rev. Plant Biol.* **2016**, *67*, 513–537. [CrossRef]
17. Pouteau, S.; Ferret, V.; Lefebvre, D. Comparison of environmental and mutational variation in flowering time in *Arabidopsis*. *J. Exp. Bot.* **2006**, *57*, 4099–4109. [CrossRef]
18. Alabadí, D.; Blázquez, M.A. Molecular interactions between light and hormone signaling to control plant growth. *Plant Mol. Biol.* **2009**, *69*, 409–417. [CrossRef]
19. Li, L.; Li, X.; Liu, Y.; Liu, H. Flowering responses to light and temperature. *Sci. China Life Sci.* **2016**, *59*, 403–408. [CrossRef]
20. Filo, J.; Wu, A.; Eliason, E.; Richardson, T.; Thines, B.C.; Harmon, F.G. Gibberellin driven growth in elf3 mutants requires *PIF4* and *PIF5*. *Plant Signal Behav.* **2015**, *10*, e992707. [CrossRef]
21. Wang, H.; Pan, J.; Li, Y.; Lou, D.; Hu, Y.; Yu, D. The DELLA-CONSTANS transcription factor cascade integrates gibberellic acid and photoperiod signaling to regulate flowering. *Plant Physiol.* **2016**, *172*, 479–488. [CrossRef] [PubMed]
22. Eriksson, S.; Böhlenius, H.; Moritz, T.; Nilsson, O. GA_4 is the active gibberellin in the regulation of LEAFY transcription and *Arabidopsis* floral initiation. *Plant Cell.* **2006**, *18*, 2172–2181. [CrossRef] [PubMed]
23. Mathew, D.; Forer, Y.; Rabinowitch, H.D.; Kamenetsky, R. Effect of long photoperiod on the reproductive and bulbing processes in garlic (*Allium sativum* L.) genotypes. *Environ. Exp. Bot.* **2011**, *71*, 166–173. [CrossRef]
24. Hao, J.H.; Su, H.N.; Zhang, L.L.; Liu, C.J.; Han, Y.Y.; Qin, X.X.; Fan, S.X. Quantitative proteomic analyses reveal that energy metabolism and protein biosynthesis reinitiation are responsible for the initiation of bolting induced by high temperature in lettuce (*Lactuca sativa* L.). *BMC Genom.* **2021**, *22*, 427. [CrossRef] [PubMed]
25. Wu, C.; Wang, M.; Cheng, Z.; Meng, H. Response of garlic (*Allium sativum* L.) bolting and bulbing to temperature and photoperiod treatments. *Biol. Open* **2016**, *5*, 507–518. [CrossRef]
26. Goldberg-Moeller, R.; Shalom, L.; Shlizerman, L.; Samuels, S.; Zur, N.; Ophir, R.; Blumwald, E.; Sadka, A. Effects of gibberellin treatment during flowering induction period on global gene expression and the transcription of flowering-control genes in *Citrus* buds. *Plant Sci.* **2013**, *198*, 46–57. [CrossRef]
27. Davière, J.M.; Achard, P. A pivotal role of DELLAs in regulating multiple hormone signals. *Mol. Plant.* **2016**, *9*, 10–20. [CrossRef]
28. Xu, Q.; Krishnan, S.; Merewitz, E.; Xu, J.; Huang, B. Gibberellin-regulation and genetic variations in leaf elongation for tall fescue in association with differential gene expression controlling cell expansion. *Sci. Rep.* **2016**, *6*, 30258. [CrossRef]
29. Achard, P.; Gusti, A.; Cheminant, S.; Alioua, M.; Dhondt, S.; Coppens, F.; Beemster, G.T. Gibberellin signaling controls cell proliferation rate in *Arabidopsis*. *Curr. Biol.* **2009**, *19*, 1188–1193. [CrossRef]
30. Rosental, L.; Still, D.W.; You, Y.; Hayes, R.J.; Simko, I. Mapping and identification of genetic loci affecting earliness of bolting and flowering in lettuce. *Theor. Appl. Genet.* **2021**, *134*, 3319–3337. [CrossRef]
31. Wang, Y.; Song, S.; Hao, Y.; Chen, C.; Ou, X.; He, B.; Zhang, J.; Jiang, Z.; Li, C.; Zhang, S.; et al. Role of *BraRGL1* in regulation of *Brassica rapa* bolting and flowering. *Hortic. Res.* **2023**, *10*, uhad119. [CrossRef] [PubMed]
32. Rieu, I.; Ruiz-Rivero, O.; Fernandez-Garcia, N.; Griffiths, J.; Powers, S.J.; Gong, F.; Linhartova, T.; Eriksson, S.; Nilsson, O.; Thomas, S.G.; et al. The gibberellin biosynthetic genes *AtGA20ox1* and *AtGA20ox2* act, partially redundantly, to promote growth and development throughout the *Arabidopsis* life cycle. *Plant J.* **2008**, *53*, 488–504. [CrossRef] [PubMed]
33. Huang, X.; Lei, Y.; Guan, H.; Hao, Y.; Liu, H.; Sun, G.; Chen, R.; Song, S. Transcriptomic analysis of the regulation of stalk development in flowering Chinese cabbage (*Brassica campestris*) by RNA sequencing. *Sci. Rep.* **2017**, *7*, 15517. [CrossRef] [PubMed]
34. Huo, Z.; Xu, Y.; Yuan, S.; Chang, J.; Li, S.; Wang, J.; Zhao, H.; Xu, R.; Zhong, F. The AP2 transcription factor *BrSHINE3* regulates wax accumulation in nonheading Chinese cabbage. *Int. J. Mol. Sci.* **2022**, *23*, 13454. [CrossRef] [PubMed]
35. Xiao, D.; Wang, H.; Basnet, R.K.; Zhao, J.; Lin, K.; Hou, X.; Bonnema, G. Genetic dissection of leaf development in *Brassica rapa* using a genetical genomics approach. *Plant Physiol.* **2014**, *164*, 1309–1325. [CrossRef]
36. Santos, A.D.; Bandeira, E.S.M.; Cunha Alves, A.A.; de Oliveira, E.J. Flowering induction in *Cassava* using photoperiod extension premature pruning and plant growth regulators. *PLoS ONE* **2023**, *18*, e0292385. [CrossRef]

37. Ueda, M.; Tanaka, A.; Sugimoto, K.; Shikanai, T.; Nishimura, Y. ChlB requirement for chlorophyll biosynthesis under short photoperiod in *Marchantia polymorpha* L. *Genome Biol. Evol.* **2014**, *6*, 620–628. [CrossRef]
38. Egorova, K.V.; Sinyavina, N.G.; Artemyeva, A.M.; Kocherina, N.V.; Chesnokov, Y.V. QTL analysis of the content of some bioactive compounds in *Brassica rapa* L. grown under light culture conditions. *Horticulturae* **2021**, *7*, 583. [CrossRef]
39. Takatsuka, H.; Umeda, M. Hormonal control of cell division and elongation along differentiation trajectories in roots. *J. Exp. Bot.* **2014**, *65*, 2633–2643. [CrossRef]
40. Andrés, F.; Porri, A.; Torti, S.; Mateos, J.; Romera-Branchat, M.; García-Martínez, J.L.; Fornara, F.; Gregis, V.; Kater, M.M.; Coupland, G. SHORT VEGETATIVE PHASE reduces gibberellin biosynthesis at the *Arabidopsis* shoot apex to regulate the floral transition. *Proc. Natl. Acad. Sci. USA* **2014**, *111*, E2760–E2769. [CrossRef]
41. Osnato, M.; Castillejo, C.; Matías-Hernández, L.; Pelaz, S. TEMPRANILLO genes link photoperiod and gibberellin pathways to control flowering in *Arabidopsis*. *Nat. Commun.* **2012**, *3*, 808. [CrossRef] [PubMed]
42. Li, Q.; Li, J.; Zhang, L.; Pan, C.; Yang, N.; Sun, K.; He, C. Gibberellins are required for dimorphic flower development in *Viola philippica*. *Plant Sci.* **2021**, *303*, 110749. [CrossRef] [PubMed]
43. Lee, D.J.; Zeevaart, J.A. Regulation of gibberellin 20-oxidase1 expression in *Spinach* by photoperiod. *Planta* **2007**, *226*, 35–44. [CrossRef] [PubMed]
44. Mutasa-Göttgens, E.S.; Qi, A.; Zhang, W.; Schulze-Buxloh, G.; Jennings, A.; Hohmann, U.; Müller, A.E.; Hedden, P. Bolting and flowering control in sugar beet: Relationships and effects of gibberellin, the bolting gene B and vernalization. *AoB Plants* **2010**, *2010*, plq012. [CrossRef] [PubMed]
45. Reeves, P.H.; Coupland, G. Analysis of flowering time control in *Arabidopsis* by comparison of double and triple mutants. *Plant Physiol.* **2001**, *126*, 1085–1091. [CrossRef] [PubMed]
46. Kong, Y.; Masabni, J.; Niu, G. Effect of temperature variation and blue and red LEDs on the elongation of arugula and mustard microgreens. *Horticulturae* **2023**, *9*, 608. [CrossRef]
47. Alabadí, D.; Gil, J.; Blázquez, M.A.; García-Martínez, J.L. Gibberellins repress photomorphogenesis in darkness. *Plant Physiol.* **2004**, *134*, 1050–1057. [CrossRef]
48. Guan, H.; Huang, X.; Zhu, Y.; Xie, B.; Liu, H.; Song, S.; Hao, Y.; Chen, R. Identification of *DELLA* genes and key stage for GA sensitivity in bolting and flowering of flowering Chinese cabbage. *Int. J. Mol. Sci.* **2021**, *22*, 12092. [CrossRef]
49. Galvão, V.C.; Horrer, D.; Küttner, F.; Schmid, M. Spatial control of flowering by DELLA proteins in *Arabidopsis thaliana*. *Development* **2012**, *139*, 4072–4082. [CrossRef]
50. Cao, Y.; Dong, Y.; Zhang, R.; Li, Q.; Peng, R.; Chen, C.; Lu, M.; Jin, X. Cucumber strigolactone receptor CsDAD2 and GA_3 interact to regulate shoot branching in *Arabidopsis thaliana* L. *Horticulturae* **2023**, *9*, 23. [CrossRef]
51. Yamaguchi, S. Gibberellin metabolism and its regulation. *Annu. Rev. Plant Biol.* **2008**, *59*, 225–251. [CrossRef] [PubMed]
52. Castro-Camba, R.; Sánchez, C.; Vidal, N.; Vielba, J.M. Interactions of gibberellins with phytohormones and their role in stress responses. *Horticulturae* **2022**, *8*, 241. [CrossRef]
53. Suo, H.; Ma, Q.; Ye, K.; Yang, C.; Tang, Y.; Hao, J.; Zhang, Z.J.; Chen, M.; Feng, Y.; Nian, H. Overexpression of *AtDREB1A* causes a severe dwarf phenotype by decreasing endogenous gibberellin levels in soybean [*Glycine max* (L.) Merr]. *PLoS ONE* **2012**, *7*, e45568. [CrossRef] [PubMed]
54. Tenhaken, R. Cell wall remodeling under abiotic stress. *Front. Plant Sci.* **2014**, *5*, 771. [CrossRef] [PubMed]
55. Alabadí, D.; Blázquez, M.A. Integration of light and hormone signals. *Plant Signal Behav.* **2008**, *3*, 448–449. [CrossRef] [PubMed]
56. Park, J.; Nguyen, K.T.; Park, E.; Jeon, J.S.; Choi, G. DELLA proteins and their interacting RING Finger proteins repress gibberellin responses by binding to the promoters of a subset of gibberellin-responsive genes in *Arabidopsis*. *Plant Cell* **2013**, *25*, 927–943. [CrossRef] [PubMed]
57. Wang, Y.; Li, B.; Li, Y.; Du, W.; Zhang, Y.; Han, Y.; Liu, C.; Fan, S.; Hao, J. Application of exogenous auxin and gibberellin regulates the bolting of lettuce (*Lactuca sativa* L.). *Open Life Sci.* **2022**, *17*, 438–446. [CrossRef]
58. Wang, W.; Hu, J.; Fang, B.; Gao, X.; Hao, C.; Mu, Y.; Feng, H.; Qu, G.; Wang, Y. *Brcd1* is associated with plant height through the gibberellin pathway in *Brassica rapa* L. *Horticulturae* **2023**, *9*, 282. [CrossRef]
59. Zhao, X.; Liu, W.; Aiwaili, P.; Zhang, H.; Xu, Y.; Gu, Z.; Gao, J.; Hong, B. PHOTOLYASE/BLUE LIGHT RECEPTOR2 regulates chrysanthemum flowering by compensating for gibberellin perception. *Plant Physiol.* **2023**, *193*, 2848–2864. [CrossRef]

horticulturae

MDPI

Article

Colored Shading Nets Differentially Affect the Phytochemical Profile, Antioxidant Capacity, and Fruit Quality of Piquin Peppers (*Capsicum annuum* L. var. *glabriusculum*)

Yamir Jiménez-Viveros and Juan Ignacio Valiente-Banuet *

Tecnologico de Monterrey, Escuela de Ingeniería y Ciencias, Epigmenio González 500, San Pablo, Santiago de Querétaro 76130, Querétaro, Mexico
* Correspondence: valiente@tec.mx; Tel.: +52-442-592-6268

Abstract: Piquin pepper fruits, a semi-domesticated wild pepper species highly valued in Mexico, currently face the threat of unsustainable harvesting practices that endanger the species. For this reason, it is necessary to establish sustainable agricultural practices for the cultivation of these peppers. Solar radiation, a critical determinant in crop production, plays a crucial role in plant development, influencing a spectrum of physiological and morphological processes, including the synthesis of phytochemicals. Our study evaluated the effect of light manipulation through colored shading nets on the phytochemical profile, antioxidant capacity, and fruit quality of semi-domesticated piquin peppers at two maturation stages: immature and mature (green and red fruits). Our hypothesis posits that these shading treatments may induce changes in these fruits' phytochemical composition and antioxidant properties, as well as quality. Our results indicate that the shading treatments and maturity stage have significant on capsaicinoid and carotenoid levels, with the highest levels observed in mature fruits. Notably, red fruits grown under black shading treatments resulted in the highest capsaicinoid levels. Carotenoid levels were higher in the black shading treatment during the first cycle, while in the second cycle, the blue shading treatment showed elevated carotenoid levels, suggesting that high irradiance conditions could reduce carotenoid contents. Although no significant differences were observed among the treatments in green fruits, in red fruits, both black and blue treatments exhibited the highest total phenolic compounds in both production cycles. Furthermore, the antioxidant capacity revealed that red fruits exhibited higher antioxidant levels than green fruits. Color analysis showed that red fruits had higher chroma and hue angle values, indicating their brighter and more intense red color than green fruits. The morphological changes in fruit width, length, and weight can be attributed to shading treatments and maturation stages. These results indicate the potential of piquin peppers to act as rich sources of bioactive compounds, emphasizing the benefits of shading as an effective strategy to improve the quality and quantity of phytochemical compounds in piquin peppers. Our findings provide substantial insights into the intricate relationship between maturation, shading treatments, and phytochemical composition, offering a path to improve the nutritional value and quality of piquin peppers.

Keywords: capsaicinoids; carotenoids; light quality; plant secondary metabolites; phenolic compounds; pungency; shading nets; solar radiation; wild peppers

1. Introduction

Piquin pepper (*Capsicum annuum* L. var. *glabriusculum)*, also known as *chiltepín*, is a semi-domesticated pepper widely distributed on the American continent. In northern Mexico, piquin peppers are considered to be of high commercial value [1] and are appreciated for their flavor, aromatic profile, and high pungency [2]. Currently, most piquin peppers are collected from wild specimens, which has drastically reduced the populations of wild pepper plants. [3] The Mexican states reported to have the highest production of piquin peppers include Nuevo León, Tamaulipas, and Coahuila in northern Mexico (Figure 1) [4].

Figure 1. Main growing regions of piquin pepper (*Capsicum annuum* L. var. *glabriusculum*) in Mexico, and location of the experimental site: CAETEC experimental station of Tecnologico de Monterrey in Pedro Escobedo, Querétaro, Mexico (20.535169 N, −100.211472 W).

Piquin peppers have a phytochemical profile associated with health benefits, including analgesic, anti-inflammatory, and anti-cancer properties, among others [5]. The bioactive compounds related to these health benefits include carotenoids, flavonoids, and phenolic compounds. These phytochemicals are biosynthesized by distinct cellular and physiological mechanisms that occur during regular fruit development and maturation. The level of accumulation is regulated by internal signals (including plant hormones) and external factors [6].

The accumulation of phytochemicals in peppers involves complex cellular and physiological mechanisms. Capsaicinoids, responsible for the pungency of peppers, are biosynthesized within the placental epidermis cells, where they are secreted to the outer cell wall, and then accumulate within specialized structures known as "blisters" [7]. Capsaicinoid biosynthesis involves enzymes such as capsaicin synthase and fatty acid synthase, stemming from the phenylpropanoid pathway. The regulation of capsaicinoid production can be mediated by hormonal signals like jasmonic acid and may also be influenced by environmental factors such as temperature and light exposure [8,9].

On the other hand, carotenoids, which contribute to the color and nutritional value of peppers, play a pivotal role as pigments and antioxidants. They are synthesized in the plastids (chromoplasts) of fruit cells using isopentenyl pyrophosphate (IPP) as a precursor, which is generated from the methylerythritol-4-phosphate (MET) pathway. Enzymes such as phytoene synthase and phytoene desaturase are involved in carotenoid biosynthesis and accumulation [10]. At the cellular level, carotenoids are crucial for capturing and transferring light energy in photosynthesis. The accumulation of carotenoids can be affected by environmental factors, including light exposure and temperature fluctuations [11].

Finally, phenolic compounds, known for their antioxidant properties, play a significant role in safeguarding plants against oxidative stress and their defense mechanisms. These compounds are produced in various plant tissues through the phenylpropanoid pathway, relying on key enzymes such as phenylalanine ammonia-lyase (PAL) and chalcone synthase [12]. At the cellular level, phenolic compounds protect cells from damage caused by free radicals and have various other functions. The accumulation of phytochemicals can increase in response to environmental factors such as light exposure and temperature variations [13].

Light intensity is one of the environmental factors that affect the phytochemical profile of peppers [13,14]. In recent years, pepper cultivation has been carried out primarily under shaded conditions to enhance horticultural productivity and fruit quality [15]. Nowadays, most pepper production occurs under protected horticultural conditions, where light conditions are manipulated with photo-selective shading nets or plastic covers [16,17]. Shading nets may be needed in regions with intense solar radiation and high temperatures [18], where excessive irradiance can result in photodamage and adverse effects on plant growth and development [19]. Shading induces changes in the physiology and biochemistry of plants [20], leading to changes in secondary metabolites [21].

Previous studies on piquin peppers have indicated that the use of black shading nets (the most used in horticulture) has a direct effect on productivity, fruit quality, and vegetative growth, and that these effects are dependent on shade levels [17]. Colored shading nets selectively filter sunlight and modify the spectral composition of light, promoting specific wavelengths [22]. We hypothesize that colored shading nets may induce changes in the phytochemical profile of peppers, thus leading to variations in the antioxidant activity and potential health benefits of piquin peppers.

The objective of this study was to determine how shading nets of different colors can influence the phytochemical profile of peppers, in particular capsaicinoids, carotenoids, and phenolic compounds in piquin pepper fruits, as well as their antioxidant activity. The effects of shading on color and fruit quality were also determined.

2. Materials and Methods

2.1. Plant Material and Experimental Site

This study was carried out at the CAETEC experimental station of Tecnológico de Monterrey in Pedro Escobedo, Qro, Mexico (20.535169 N, −100.211472 W) during the 2021 and 2022 production cycles (Figure 1). This location has a tropical and subtropical steppe climate (BSk) according to the Köppen climate classification system [23]. The experimental site has a vertisol soil type according to the FAO/UNESCO soil classification system [24]. Analysis of soil particles indicates that the study was conducted in a clay loam soil containing 38.2% sand, 30.02% silt, and 31.78% clay (Fertilab, 2018. Laboratorio de Nutrición Vegetal Celaya, Guanajuato).

The piquin pepper plants (*Capsicum annuum* L. var. *glabriusculum*) used in this study were obtained from eight selection cycles of a wild ecotype originally from San Fernando, Tamaulipas, Mexico. Before the study, the experimental plants had been selected based on their productivity and phenological attributes.

The seeds of piquin peppers have naturally low germination rates, which makes it necessary to apply pre-germination treatments (Figure 2). These treatments consisted of seed imbibition in a solution with 5000 ppm of gibberellic acid (Cyto-Gibb® CbM, Tlalnepantla de Baz, Mexico) for 24 h, with constant stirring at room temperature. After the treatments, the seeds were planted in trays containing peat moss as container media and placed in moist conditions in a greenhouse. The planting dates were 15 November 2020 and 1 December 2021 for the 2021 and 2022 cycles, respectively. Seedlings were transplanted approximately 125 d after sowing to the experimental site when they reached an average size of 10–15 cm. The seedlings were planted at 90 cm intervals under the shading treatments. Plants were irrigated using a nutrient solution to provide the necessary elements for plant growth. Irrigation was carried out three times a day for five minutes every day. The nutrient solution used for irrigation consisted of 15 mM·L^{-1} nitrates, 1 mM·L^{-1} ammonium, 1.5 mM·L^{-1} phosphates, 8 mM·L^{-1} potassium, 4 mM·L^{-1} calcium, 2 mM·L^{-1} magnesium, and 3 mM·L^{-1} sulfur. The electrical conductivity of the solution was 1.5 dS·m^{-1}, and the pH was set at 6.0.

Air temperatures were recorded during the production cycles in each treatment using a data logger (RC-51H waterproof USB temperature humidity data logger, Elitech, San Jose, CA, USA). The data loggers were programmed to record temperature data once every 30 min.

Light transmission at the canopy level was quantified for each color shade treatment in terms of photosynthetically active radiation (PAR, $\mu mol\ m^{-2}\ s^{-1}$) using the LI-190R quantum sensor (LI-COR, Lincoln, NE, USA). These measurements were conducted at two time points at 12:00 and 14:00 h.

(a)

(b)

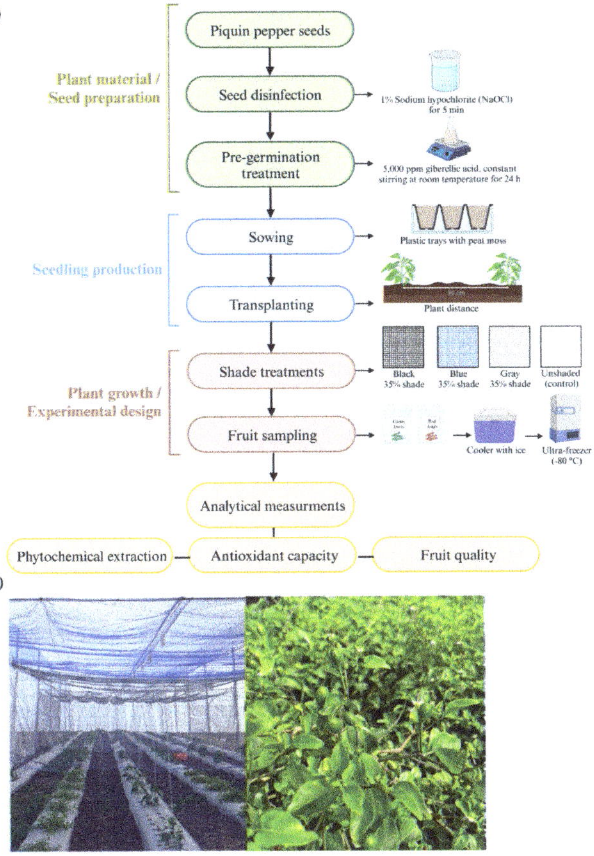

Figure 2. (**a**) Methodology flowchart: assessing the effects of shading treatment on phytochemical profile, antioxidant capacity, and fruit quality of piquin pepper fruits (*Capsicum annuum* L. var. *glabriusculum*) at two maturation stages. (**b**) Pictures of experimental setup and piquin pepper plants with fruits.

2.2. Experimental Design

The study was conducted using a factorial experimental design. The treatments were applied by growing the plants under black, blue, and gray shading nets. The nets used were high-density polyethylene (HDPE) with 35% light interception (Eurosol 54, EURAM, Santiago, Chile). An additional unshaded treatment was used as the control. The nets were placed at 3 m above the ground, covering all sides of the structure. The experimental units consisted of four randomly selected plants per treatment. A 200 g fruit sample was obtained from the plants. The sample fruits were collected at two different maturation stages: immature (green) and mature (red); homogeneity in size and color was ensured within the samples. Fruit collection was conducted during the months of June for green fruits, approximately 35–40 days after flowering (DAF), and in August for mature (red) fruits, approximately 70–75 DAF in both cycles. All analytical measurements were tested in triplicate.

2.3. Sample Preparation

After harvest, the peppers were placed in labeled polyethylene bags and stored in a cooler with ice to keep the fruits fresh. Once in the laboratory, the fruits were placed initially in a −20 °C freezer, and then in an ultra-freezer at −80 °C until further analysis.

Prior to phytochemical extractions, the moisture content of the piquin pepper fruits was determined at both maturation stages using a convection oven (Binder ED, Tuttlingen, Germany). A total of 20 fruits from each category were individually weighed on an analytical balance (Mettler Toledo, ME54E, Columbus, OH, USA) to record initial weights. All fruits were placed in the oven at 70 °C until constant weight, demonstrated by minimal mass fluctuations over time. The fruits were subsequently allowed to cool in a desiccator. After cooling, the fruits were weighed. The moisture content was calculated for each group using the following formula: Moisture Content (%) = [(Initial Weight − Final Weight) × Initial Weight^{-1}] × 100. This method ensures an accurate determination of moisture content, which was used to report all phytochemical results on a dry weight basis (DWB).

Before the extraction procedures, all fruits were washed with water and soap and rinsed with distilled water. The analysis of capsaicinoids and the total phenolic compounds were determined using fresh fruits. For carotenoid analysis, fruits were dried in a convection oven at 65 °C for 24 h. After drying, the seeds were removed and the placenta and pericarp were pulverized with an electric mill (Krups GX4100®®, Mexico City, Mexico). The extraction procedures for the different phytochemicals varied depending on the compound of interest.

2.4. HPLC Analysis for Capsaicinoid Content

The extraction procedure used to determine capsaicinoids used 2 g of fresh fruits that included pericarp, placenta, and seeds. Samples were homogenized using diatomaceous earth (1:1 w/w) and ground using a porcelain mortar. The extraction solvent was 6 mL of 100% methanol. The resulting mixture was sonicated (Ultrasonic Cleaner 8890, Cole-Parmer, Niles, IL, USA) for 15 min and then centrifuged (Centrifuge Multifuge X1R, Thermo Fisher, Walthman, MA, USA) at 22,830 times gravity (× *g*) at 4 °C for 5 min. The supernatant was filtered using Whatman N° 2 paper.

Capsaicinoid analyses were carried out via HPLC using the method described by Wahyuni et al. (2011) [25], with minor modifications. The HPLC system was an Agilent Technologies 1200 series with a UV–Vis detector. Extracts were filtered through a 0.2 µm PTFE membrane filter into a 1.5 mL amber vial. For each HPLC determination, 10 µL samples were injected in triplicate. Compounds were separated using an analytical column (ZORBAX Eclipse XDB-C18, 4.6 × 150 mm, 5 Micron, Agilent, Santa Clara CA, USA). The mobile phases consisted of formic acid and ultrapure water (1:10^3, v/v eluent A), and formic acid and acetonitrile (1:10^3, v/v, eluent B). The gradient applied started at 25% B for 5 min and increased linearly to 75% B for 10 min, and preequilibrated for 2 min to the initial conditions before the next injection. The column temperature was set at 40 °C, and the flow rate was 1 mL·min^{-1}. Capsaicinoids were detected at a 280 nm wavelength and quantified using external standards for capsaicin (CAP) and dihydrocapsaicin (DHC) from Sigma (St. Louis, MO, USA) at different concentrations.

2.5. HPLC Analysis for Carotenoid Content

The extraction process for carotenoid determinations used 0.5 g of peppers (dry weight basis or DW) that included the pericarp and placenta. Then, 10 mL of chloroform and methanol (1:1, v/v) was added as the extraction solvent. The resulting mixture was sonicated for 30 min, followed by centrifugation at 7000× *g* for 15 min. The entire process was carried out twice under dark conditions and at room temperature to ensure maximum carotenoid extraction. The supernatant was filtered using Whatman N° 2 paper. The solvent of the extracts was evaporated using a SpeedVac concentrator (Thermo Scientific, San Jose, CA, USA) and reconcentrated by adding 2 mL of the same solvent.

The analyses of carotenoids were performed via HPLC using the methods described by Mínguez-Mosquera and Hornero-Méndez (1993) [26] and Blanco Ríos et al. (2013) [27], with minor modifications. The HPLC system used was the Agilent Technologies 1200 series equipped with a UV–Vis detector. Extracts were filtered through a 0.2 µm PTFE membrane filter into a 1.5 mL amber vial. For each determination, 20 µL of sample was injected in triplicate for HPLC analysis. Compounds were separated using an analytical column (ZORBAX Eclipse XDB-C18, 4.6 × 150 mm, 5 Micron, Agilent, Santa Clara, CA, USA). The mobile phases consisted of acetone (eluent A) and ultrapure water (eluent B). The gradient applied started at 25% B for 5 min, increased linearly to 75% B for 10 min, and preequilibrated for 2 min to the initial conditions before the next injection. The temperature in the column was maintained at 25 °C, and the flow rate was set at 1.7 mL·min^{-1}. Carotenoid levels were detected at a wavelength of 450 nm and quantified using an external standard for β-carotene from Sigma (St. Louis, MO, USA) at different concentrations.

2.6. Total Phenolic Compounds

The total phenolic compounds were determined spectrophotometrically using the Folin–Ciocalteu method [28]. Compounds were extracted by adding 10 mL of a solution of methanol and water (70:30 v/v) to 5 g of ground fresh fruits. This mixture was sonicated for 30 min and centrifuged at 7000× *g* for 15 min. This process was performed twice to maximize the compound extraction. All extracts were stored at −20 °C until further analysis.

The Folin–Ciocalteu assay was performed in a 96-microwell plate. In each microwell, 20 µL of sample was added, followed by 50 µL of 0.5 N of the Folin–Ciocalteu reagent and 150 µL of water. After five minutes of incubation at room temperature, the reaction was neutralized with 50 µL of sodium carbonate (Na_2CO_3 20%, p·v^{-1}). Subsequently, the mixture was incubated for 2 h at room temperature under dark conditions. After this incubation time, the absorbances were read at 765 nm using a microplate absorbance spectrophotometer (xMark^{TM}, Bio-Rad, Hercules, CA, USA).

The spectrophotometric readings were compared against a Gallic acid standard. The calibration curve was obtained using dilutions of Gallic acid with concentrations ranging from 0 to 1000 µM. All samples were measured in triplicate and the final concentration was expressed as milligrams of gallic acid equivalents (GAE) on a dry weight basis.

2.7. Antioxidant Capacity

A comprehensive analysis of the antioxidant capacity of piquin pepper fruits was determined using in vitro assays comparing the antioxidant effects of the fruits at two different maturation levels using the ABTS and DPPH methods [29]. These methods yield complementary information about the antioxidant properties of the samples, and their combined results allow for a more complete determination of the overall antioxidant capacity. The same extracts obtained for the phenolic compound analysis were employed to assess antioxidant activity.

2.7.1. ABTS Method

The ABTS method measures the Trolox equivalent antioxidant capacity (TEAC), which compares the antioxidant capacity to cleave the radical cation of ABTS and Trolox [30]. The ABTS stock solution was obtained by reacting 7 mMol·L^{-1} and 2.45 mMol·L^{-1} of potassium persulfate after incubation in the dark for 16 h. The stock solution was subsequently diluted in ethanol to an absorbance of 0.8 ± 0.1 at 734 nm. Trolox standard solutions were prepared in methanol from 0 to 700 µmol·L^{-1} and assayed under the same conditions. In each well of a 96-microplate plate, 200 µL of reagent and 20 µL of the sample extracts were added and incubated for 6 min with constant agitation. Each sample was assessed in triplicate. The measurements were taken at 734 nm using xMark™ Microplate Absorbance Spectrophotometer (xMark^{TM}, Bio-Rad, Hercules, CA, USA). The calibration curve for the ABTS method was estimated using Trolox standard solutions from 0 to 700 µmol·L^{-1}. These solutions were prepared in methanol and assayed under the same conditions. Based on these solutions, the calibration curve was estimated as The TEAC was reported as $y = 0.0015x + 0.0426$, $R^2 = 0.9941$. Trolox equivalents (mM TE·g^{-1} DW). These units quantify the antioxidant capacity of the sample based on its ability to neutralize the ABTS radicals relative to Trolox.

2.7.2. DPPH Method

The DPPH method is based on the radical unpaired electron yield of an antioxidant substance, in which DPPH is demoted from a blue–purple color to light yellow [31]. For this assay, a stock solution of 125 µM of DPPH (1,1-diphenyl-2-picrylhydrazyl) was prepared. For the assay, 20 µL of sample extracts and 200 µL of DPPH were added to a well in a 96-microwell plate and mixed; analysis was carried out in triplicate. The plaque was stored in the dark for 90 min at room temperature. After incubation, DPPH stock solution was added as a control to the plaque. Absorbance readings were taken at 520 nm using an xMark™ Microplate Absorbance Spectrophotometer (xMark™, Bio-Rad, Hercules, CA, USA). Scavenging was expressed in the mg of dry sample needed to decolorate 50% of the reagent and as Trolox equivalents (mM TE·g^{-1} DW). The calibration curve for the DPPH method was estimated in a similar way to the ABTS method. The standard curve equation obtained for the ABTS method was $y = 0.0015x + 0.0622$, $R^2 = 0.9966$).

2.8. Fruit Quality

2.8.1. Analysis of Color

The colorimetric determinations of piquin pepper fruits were carried out using a Konica Minolta spectrophotometer (CM-5, Ramsey, NJ, USA). Ten fruits per treatment were placed in a glass petri dish, and the measurements were taken using the spectrophotometer in triplicate. Color results were reported using the CIE L*a*b* (CIELAB) color space parameters, as these parameters allow an accurate description of the color of fruits [32]. In the CIELAB parameters, L* represents the lightness value, ranging from 0 (black) to 100 (white), a* represents the chromaticity value between green (−) and red (+), and b* represents the chromaticity value between blue (−) and yellow (+) [33].

2.8.2. Morphological Analysis

The morphological analyses of peppers were conducted on samples of ten individual fruits per treatment. Individual fruit sizes (width and height) were determined using a digital vernier. Additionally, the weight of ten fruits was determined using an analytical balance (Mettler Toledo, ME54E, Columbus, OH, USA).

2.9. Statistical Analysis

The results were analyzed using a one-way analysis of variance (ANOVA) to determine the significant differences between treatments. Significance levels were set at $p < 0.05$ throughout the study. When significant differences were detected, the means were separated using Tukey's Honestly Significant Difference (HSD) test at a significance level of $p < 0.05$. The statistical analyses were performed using Minitab Statistical Software (version 21.4). All results are presented as the mean ± standard deviation (SD) of three replicates.

3. Results and Discussion

3.1. Temperature

The effect of the shading treatments on air temperature (Figure 3) indicated that the highest temperatures were registered in the unshaded treatments in comparison to the rest of the treatments. The largest differences in mean temperature occurred in the middle of the day but were less than 4 °C across treatments. Shading treatments can effectively reduce the air temperature and elevate air humidity. Shading nets represent an impactful approach to creating an optimal environment for crop cultivation, leading to improved quality and increased crop productivity in regions with high levels of solar radiation [34].

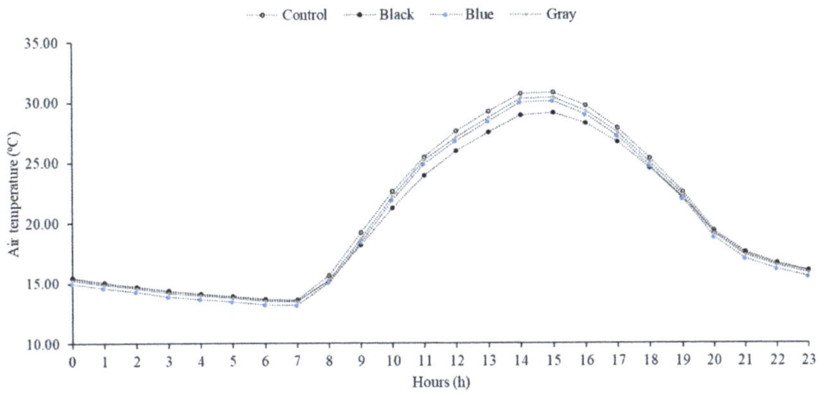

Figure 3. Mean hourly air temperature (°C) for each treatment during the production cycles of piquin peppers (*Capsicum annuum* L. var. *glabriusculum*) in Querétaro, México.

3.2. Photosynthetically Active Radiation (PAR)

All colored shading nets exhibited a reduction in light transmittance compared to the control (Figure 4). The black shading treatment reduced transmittance by an average of 31%, while the blue and gray nets showed reductions of 27% and 25%, respectively. The differences in variability in the modulation of light transmission may have implications for photosynthesis and plant development. The microclimatic impact of colored shading nets on the reduction in photosynthetically active radiation (PAR) has the potential to significantly impact the key physiological processes determining fruit yield and quality in crops, including photosynthesis and carbon allocation [35]. The variations in PAR availability observed among the different nets stem from the color of threads, which alters the proportion of diffuse light compared to the total light transmitted under nets [36].

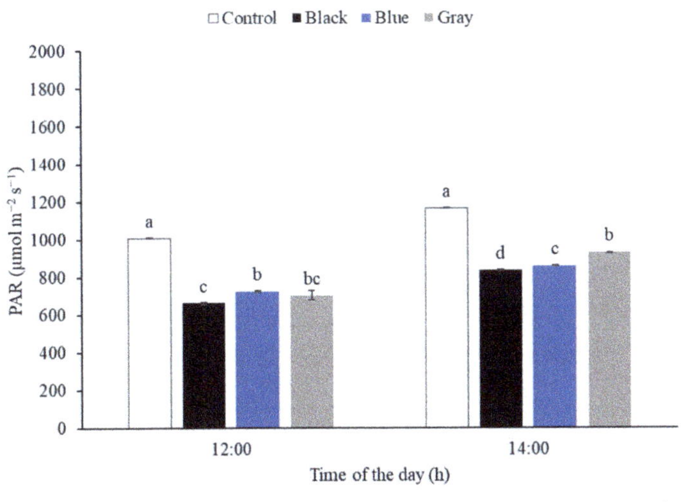

Figure 4. Mean solar radiation as photosynthetically active radiation (PAR, $\mu mol \cdot m^{-2} \cdot s^{-1}$) measured in experimental treatments at two different times of the day. Effect of the black, blue, and grey shading treatments on light transmission compared to the control (unshaded treatment). Error bars represent standard error of replicates. Values with the same letters are statistically similar. Tukey's mean comparison test (a = 0.05).

3.3. Capsaicinoid Content of Piquin Peppers

The effect of the different shading nets on the contents of capsaicinoids in piquin peppers showed significant differences, and the results varied depending on the shading treatments and the maturation stage of the peppers. Capsaicinoid levels increased in red fruits of both production cycles (Figure 5). An increase in capsaicinoids during maturation has been reported in other peppers, where capsaicinoids start to accumulate in the early stages of fruit development, followed by an increase during fruit maturation [37]. The results of the second production cycle showed a similar increase in capsaicinoids, but the final contents were higher in all treatments compared to the first cycle.

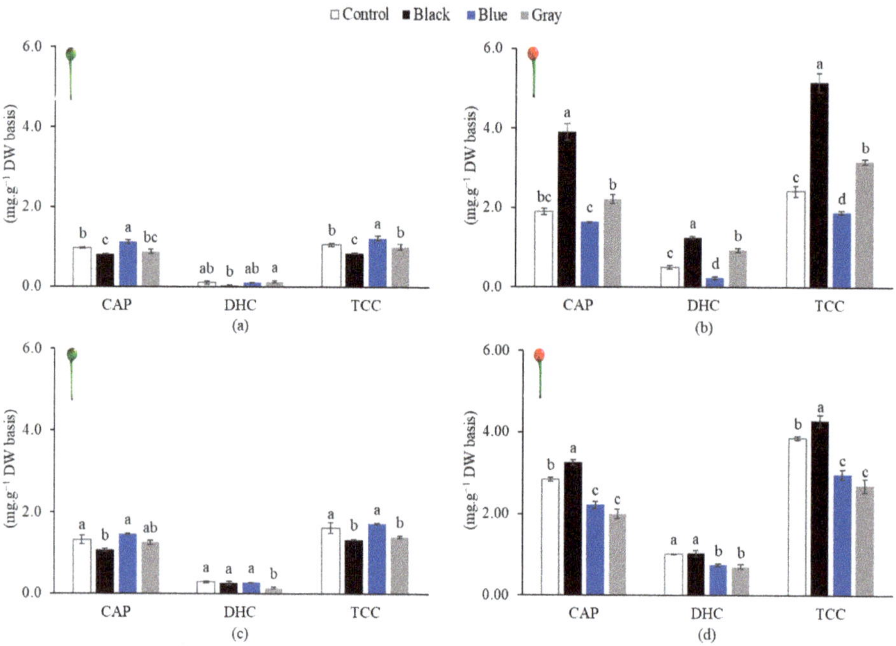

Figure 5. Mean capsaicinoid contents: capsaicinoids (CAPs), dihydrocapsaicin (DHC), and total capsaicinoid content (TCC) (mg·g^{-1} DW basis) of green and red fruits of piquin peppers grown under colored shading net treatments in Queretaro, Mexico, across two production cycles (2021–2022). (**a**) Green fruits, first production cycle (2021). (**b**) Red fruits, first production cycle (2021). (**c**) Green fruits, second production cycle (2022). (**d**) Red fruits, second production cycle (2022). Error bars represent standard error of replicates. Values with the same letters are statistically similar. Tukey's mean comparison test (a = 0.05).

The immature fruits grown under the blue shade treatment had the highest capsaicinoid contents in both production cycles. However, as the fruits matured, the black shading treatment presented the highest capsaicinoid contents, measured as capsaicin (CAP), dihydrocapsaicin (DHC), and the total contents of capsaicinoids (TCC). These results were significantly higher than the control and the other shading treatments. These results are consistent in both cycles and indicate that the black shading treatment increased the capsaicinoid content of piquin peppers in the red maturation stage. Black shading nets reduce the light intensity reaching the plant canopy, creating a shaded environment, which, in turn, may influence the expression of genes involved in capsaicinoid biosynthesis [9,14].

As mentioned, the capsaicinoid levels were higher in mature fruits than in green fruits, confirming previous reports in which the capsaicinoid contents of piquin peppers increased as the fruit matured. Nonetheless, the observed differential effect of the colored shading treatments on capsaicinoid levels [17] could be related to the modulation of gene expression associated with capsaicinoid biosynthesis, where light plays a role in the expression of the capsaicin synthase gene (*CS*). The promoter regions of the *CS* gene contain light-responsive motifs, indicating that light can influence its expression [9]. In a similar study on *C. annuum* 'Star Flame' and 'Fire flame,' the interaction of reduced light intensity using different colored shading nets at different harvest times significantly impacted capsaicinoid contents [38]. The reduced light intensity and modified light qualities caused by colored shading nets might influence the gene expression responsible for capsaicin production [39,40]. Nonetheless, the higher capsaicinoid contents associated with green nets could also be related to higher temperatures [41].

Overall, light intensity and changes in its spectrum seem to be critical environmental factors that influence capsaicinoid accumulation in pepper plants. The duration of light exposure may also impact the synthesis of capsaicinoids, leading to variations in the pungency levels of the different pepper cultivars.

3.4. Carotenoid Content of Piquin Peppers

The analysis of carotenoid content, reported as β-carotene equivalents, showed significant differences between shading treatments for green and red fruits in both production cycles (Figure 6). In the first production cycle, the black shading treatment resulted in the highest carotenoid content for green and red fruits, while the gray treatment had the lowest carotenoid content for red fruits. In the second production cycle, the blue shading treatment caused the highest carotenoid content for red fruits.

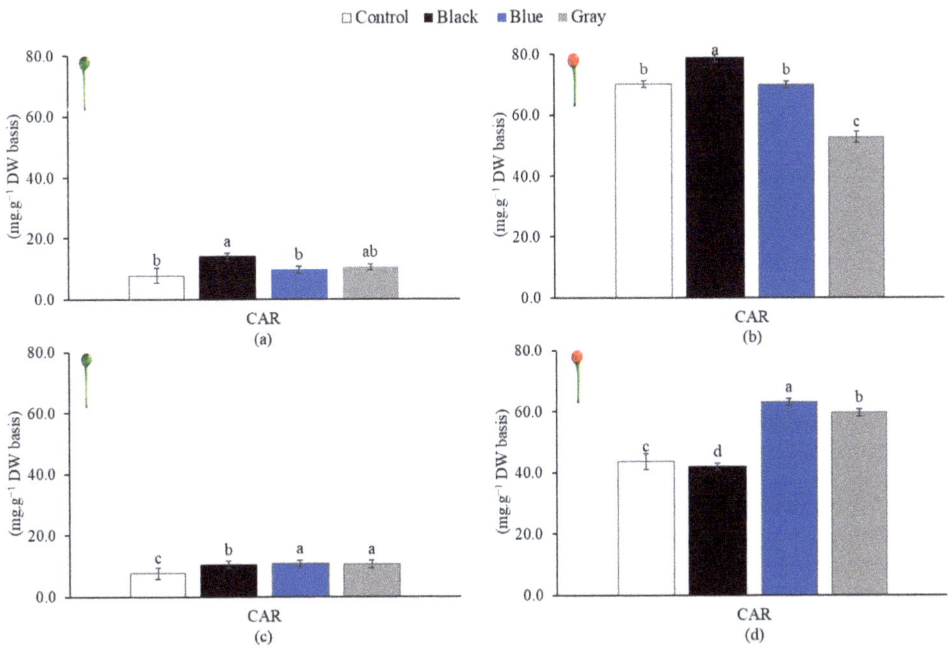

Figure 6. Mean carotenoid contents (β-carotene) (CAR, mg·g^{-1} DW basis) of green and red fruits of piquin peppers grown under colored shading net treatments in Queretaro, Mexico, across two production cycles (2021–2022). (**a**) Green fruits, first production cycle (2021). (**b**) Red fruits, first production cycle (2021). (**c**) Green fruits, second production cycle (2022). (**d**) Red fruits, second production cycle (2022). Error bars represent standard error of replicates. Values with the same letters are statistically similar. Tukey's mean comparison test (a = 0.05).

The maturation process of peppers brings about increased carotenoids in *Capsicum* fruits. In this process, chloroplasts differentiate into chromoplasts in the epicarp of the fruit, which leads to an accumulation of carotenoids. Increased carotenoids contribute to a change in the coloration of peppers, in which more than thirty types of carotenoids are involved [42–44]. In piquin peppers, this process starts at the initial stages of fruit development, where fruits exhibit a green color, indicating their immature stage. As they progress to the breaker stage, their color transitions to purple or orange, and finally, at full maturity, piquin peppers acquire a vibrant red color [1,45].

In our results, the shaded treatments affected the carotenoid content of piquin peppers. The black and blue shading treatments led to higher carotenoid contents than the unshaded control. Our findings coincide with previous studies on other *Capsicum* fruits that showed a significant increase in the carotenoid content in the shaded treatments compared to the unshaded conditions [46,47].

When bell pepper plants were grown in shaded conditions, there was a reduction in the exposure of plants to light. Reduced light intensity can increase carotenoid levels in sweet peppers grown under black nets, particularly β-carotene and lycopene [48]. In another cultivar of sweet pepper, unshaded treatments produced over 50% lower carotenoid levels than those grown under white or colored nets [49]. Nonetheless, the optimal shading treatment for increasing carotenoid contents may vary based on the specific cultivar and prevailing environmental conditions during production [50].

The reduced carotenoid contents in the unshaded nets (control treatment) could be related to the rapid destruction of carotenoids by high-intensity illumination [51]. Thus, the use of shade nets reduces light-intensity stress and protects the leaves from thylakoid damage caused by high irradiance [52]. Increased carotenoids enhance the nutritional value of peppers and enrich the vibrant colors, with a positive impact on fruit quality.

3.5. Total Phenolic Compounds of Piquin Peppers

The general effects of the different shading net treatments on the total phenolic compounds (TPC) of piquin peppers indicate that the shading treatments did not significantly affect the phenolic contents of the immature (green) fruits (Figure 7). In mature fruits, there was an increase in TPC compared to immature fruits in both production cycles. For mature fruits in the first production cycle, the control and black treatments showed slightly higher phenolic contents than the other treatments. In the second production cycle, the grey treatment showed a slightly higher phenolic content than the control and black treatments. Higher TPC in mature fruits has been previously described during the maturation process of habanero peppers and other *Capsicum* cultivars [53,54].

Our results indicate that the different shading net treatments affected the production of TPC as the piquin peppers reached maturity, but did not have a significant effect on immature fruits. Nonetheless, the specific effects of the colored shading treatments on the TPC of mature fruits are not clear, as the results varied from the 2021 to the 2022 cycle. In other shading studies using cultivars of sweet peppers, the unshaded, white, and pearl shade treatments increased the phenolic compounds and antioxidant activity, whereas black shade nets caused a reduction in phenolic compounds [15,50]. The higher phenolic compounds in the unshaded environments could be attributed to higher irradiance, which triggers a stress response in the plant, leading to the production of higher levels of phenolic compounds as a protective mechanism. Phenolic compounds can act as antioxidants and protect the plant from oxidative damage caused by increased irradiance. UV radiation is a stress factor that stimulates the biosynthesis of flavonoids and phenolic compounds, resulting in higher contents as a response to UV radiation [55].

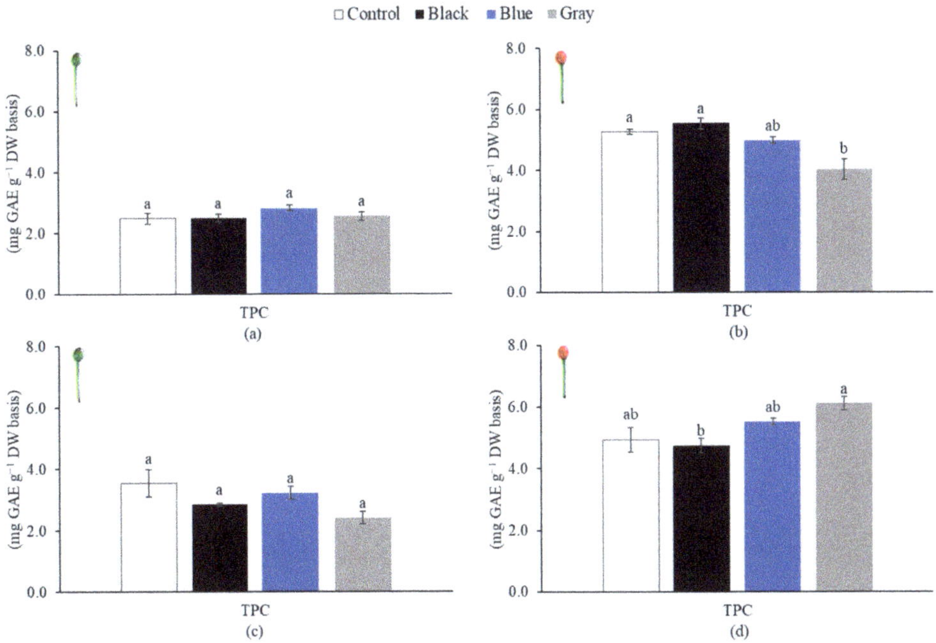

Figure 7. Mean total phenolic compounds (TPC, mg GAE·g^{-1} DW basis) of green and red fruits of piquin peppers grown under colored shading net treatments in Queretaro, Mexico, across two production cycles (2021–2022). (**a**) Green fruits, first production cycle (2021). (**b**) Red fruits, first production cycle (2021). (**c**) Green fruits, second production cycle (2022). (**d**) Red fruits, second production cycle (2022). Error bars represent standard error of replicates. Means with the same letters are statistically similar. Tukey's mean comparison test (a = 0.05).

3.6. Antioxidant Capacity

The antioxidant capacity of immature (green) and mature (red) peppers determined using the ABTS method showed that the blue treatment caused the highest antioxidant capacity at both maturation stages, followed by the gray and black shades (Figure 8). The control treatment had the lowest antioxidant capacity in both maturation stages. Comparable results were obtained in the second production cycle, in which the blue and gray treatments also exhibited a higher antioxidant capacity than the control fruits in both maturation stages. The black shade treatment also increased the antioxidant capacity in immature fruits in both production cycles. However, the differences in antioxidant capacity were not statistically significant in the mature fruits of the second production cycle, and the black shading treatment showed a reduction in antioxidant capacity compared to the control group.

The determination of antioxidant capacity by DPPH showed a similar trend, with the blue shade net treatment having the highest antioxidant capacity in most cases; the control treatment had the lowest antioxidant capacity. However, some differences between the results from the ABTS and DPPH could be identified, with some treatments showing different antioxidant capacity levels depending on the method. These methods served as complementary assays to explore a wide range of antioxidant compounds; ABTS can estimate both hydrophilic and lipophilic radicals [56], while DPPH is particularly sensitive to lipophilic radicals [57]. Antioxidants in peppers predominantly arise from a combination of hydrophilic compounds, mainly phenolic compounds, and lipophilic compounds such as carotenoids [43,58].

During maturation, the red fruits generally had a higher antioxidant capacity than the green fruits, as expected. The antioxidant capacity and ascorbic acid content significantly increased in different pepper cultivars during the growth and maturation of the fruits; the highest levels were found in the last stage of maturity [53,54]. This difference was more pronounced in the DPPH results, where the differences between green and red fruits were higher than in the ABTS results.

Our results indicate that the shading treatments had an enhancing effect on the antioxidant capacity of the piquin peppers, particularly the blue shade treatment. The specific effects varied depending on the shade color and fruit maturation stage. The results confirmed the potential health benefits of consuming piquin peppers, particularly when they are fully ripe. Previous results regarding piquin peppers have reported that differences in antioxidant capacity are a result of more than 32 compounds identified, mainly phenolic compounds that contribute to the free radical scavenging of fruits [59]. The antioxidant capacity of *Capsicum* fruits is directly related to the total phenolic compounds [60].

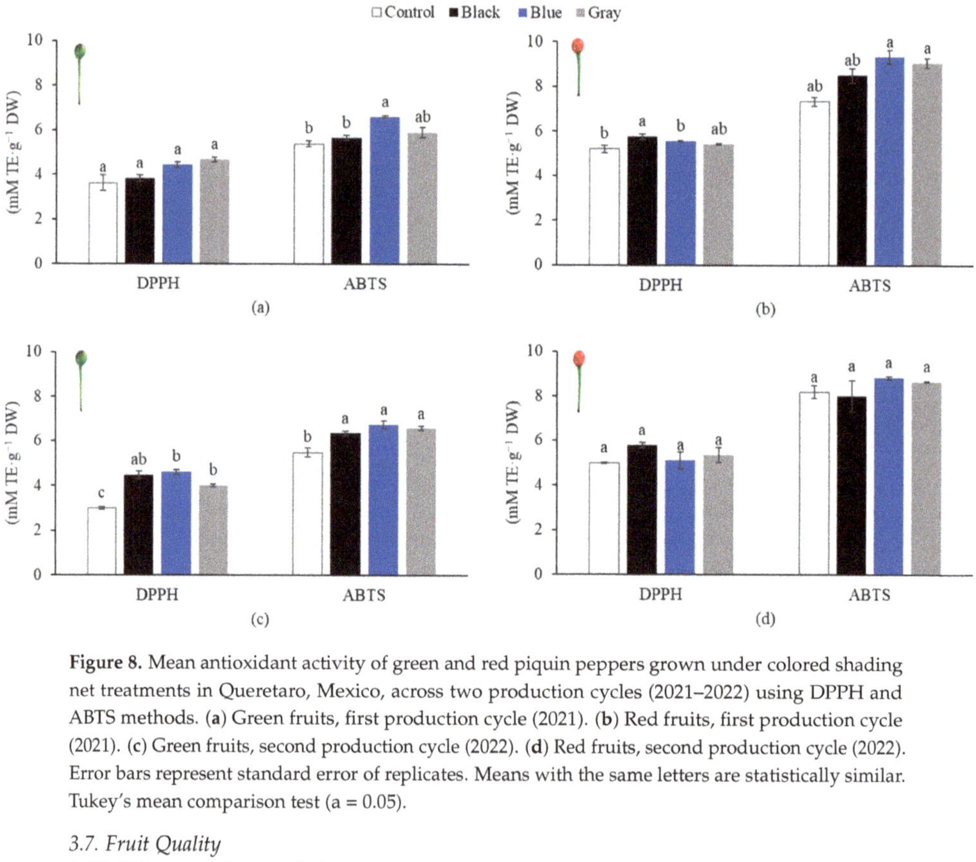

Figure 8. Mean antioxidant activity of green and red piquin peppers grown under colored shading net treatments in Queretaro, Mexico, across two production cycles (2021–2022) using DPPH and ABTS methods. (**a**) Green fruits, first production cycle (2021). (**b**) Red fruits, first production cycle (2021). (**c**) Green fruits, second production cycle (2022). (**d**) Red fruits, second production cycle (2022). Error bars represent standard error of replicates. Means with the same letters are statistically similar. Tukey's mean comparison test (a = 0.05).

3.7. Fruit Quality

3.7.1. Effects on Pepper Color

The color of piquin peppers was significantly affected by the shading treatments in both production cycles (Tables 1 and 2). In the first cycle, the control fruits had a lighter color (the highest L* value) for both immature and red peppers in relation to the other treatments. In immature peppers, the black treatment had the highest a* and b* values, indicating a more intense green color. The blue treatment had the highest a* and b* values for mature peppers, indicating a more intense red color. Color values in fruits may be used as predictors of pigment concentrations, and the reduced a* values observed in unshaded treatments can be related to a reduction in carotenoids [50].

Table 1. Mean colorimetric values of green and red piquin peppers grown under colored shading net treatments in Queretaro, Mexico, during the first production cycle (2021).

Treatment	Green Fruits			Red Fruits		
	L*	a*	b*	L*	a*	b*
Control	30.88 ± 2.32 a	−8.67 ± 5.09 a	31.20 ± 2.28 a	28.51 ± 0.88 a	43.69 ± 4.92 a	38.24 ± 1.72 b
Black	30.26 ± 2.69 a	−15.45 ± 21.48 a	28.04 ± 3.07 b	27.64 ± 2.42 a	43.39 ± 1.40 a	38.60 ± 3.07 b
Blue	29.33 ± 1.57 a	−9.98 ± 0.38 a	28.46 ± 1.83 b	28.83 ± 1.60 a	45.00 ± 1.07 a	40.50 ± 2.12 ab
Gray	30.16 ± 1.58 a	−10.24 ± 0.31 a	31.73 ± 1.85 a	28.88 ± 1.58 a	44.55 ± 1.07 a	42.06 ± 2.92 a

L*: indicates lightness, a*: is the red/green coordinate, and b*: is the yellow/blue coordinate. ± Standard deviation (n = 4). Means with the same letters are statistically similar. Tukey's mean comparison test (a = 0.05).

Table 2. Mean colorimetric values of green and red piquin peppers grown under colored shading net treatments in Queretaro, Mexico., during the second production cycle (2022).

Treatment	Green Fruits			Red Fruits		
	L*	a*	b*	L*	a*	b*
Control	28.64 ± 1.78 a	−7.43 ± 1.60 a	40.47 ± 2.15 a	36.77 ± 2.29 a	35.02 ± 2.23 a	20.84 ± 1.97 a
Black	24.81 ± 1.59 b	−9.03 ± 0.94 b	35.57 ± 2.70 bc	38.06 ± 0.60 a	34.00 ± 1.40 a	19.22 ± 1.63 a
Blue	23.73 ± 2.20 b	−10.29 ± 0.78 c	36.90 ± 1.61 b	35.30 ± 8.86 a	21.49 ± 19.67 b	22.36 ± 7.29 a
Gray	24.69 ± 2.56 b	−9.82 ± 0.38 bc	34.61 ± 2.28 c	38.19 ± 1.16 a	34.17 ± 0.77 a	20.25 ± 1.10 a

L*: indicates lightness, a*: is the red/green coordinate, and b*: is the yellow/blue coordinate. ± Standard deviation (n = 4). Means with the same letters are statistically similar. Tukey's mean comparison test (a = 0.05).

In the second production cycle, the control fruits also had the highest L* value for both maturation stages, indicating a lighter color than in the other treatments. The black treatment had a more intense red color (highest a* value) for mature peppers, while for green peppers, the blue treatment caused a more intense yellowish-green color (highest b* value). Consistent findings have been observed in bell peppers, where the L* value was highest in unshaded treatments, while the a* value was the lowest in the unshaded treatments and reached the highest values under black and red shading nets [50]. The intense red color may be related to the higher concentration of carotenoids and anthocyanins, which are responsible for the red coloration of mature fruits [38]. Our results suggest that the differential effects of the colored shade treatments on pepper color could be related to an effect on the pigment biosynthesis (particularly carotenoids), but it is not clear whether the treatments affected chlorophyll degradation, a process that occurs during fruit maturation [61].

Nonetheless, our results indicate that the different shading treatments affected the color of piquin peppers, with some treatments resulting in more intense and vibrant colors than others. These findings could be useful for understanding the factors that influence the color of piquin peppers and to develop commercial strategies to improve color and fruit quality.

3.7.2. Size and Weight

In general, the immature (green) fruits were longer and narrower than the mature fruits (Tables 3 and 4). During fruit maturation, peppers undergo a highly intense metabolism, the emission of volatile compounds associated with fruit respiration, and changes in their cellular structure, leading to water loss, thus becoming denser [53,62]. This can result in a reduction in overall size, even as the fruit reaches its full flavor and nutrient potential. Additionally, size reduction may also be influenced by genetic factors [62], environmental conditions [48], and cultural practices [63].

Table 3. Mean morphological analysis, width, length, and weight of green and red piquin peppers grown under colored shading net treatments in Queretaro, Mexico, during the first production cycle (2021).

Treatment	Green Fruits			Red Fruits		
	Fruit Width	Fruit Length	Weight	Fruit Width	Fruit Length	Weight
Control	8.73 ± 0.32 ab	10.22 ± 0.40 ab	2683.83 ± 344.82 a	8.84 ± 0.33 a	9.98 ± 0.40 a	2553.73 ± 180.66 a
Black	8.91 ± 0.24 a	10.81 ± 0.49 a	2883.33 ± 336.51 a	8.33 ± 0.42 bc	9.57 ± 0.58 a	2242.59 ± 423.48 a
Blue	8.23 ± 0.35 c	10.42 ± 0.80 ab	2711.35 ± 319.59 a	8.62 ± 0.33 ab	10.01 ± 0.57 a	2348.14 ± 428.44 a
Gray	8.39 ± 0.26 bc	10.01 ± 0.38 b	2802.46 ± 306.35 a	8.14 ± 0.41 c	9.99 ± 0.57 a	2731.55 ± 411.68 a

Width and length, average value (mm) ± Standard deviation (n = 10). Weight, average value (mg) ± Standard deviation (n = 10). Means with the same letters are statistically similar. Tukey's mean comparison test (a = 0.05).

Table 4. Mean morphological analysis, width, length, and weight of green and red piquin peppers grown under colored shading net treatments in Queretaro, Mexico, during the second production cycle (2022).

Treatment	Green Fruits			Red Fruits		
	Fruit Width	Fruit Length	Weight	Fruit Width	Fruit Length	Weight
Control	8.56 ± 0.28 a	10.32 ± 0.43 a	2977.50 ± 283.44 a	7.93 ± 0.19 b	9.54 ± 0.20 b	2924.50 ± 162.81 b
Black	8.47 ± 0.27 a	10.02 ± 0.34 a	2852.78 ± 187.14 ab	8.54 ± 0.17 a	9.84 ± 0.41 ab	3046.45 ± 232.93 b
Blue	8.02 ± 0.33 b	8.91 ± 0.31 b	2484.24 ± 107.96 b	8.40 ± 0.15 a	10.21 ± 0.33 a	3375.51 ± 46.85 a
Gray	8.24 ± 0.38 ab	9.26 ± 0.33 b	2586.40 ± 154.60 ab	8.37 ± 0.26 a	10.27 ± 0.50 a	2555.88 ± 99.13 c

Width and length, average value (mm) ± Standard deviation (n = 10). Weight, average value (mg) ± Standard deviation (n = 10). Means with the same letters are statistically similar. Tukey's mean comparison test (a = 0.05).

The colored shade treatments had a significant effect on the morphology of piquin peppers. While the control and black shade treatments produced the widest fruits in both production cycles, the blue shade treatment caused the thinnest fruits. In terms of fruit length, the black and the blue shade treatment caused the longest fruits in the first and second cycles, respectively. Our results seem to concur with previous studies on piquin peppers that indicate that intermediate black shading (50% shade) increases yield in terms of fruit size and the number of fruits [17].

Previous studies have reported that the use of shading nets can increase the size of *Capsicum* fruits [64], although the specific conditions and types of nets used were different. In sweet peppers, the use of red and pearl shade nets favored fruit growth, which resulted in an increased fruit yield, producing fruits with a thicker pericarp [16]. As for fruit weight, the blue shade treatment had the heaviest fruits in the second production cycle after maturation.

4. Conclusions

Our results highlight the effects of the different color shading treatments on the capsaicinoids, carotenoids, and phenolic compounds of piquin pepper fruits at two maturation stages. The maturation process naturally increased the compound levels and antioxidant capacity, as mature red fruits exhibited the highest levels of phytochemicals and showed a greater antioxidant capacity than immature green fruits. Black shading was determined to boost capsaicinoid content, while black and blue shading led to an increase in carotenoid levels. The use of black and gray shading resulted in elevated phenolic compound levels. Notably, no significant variations in antioxidant capacity were observed in the different treatments. Morphological attributes such as fruit size and weight were affected by both colored shading and fruit maturity. These findings emphasize the potential health benefits of mature red piquin peppers, suggesting their utility in the development of functional foods, nutraceuticals, and pharmaceuticals. Shading can be considered a promising technique to enhance the phytochemical contents of piquin pepper fruits. Our results suggest promising avenues for future research and applications. In the agricultural context, we recommend the further exploration of innovative light management techniques to increase the phytochemical content of piquin or other peppers. Given the limited existing agricultural practices in this area, our work opens the door to a wide range of possibilities for researchers and farmers engaged in the cultivation of piquin peppers.

Author Contributions: Conceptualization, Y.J.-V. and J.I.V.-B.; methodology, Y.J.-V. and J.I.V.-B.; formal analysis, Y.J.-V.; investigation, Y.J.-V.; resources, J.I.V.-B.; data curation, Y.J.-V.; writing—original draft preparation, Y.J.-V.; writing—review and editing, J.I.V.-B.; visualization, Y.J.-V.; supervision, J.I.V.-B.; project administration, J.I.V.-B.; funding acquisition, J.I.V.-B. All authors have read and agreed to the published version of the manuscript.

Funding: This research was funded by the Consejo Nacional de Humanidades, Ciencia y Tecnología (CONAHCyT Spanish acronym) grant number 624964 and the APC was funded by Tecnologico de Monterrey.

Data Availability Statement: The data presented in this study are available on request from the corresponding author due to laboratory policy.

Acknowledgments: The authors thank CONAHCyT and Tecnologico de Monterrey for providing the Ph.D. scholarship for Yamir Jiménez-Viveros (624964). In addition, they would like to thank Anaberta Cardador for her valuable insights for the analytical determinations, and to CAETEC for field support during the field experiments.

Conflicts of Interest: The authors declare no conflict of interest.

References

1. Mares-Quiñones, M.D.; Valiente-Banuet, J.I. Horticultural aspects for the cultivated production of piquin peppers (*Capsicum annuum* L. var. *Glabriusculum*) a review. *HortScience* **2019**, *54*, 70–75. [CrossRef]
2. Rodríguez-Del Bosque, L.A. Preferencia del consumidor por el chile piquín en comparación con otros chiles en el noreste de México. *Rev. Chapingo Ser. Hortic.* **2005**, *11*, 279–281. Available online: https://www.redalyc.org/articulo.oa?id=60911214 (accessed on 5 October 2023).
3. Coronado-García, M.A.; Córdova-Yánez, A.; García-Porchas, M.; Santiago-Hernández, V.G.; Vásquez Navarro, R.A. Estrategias de mercado para productos elaborados a base de chiltepín en la sierra de Sonora. *Rev. Mex. Agroneg.* **2013**, *32*, 359–370.
4. Rodríguez-del Bosque, L.A. Producción intensiva de chile piquín en el norte de Tamaulipas. Ficha tecnológica por sistema producto. *INIFAP* **2008**. Available online: http://www.inifapcirne.gob.mx/Biblioteca/Publicaciones/536.pdf (accessed on 6 October 2023).
5. Meckelmann, S.W.; Riegel, D.W.; van Zonneveld, M.; Ríos, L.; Peña, K.; Mueller-Seitz, E.; Petz, M. Capsaicinoids, flavonoids, tocopherols, antioxidant capacity and color attributes in 23 native peruvian chili peppers (*Capsicum* spp.) grown in three different locations. *Eur. Food Res. Technol.* **2014**, *240*, 273–283. [CrossRef]
6. Kumar, A.; Kumar, S.; Anju, T.; Ramchiary, N. Genetic, epigenetic, and hormonal regulation of fruit development and ripening in *Capsicum* L. species. *Annu. Plant Rev. Online* **2021**, *4*, 295–356. [CrossRef]
7. Aza-González, C.; Núñez-Palenius, H.G.; Ochoa-Alejo, N. Molecular biology of capsaicinoid biosynthesis in chili pepper (*Capsicum* spp.). *Plant Cell Rep.* **2011**, *30*, 695–706.
8. Arce-Rodríguez, M.L.; Ochoa-Alejo, N. An R2R3-MYB transcription factor regulates capsaicinoid biosynthesis. *Plant Physiol.* **2017**, *174*, 1359–1370. [CrossRef]
9. Naves, E.R.; de Ávila Silva, L.; Sulpice, R.; Araújo, W.L.; Nunes-Nesi, A.; Peres, L.E.P.; Zsögön, A. Capsaicinoids: Pungency beyond *Capsicum*. *Trends Plant Sci.* **2019**, *24*, 109–120. [CrossRef]
10. Llorente, B.; D'Andrea, L.; Ruiz-Sola, M.A.; Botterweg, E.; Pulido, P.; Andilla, J.; Loza-Alvarez, P.; Rodriguez-Concepcion, M. Tomato fruit carotenoid biosynthesis is adjusted to actual ripening progression by a light-dependent mechanism. *Plant J.* **2016**, *85*, 107–119. [CrossRef]
11. Bian, Z.H.; Yang, Q.C.; Liu, W.K. Effects of light quality on the accumulation of phytochemicals in vegetables produced in controlled environments: A review. *J. Sci. Food Agric.* **2015**, *95*, 869–877.
12. Lemos, V.C.; Reimer, J.J.; Wormit, A. Color for life: Biosynthesis and distribution of phenolic compounds in pepper (*Capsicum annuum*). *Agriculture* **2019**, *9*, 81. [CrossRef]
13. Ncise, W.; Daniels, C.W.; Nchu, F. Effects of light intensities and varying watering intervals on growth, tissue nutrient content and antifungal activity of hydroponic cultivated *Tulbaghia violacea* L. under greenhouse conditions. *Heliyon* **2020**, *6*, e03906. [CrossRef]
14. Jiménez-Viveros, Y.; Núñez-Palenius, H.G.; Fierros-Romero, G.; Valiente-Banuet, J.I. Modification of light characteristics affect the phytochemical profile of peppers. *Horticulturae* **2023**, *9*, 72. [CrossRef]

15. Mashabela, M.N.; Selahle, K.M.; Soundy, P.; Crosby, K.M.; Sivakumar, D. Bioactive compounds and fruit quality of green sweet pepper grown under different colored shade netting during postharvest storage. *J. Food Sci.* **2015**, *80*, H2612–H2618. [CrossRef]
16. Ilić, Z.S.; Milenković, L.; Šunić, L.; Barać, S.; Mastilović, J.; Kevrešan, Ž.; Fallik, E. Effect of shading by coloured nets on yield and fruit quality of sweet pepper. *Zemdirbyste* **2017**, *104*, 53–62. [CrossRef]
17. Valiente-Banuet, J.I.; Gutierrez-Ochoa, A. Effect of irrigation frequency and shade levels on vegetative growth, yield, and fruit quality of piquin pepper (*Capsicum annuum* L. Var. *glabriusculum*). *HortScience* **2016**, *51*, 573–579. [CrossRef]
18. Mohawesh, O.; Albalasmeh, A.; Deb, S.; Singh, S.; Simpson, C.; Alkafaween, N.; Mahadeen, A. Effect of colored shading nets on the growth and water use efficiency of sweet pepper grown under semi-arid conditions. *Horttechnology* **2022**, *32*, 21–27. [CrossRef]
19. Sivakumar, D.; Jifon, J.; Soundy, P. Spectral quality of photo-selective shade nettings improves antioxidants and overall quality in selected fresh produce after postharvest storage. *Food Rev. Int.* **2018**, *34*, 290–307. [CrossRef]
20. Santana, J.Q.; Balbino, M.A.; Tavares, T.R.; Bezerra, R.S.; Farias, J.G.; Ferreira, R.C. Effect of photoselective screens in the development and productivity of red and yellow sweet pepper. *Acta Hortic.* **2012**, *956*, 493–500. [CrossRef]
21. Stamps, R.H. Use of colored shade netting in horticulture. *HortScience* **2009**, *44*, 239–241. [CrossRef]
22. Castellano, S.; Mugnozza, G.S.; Russo, G.; Briassoulis, D.; Mistriotis, A.; Hemming, S.; Waaijenberg, D. Plastic nets in agriculture: A general review of types and applications. *Appl. Eng. Agric.* **2008**, *24*, 799–808. [CrossRef]
23. Britannica. Tropical and Subtropical Steppe Climate, Deserts, arid Regions, Semi-Arid. Available online: https://www.britannica.com/science/tropical-and-subtropical-steppe-climate (accessed on 11 July 2023).
24. INEGI. Available online: https://www.inegi.org.mx/app/biblioteca/ficha.html?upc=702825293147 (accessed on 11 July 2023).
25. Wahyuni, Y.; Ballester, A.R.; Sudarmonowati, E.; Bino, R.J.; Bovy, A.G. Metabolite biodiversity in pepper (*Capsicum*) fruits of thirty-two diverse accessions: Variation in health-related compounds and implications for breeding. *Phytochemistry* **2011**, *72*, 1358–1370. [CrossRef]
26. Mínguez-Mosquera, M.I.; Hornero-Méndez, D. Separation and quantification of the carotenoid pigments in red peppers (*Capsicum annuum* L.), paprika, and oleoresin by reversed-phase HPLC. *J. Agric. Food Chem.* **1993**, *41*, 1616–1620. [CrossRef]
27. Blanco-Ríos, A.K.; Medina-Juárez, L.Á.; González-Aguilar, G.A.; Gámez-Meza, N. Antioxidant activity of the phenolic and oily fractions of different sweet bell peppers. *J. Mex. Chem. Soc.* **2013**, *57*, 137–143. [CrossRef]
28. Singleton, V.L.; Orthofer, R.; Lamuela-Raventós, R.M. Analysis of total phenols and other oxidation substrates and antioxidants by means of folin-ciocalteu reagent. *Methods Enzymol.* **1999**, *299*, 152–178. [CrossRef]
29. Rajurkar, N.; Hande, S.M. Estimation of phytochemical content and antioxidant activity of some selected traditional Indian medicinal plants. *Indian J. Pharm. Sci.* **2011**, *73*, 146. [CrossRef]
30. Re, R.; Pellegrini, N.; Proteggente, A.; Pannala, A.; Yang, M.; Rice-Evans, C. Antioxidant activity applying an improved ABTS radical cation decolorization assay. *Free Radic. Biol. Med.* **1999**, *26*, 1231–1237. [CrossRef]
31. Fukumoto, L.R.; Mazza, G. Assessing Antioxidant and Prooxidant Activities of Phenolic Compounds†. *J. Agric. Food Chem.* **2000**, *48*, 3597–3604. [CrossRef]
32. Rhim, J.W.; Hong, S.I. Effect of Water Activity and Temperature on the Color Change of Red Pepper (*Capsicum annuum* L.) Powder. *Food Sci. Biotechnol.* **2011**, *20*, 215–222. [CrossRef]
33. Ibraheem, N.A.; Hasan, M.M.; Khan, R.Z.; Mishra, P.K. Understanding color models: A review. *ARPN J. Sci. Technol.* **2012**, *2*, 265–275. [CrossRef]
34. Ahemd, H.A.; Al-Faraj, A.A.; Abdel-Ghany, A.M. Shading greenhouses to improve the microclimate, energy and water saving in hot regions: A review. *Sci. Hortic.* **2016**, *201*, 36–45. [CrossRef]
35. Olivares-Soto, H.; Bastías, R.M. Photosynthetic efficiency of apples under protected shade nets. *Chil. J. Agric. Res.* **2018**, *78*, 126–138. [CrossRef]
36. Bastías, R.M.; Boini, A. Apple Production under Protective Netting Systems. In *Apple Cultivation—Recent Advances*, 1st ed.; Küden, A., Ed.; IntechOpen: London, UK, 2022; pp. 91–102.
37. Fayos, O.; De Aguiar, A.C.; Jiménez-Cantizano, A.; Ferreiro-González, M.; Garcés-Claver, A.; Martínez, J.; Mallor, C.; Ruiz-Rodríguez, A.; Palma, M.; Barroso, C.G.; et al. Ontogenetic variation of individual and total capsaicinoids in malagueta peppers (*Capsicum Frutescens*) during fruit maturation. *Molecules* **2017**, *22*, 736. [CrossRef] [PubMed]
38. Nagy, Z.; Daood, H.; Neményi, A.; Ambrózy, Z.; Pék, Z.; Helyes, L. Impact of shading net color on phytochemical contents in two chili pepper hybrids cultivated under greenhouse conditions. *Korean J. Hortic. Sci. Technol.* **2017**, *35*, 418–430. [CrossRef]
39. Agyemang, D.S.; Nagy, Z.; e Souza, C.S.; Pék, Z.; Neményi, A.; Helyes, L. Effect of net shading technology on the yield quality and quantity of chilli pepper under greenhouse cultivation. *Acta Agrar. Debr.* **2021**, *1*, 5–9. [CrossRef]
40. Pacheco, F.V.; Alvarenga, I.C.A.; Junior, P.M.R.; Pinto, J.E.B.P.; Avelar, R.d.P.; Alvarenga, A.A. Growth and production of secondary compounds in monkey-pepper (*Piper Aduncum* L.) leaves cultivated under altered ambient light. *Aust. J. Crop Sci.* **2014**, *8*, 1510–1516. [CrossRef]
41. Jeeatid, N.; Techawongstien, S.; Suriharn, B.; Bosland, P.W.; Techawongstien, S. Light intensity affects capsaicinoid accumulation in hot pepper (*Capsicum Chinense* Jacq.) cultivars. *Hortic. Environ. Biotechnol.* **2017**, *58*, 103–110. [CrossRef]
42. Batiha, G.E.S.; Alqahtani, A.; Ojo, O.A.; Shaheen, H.M.; Wasef, L.; Elzeiny, M.; Ismail, M.; Shalaby, M.; Murata, T.; Zaragoza-Bastida, A.; et al. Biological properties, bioactive constituents, and pharmacokinetics of some *Capsicum* spp. and capsaicinoids. *Int. J. Mol. Sci.* **2020**, *21*, 5179. [CrossRef]

43. Hassan, N.M.; Yusof, N.A.; Yahaya, A.F.; Rozali, N.N.M.; Othman, R. Carotenoids of *Capsicum* fruits: Pigment profile and health-promoting functional attributes. *Antioxidants* **2019**, *8*, 469. [CrossRef]
44. Rodríguez-Rodríguez, E.; Sánchez-Prieto, M.; Olmedilla-Alonso, B. Assessment of carotenoid concentrations in red peppers (*Capsicum annuum*) under domestic refrigeration for three weeks as determined by HPLC-DAD. *Food Chem. X* **2020**, *6*, 100092. [CrossRef]
45. Salinas Hernández, M.; Ma, R.; Liévano, L.; Andrés, E.; Jiménez, P. Caracterización morfológica y cambios durante la vida postcosecha de cuatro tipos de chile amashito (*Capsicum annuum* L.) variedad *glabriusculum* (Dunal) Heiser & Pickersgill. *Rev. Iberoam. Tecnol. Postcosecha* **2010**, *11*, 92–100. Available online: https://www.redalyc.org/articulo.oa?id=81315093012 (accessed on 17 August 2023).
46. Ambrózy, Z.; Daood, H.; Nagy, Z.; Darázsi Ledó, H.; Helyes, L. Effect of net shading technology and harvest times on yield and fruit quality of sweet pepper. *Appl. Ecol. Environ. Res.* **2016**, *14*, 99–109. [CrossRef]
47. Castillejo, N.; Martínez-Zamora, L.; Artés-Hernández, F. Postharvest UV radiation enhanced biosynthesis of flavonoids and carotenes in bell peppers. *Postharvest Biol. Technol.* **2022**, *184*, 111774. [CrossRef]
48. Ilić, Z.S.; Fallik, E. Light quality manipulation improves vegetable quality at harvest and postharvest: A review. *Environ. Exp. Bot.* **2017**, *139*, 79–90. [CrossRef]
49. Ombódi, A.O.; Pék, Z.; Szuvandzsiev, P.; Lugasi, A.; Ledóné Darázsi, H.; Helyes, L. Effect of coloured shade nets on some nutritional characteristics of a kapia type pepper grown in plastic tunnel. *Columella J. Agric. Environ. Sci.* **2016**, *3*, 25–33. [CrossRef]
50. Díaz-Pérez, J.C.; St. John, K.; Kabir, M.Y.; Alvarado-Chávez, J.A.; Cutiño-Jiménez, A.M.; Bautista, J.; Gunawan, G.; Nambeesan, S.U. Bell Pepper (*Capsicum Annum* L.) under colored shade nets: Fruit yield, postharvest transpiration, color, and chemical composition. *HortScience* **2020**, *55*, 181–187. [CrossRef]
51. Horváth, G.; Kissimon, J.; Faludi-Dániel, Á. effect of light intensity on the formation of carotenoids in normal and mutant maize leaves. *Phytochemistry* **1972**, *11*, 183–187. [CrossRef]
52. Sandmann, G.; Kuhn, M.; Böger, P. Carotenoids in photosynthesis: Protection of D1 degradation in the light. *Photosynth. Res.* **1993**, *35*, 185–190. [CrossRef]
53. Howard, L.R.; Talcott, S.T.; Brenes, C.H.; Villalon, B. Changes in phytochemical and antioxidant activity of selected pepper cultivars (*Capsicum* species) as influenced by maturity. *J. Agric. Food Chem.* **2000**, *48*, 1713–1720. [CrossRef]
54. Ionică, M.E.; Nour, V. Bioactive compounds and antioxidant activity of hot pepper fruits at different stages of growth and ripening. *J. Appl. Bot. Food Qual.* **2017**, *90*, 232–237. [CrossRef]
55. Angmo, P.; Dolma, T.; Phuntsog, N.; Chaurasia, O.P.; Stobdan, T. Effect of shading and high temperature amplitude on yield and phenolic contents of greenhouse capsicum (*Capsicum annuum* L.). *J. Biol. Pharm.* **2021**, *4*, 30–39. [CrossRef]
56. Arnao, M.B.; Cano, A.; Acosta, M. The hydrophilic and lipophilic contribution to total antioxidant activity. *Food Chem.* **2001**, *73*, 239–244. [CrossRef]
57. Nagarajan, J.; Ramanan, R.N.; Raghunandan, M.E.; Galanakis, C.M.; Krishnamurthy, N.P. Carotenoids. In *Nutraceutical and Functional Food Components: Effects of Innovative Processing Techniques*; Academic Press: Cambridge, MA, USA, 2017; pp. 259–296. [CrossRef]
58. Mercy, E.R.; David, U. Potential health benefits of conventional nutrients and phytochemicals of Capsicum peppers. *Pharm. Pharmacol. Int. J.* **2018**, *6*, 62–69. [CrossRef]
59. Del Rocio Moreno-Ramírez, Y.; Martínez-Ávila, G.C.G.; González-Hernández, V.A.; Castro-López, C.; Torres-Castillo, J.A. Free radical-scavenging capacities, phenolics and capsaicinoids in wild piquin chili (*Capsicum annuum* var. *glabriusculum*). *Molecules* **2018**, *23*, 2655. [CrossRef]
60. Castro-Concha, L.A.; Tuyub-Che, J.; Moo-Mukul, A.; Vazquez-Flota, F.A.; Miranda-Ham, M.L.; Bekatorou, A.; Tariq, A.; Tripathi, N.K. Antioxidant capacity and total phenolic content in fruit tissues from accessions of *Capsicum Chinense* Jacq. (Habanero pepper) at different stages of ripening. *Sci. World J.* **2014**, *2014*, 809073. [CrossRef]
61. Pola, W.; Sugaya, S.; Photchanachai, S. Color development and phytochemical changes in mature green chili (*Capsicum annuum* L.) exposed to red and blue light-emitting diodes. *J. Agric. Food Chem.* **2020**, *68*, 59–66. [CrossRef]
62. Chaki, M.; Álvarez De Morales, P.; Ruiz, C.; Begara-Morales, J.C.; Barroso, J.B.; Corpas, F.J.; Palma, J.M. Ripening of Pepper (*Capsicum annuum*) Fruit is characterized by an enhancement of protein tyrosine nitration. *Ann. Bot.* **2015**, *116*, 637. [CrossRef] [PubMed]
63. Ombódi, A.; Pék, Z.; Szuvandzsiev, P.; Taskovics, Z.T.; Koházi-Kis, A.; Kovács, A.; Darázsi, H.L.; Helyes, L. Effects of external coloured shade nets on sweet peppers cultivated in walk-in plastic tunnels. *Not. Bot. Horti Agrobot. Cluj-Napoca* **2015**, *43*, 398–403. [CrossRef]
64. Díaz-Pérez, J.C. Bell pepper (*Capsicum annum* L.) crop as affected by shade level: Fruit yield, quality, and postharvest attributes, and incidence of phytophthora blight (caused by *Phytophthora capsici* Leon.). *HortScience* **2014**, *49*, 891–900. [CrossRef]

horticulturae

MDPI

Article

Quantifying the Effect of Light Intensity Uniformity on the Crop Yield by Pea Microgreens Growth Experiments

László Balázs *, Gergő Péter Kovács, Csaba Gyuricza, Petra Piroska, Ákos Tarnawa and Zoltán Kende

Institute of Agronomy, Hungarian University of Agriculture and Life Sciences, 2100 Gödöllő, Hungary; kovacs.gergo.peter@uni-mate.hu (G.P.K.); gyuricza.csaba@uni-mate.hu (C.G.); piroska.petra@phd.uni-mate.hu (P.P.); tarnawa.akos@uni-mate.hu (Á.T.); kende.zoltan@uni-mate.hu (Z.K.)
* Correspondence: balazs.laszlo@uni-mate.hu

Abstract: Differences in individual plant growth are affected by the spatial variation of light intensity, reducing the homogeneity of microgreen crops. Identifying the tradeoffs between light uniformity and crop quality is challenging due to the confounding effect of nonuniform illuminance with other noise factors. This study presents the results of hydroponic pea (*Pisum sativum*, L.) growth experiments aimed at quantifying the effect of photon irradiance variations. By adjusting the power of LED luminaires, we established one uniformly illuminated zone and two non-uniformly illuminated zones. Germinated seeds with 6 cm-long radicles were transplanted to cultivation trays with known light intensity in predetermined positions. Plants were cut 12 days after the start of light treatment and measured for fresh weight and shoot height. Our findings revealed no significant difference between the crop yield on trays having the same average PPFD but different light uniformity. However, correlation analysis of individual measurement data showed that local PPFD differences explained 31% of the fresh weight variation, and the rest was attributed to noise in the germination and growth processes. We also discuss the implications of our findings for the design and optimization of vertical farms.

Keywords: vertical farm; plant factory; PPFD; photon irradiance; LED

1. Introduction

Growing young edible vegetables, collectively known as microgreens, have gained popularity recently and have become one of the fastest-growing segments of indoor vertical farming [1]. Microgreens are young seedlings of leafy vegetables and herbs that are favored in new culinary trends due to their unusual appearance, bright color, intense flavor, crisp texture, and unique nutrient profile [2]. A huge number of species can be consumed as microgreens [3]. By value, *Brassicaceae* microgreens dominate the global market, led by broccoli at 15%, followed by arugula at 9% [1]. Edible plants originally cultivated for seeds and not for shoots, like peas, beans, cereals, and sunflowers, are also popular microgreens and are cultivated in large quantities.

Microgreens containing high levels of carotenoids, chlorophylls, and organic acids are associated with several health benefits, including anti-diabetic and anticholinergic activity, and are recommended as a functional food for a daily diet [4]. Microgreens are harvested at an immature growth stage, shortly after the full development of cotyledons and at the emergence of the first true leaves [5]. Depending on the species, the time between seeding and harvest is between 1 and 2 weeks [6]. The short cultivation cycle, high seeding density, low shoot height, and high market value make microgreens an attractive crop for vertical farming.

Indoor vertical farms use LED lighting as their sole source of light, giving growers complete control over the environmental factors affecting plant growth. This allows for year-round production in any location, close to consumers [7]. However, the profitability of microgreen production is hindered by two major factors: the market value of the fresh produce and the cost of operating the vertical farm. Since both factors are heavily influenced by lighting, optimizing lighting conditions for the plants becomes a critical challenge in vertical farming. To maximize space utilization, horizontal cultivation layers are densely packed with plants, and LED lights are mounted near the plant canopy [8]. The short separation distance between luminaires and the canopy can reduce photon irradiance uniformity, leading to spatial variation in plant growth. The plants in the middle of cultivation trays tend to grow taller and accumulate more biomass, while those on the edges are smaller and lighter. Although this center and edge effect is often attributed to uneven light distribution [9], other microenvironmental factors, such as airflow or genetic differences between individual seeds, can sometimes mask the effects of nonuniform light distribution.

The lighting conditions play a significant role in the growth of microgreens and are typically evaluated based on the horizontal photosynthetic photon flux density (PPFD) at the canopy level [10]. The light intensity has been found to impact both the yield and quality of microgreens [11]. At the early stages of plant development, photosynthesis by the cotyledon is a crucial process [12,13], which in turn influences the rate of subsequent seedling development [14,15].

In the present practice, one PPFD value is provided to describe lighting conditions in a vertical farm, though horticultural lighting guidelines recommend measuring PPFD values at several representative points of the working area and reporting both mean and standard deviations [16]. The spatial variations across the illuminated plane are characterized by the photon irradiance uniformity (U_o), defined as the quotient of the minimum reading and the average of data points. Another uniformity metric is diversity, (U_d), defined as the minimum to maximum ratio [10,17].

The measurement of PPFD is limited to the photosynthetically active radiation (PAR), which spans from 400 nm to 700 nm. However, studies have shown that relying solely on this range to evaluate photosynthetic activity has its limitations. Far-red photons interact with shorter-wavelength radiation, resulting in a synergistic effect that contributes to photosynthesis [18]. Consequently, pushing the upper limit of the PAR range to 750 nm has been recommended [19].

The intensity and the spectrum of light both determine the growth rate and the phytochemical content of microgreens [3]. In horticulture, most commercial lighting equipment comprises monochromatic blue, red, and far-red LED chips. The red and blue (R/B) or red and far-red (R/FR) photon irradiance ratios characterize the spectral distribution of incident radiation [20]. Plants grown under extremely low or high R/B ratios exhibit physiological disorders, and a balance between the photon irradiance ratios of various wavebands should be set to ensure proper conditions for plant development [21]. Many studies have investigated the optimal R/B or R/FR ratios for indoor crop production [22–26].

The various environmental factors that affect plant growth, such as lighting parameters, temperature, humidity, and carbon dioxide concentration of air, as well as the composition of the nutrient solution, all interact with one another. To achieve optimal crop yield and quality, it is crucial to control and optimize all these parameters. However, measuring the light response curve of plant growth in a multidimensional parameter space requires significant experimental efforts.

High-speed automated procedures have been developed for 3D characterization of the lighting environment [27–30] as well as for quick phenotyping and monitoring of plant growth [9,15,30,31]. With the aid of a high-throughput experimental unit, one can efficiently screen a vast array of parameter settings and extract the transfer function linking growth traits to environmental parameters. However, transitioning the experimental light response functions into commercial production poses a challenge due to the presence of numerous unknown noise factors.

Our approach was to carry out the lighting experiments under the conditions of commercial production. We created a gradient in the lighting conditions by controlling the luminaires of the vertical farm. The objectives of the present study were to quantify the effect of spatial photon irradiance variations on the growth traits of pea microgreens and retrieve the transfer function between the fresh weight of individual seedlings and the local PPFD values to be used for the optimization of commercial microgreen production.

2. Materials and Methods

Plant growth tests were carried out in a climate-controlled container farm of the Hungarian University of Agriculture designed to be a scalable cultivation unit of a larger plant factory. Fans positioned in the middle and at the end of each shelf maintained constant airflow over the canopy to minimize spatial differences in temperature, humidity, and CO_2 concentration. Air parameters were checked by an ALMEMO 2590 measuring instrument (Ahlborn Mess- und Regelungstechnik GmbH, Holzkirchen, Germany). Leaf surface temperature was recorded by an infrared thermometer (AHG Wahtsmuth & Krogmann mbH, Hamburg, Germany). The average temperature and relative humidity during the tests were kept within the 20 ± 2 °C and 75 ± 5% range, respectively. In our vertical farm, we did not apply CO_2 injection. The CO_2 concentration varied between 400 ppm and 600 ppm throughout the experiments. The lowest value was measured during the light period when the photosynthesis consumed carbon dioxide, and the highest concentration was detected at the end of the dark period. The carbon dioxide level increased temporarily up to 1200 ppm in the presence of a human operator, but the concentration difference between different points of the vertical farm was less than 10%. The experiments comprised two steps, starting with germination in the dark, followed by seedling development under three different light treatments. Pea (*Pisum sativum* L., cv. Kleine Rheinländerin) seeds were obtained from Royal Sluis (Enkhuizen, Holland). Seeds were soaked in distilled water for 24 h and then placed on perforated stainless-steel sheets with round holes of 5 mm diameter. The plantation distance between adjacent seeds was 4 cm. Metal sheets with the seeds were placed into a germination box, ensuring saturated moisture at the 20 °C ambient temperature.

After 7 days, germinated seeds with longer than 6 cm radicles were transplanted to plastic cultivation trays with 5 mm-diameter holes arranged in a 12 × 7 array with a 4 cm grid size. The top view of the plant arrangement, along with the definition of the directions, is shown in Figure 1. The x-axis is parallel to the line of LED luminaires, whereas the y-axis is perpendicular to the LED pairs. The LED luminaires were positioned above the first and twelfth rows. In the analysis, the row numbers on the y-axis and column numbers on the x-axis were used for the identification of seedlings' positions. The picture of the seedlings transplanted to the cultivation tray is shown in Figure 2a, and the seedlings prior to harvest are shown in Figure 2b.

The radicle was immersed into a nutrient solution mixed from a three-component commercial formula (Dutch Formula Grow, Advanced Hydroponics of Holand, 1-Grow: 2 mL/L, 2-Bloom: 2 mL/L, 3-Micro: 1 mL/L). During the entire experiment, the electrical conductivity and the pH of the nutrient solutions were in the range of 1.50 ± 0.05 mS/cm and 6.9 ± 0.1, respectively, measured by a universal measuring instrument (Combi 5000, STEP Systems GmbH, Nürnberg, Germany).

The schematic diagram of the experimental set-up is shown in Figure 3. Plant growth was carried out simultaneously at three different positions of the vertical farm in three cultivation containers coded as A, B1, and B2. The dimensions of the containers were 60 cm × 40 cm × 7.5 cm. The plants on A, B1, and B2 were exposed to three different lighting conditions determined by the position of the trays relative to the LEDs as well as the power of the luminaires. Two layers of the vertical farm were used in the experiments. In levels A and B, the separation distances between the LED luminaires and the planes of the cultivation trays were 45 cm and 21 cm, respectively. In each level, three pairs of 120 cm long variable-spectrum LED luminaires equipped with secondary optics ensuring an 80° beam angle (Hortiled Multi 4DIM, Hortilux, Den Haag, The Netherlands) were mounted to the edges of the shelves, covering the 3.6 m length of the shelving unit. The photon intensity distribution diagram of the 80° beam angle luminaires is shown in Figure S1.

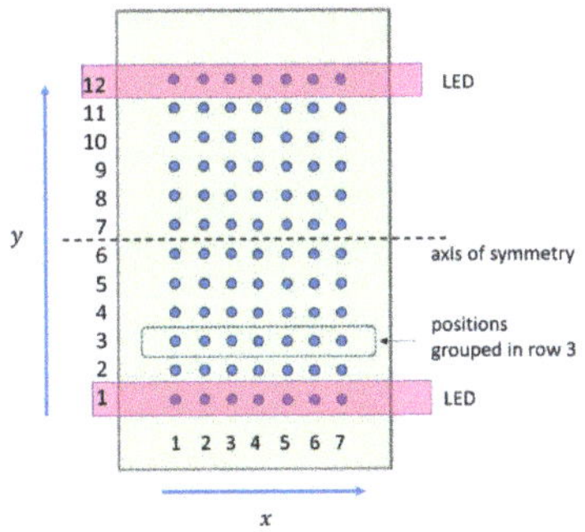

Figure 1. Schematics of the plant arrangement on a cultivation tray. Seedlings were arranged in 12 rows and 7 columns. The mesh size of the 12 × 7 grid was 4 cm in both the x and y directions. The dotted line shows the axis of symmetry of the irradiance created by the pair of LED luminaires shown as purple rectangles. The dotted box shows positions in row 3 from which the related row average was calculated in data analysis.

(a) (b)

Figure 2. Individual plants on the cultivation tray at the (**a**) transplantation; and (**b**) harvest.

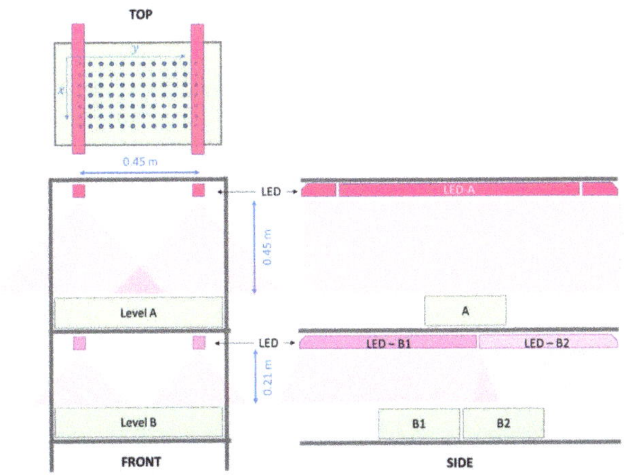

Figure 3. Top, front, and side view of the experimental vertical farm. The distance between the LED luminaires and the cultivation trays on levels A and B were 45 cm and 21 cm, respectively. The PPFD distribution was tailored by adjusting the power of LED luminaires (LED-B1, LED-B2, LED-A) and the position of trays B1, B2, and A. The color saturation of the stripes representing the LEDs indicates the relative luminaire power: LED-A = 59.5% > LED-B1 = 41.7% >> LED-B2 = 5%.

Our objective was to establish a highly uniform PPFD distribution across tray A and a non-uniform distribution on tray B1, having the same average PPFD as tray A. The light distribution across the plane of B2 was designed to cover a broad intensity range, from the high values on the left corners going down to the extremely low values on the right side. To achieve this goal, the power of the luminaires coded as LED-A, LED-B1, and LED-B2 in Figure 3 was set at 59.5%, 41.7%, and 5% of the nominal value. The on and off times as well as the power of each color channel of the luminaires were set by a DALI (Digitally Addressable Lighting Interface) controller (DLC-02 DALI Digital Lighting Controller, Mean-Well, Taiwan). Only the light intensity was tailored at the three locations of the test; the relative spectral distribution of irradiance, i.e., the power ratio of the color channels, was held constant. The spectral irradiance values were measured with 5 nm resolution at every position of the cultivation trays using a handheld spectroradiometer (Mavospec Base, GOSSEN Foto- und Lichtmesstechnik GmbH, Nürnberg, Germany).

The plant growth was carried out at a constant 20 °C temperature. The photoperiod was 16 h per day in each light treatment. Shoots were cut 12 days after transplantation and measured for length and fresh weight. A picture of seedlings prior to harvest is shown in Figure 2b. The weight of the individual seedlings was measured by a precision balance (Kern EMB 200-3, Kern & Sohn GmbH, Balingen, Germany).

Statistical analysis of measurement data was carried out using normality, Kruskall–Wallis, and Levene's tests from the statistical module of the SciPy [32] open-source Python package. Significance levels were set at $p < 0.05$ throughout the data analysis.

3. Results

The objective of the experiment was to compare the growth traits of pea seedlings cultivated under three different light-intensity distributions while keeping the light spectrum constant across all cultivation trays. Table 1 summarizes the main statistical parameters and uniformity measures (U_o, U_d) determined for A, B1, and B2. The columns of Table 1 contain both the lighting-related information (PPFD, photon irradiances in the B, G, R, and FR wavebands, as well as the R/B ratio) and the shoot fresh weight (FW). The D'Agostino-Pearson normality test [33] was carried out for all measured datasets. The null hypothesis was that the sample came from a normal distribution. The p-value of the normality test is also listed for each distribution in Table 1. The $p < 0.05$ values indicate cases where the normality assumption can be rejected with 95% confidence. By checking the rows of p-values in Table 1, it is obvious that most of the data are from non-normal distributions. The ANOVA method generally used in data analysis requires samples with normal distributions as input; therefore, non-parametric hypothesis tests were used for the comparison of data. The Kruskal–Wallis test is a non-parametric version of ANOVA, testing the medians rather than the means of the samples. In our analysis, we assumed that conclusions drawn on medians were valid for the mean values as well. In Section 3.1, we provide the quantitative measures of the lighting environments for the three cultivation trays. In Section 3.2, the individual shoot weight data are presented and compared with light measurements.

3.1. Characterization of the Lighting Environment

The four color channels of the LED luminaires covered the four adjacent wavebands generally used for the characterization of LED-based horticultural lighting [20]: blue (B): 400–499 nm, green (G): 500–599 nm, red (R): 600–699 nm, and far-red (FR): 700–800 nm. A representative example of the measured irradiance spectra is depicted in Figure 4, exhibiting four peaks corresponding to the four types of LEDs built into the luminaires: blue peak emission at 450 nm, deep red at 660 nm, far-red at 730 nm, and a broad peak extending over the green waveband corresponding to the phosphor emission of the white LEDs. In this particular spectrum, the R/B and R/FR ratios were 2.5 and 3.6, respectively.

Table 1. Photon irradiance parameters were measured for the trays A, B1, and B2. The unit of PPFD, B, G, R and FR is μmol m^{-2} s^{-1}. FW: shoot fresh weight in grams. R/B denotes the ratio of red and blue photon irradiances. Uo is the overall uniformity defined as the quotient of the minimum and the average of the distribution. The *p*-value of the normality test is denoted by *p*.

Tray	Parameter	PPFD	B	G	R	FR	R/B	FW
A	Minimum	234.90	59.12	22.55	152.98	45.92	2.42	0.55
	Average	248.91	63.68	23.75	161.47	49.34	2.54	2.35
	Maximum	269.37	69.06	25.32	175.30	51.90	2.65	4.55
	U_o	0.94	0.93	0.95	0.95	0.93	0.96	0.23
	U_d	0.87	0.86	0.89	0.87	0.88	0.91	0.12
	p	0.009	0.002	0.076	0.013	0.138	0.715	0.12
B1	Minimum	98.62	25.98	11.01	61.63	17.28	2.22	0.23
	Average	232.48	59.41	22.05	151.02	41.27	2.51	2.31
	Maximum	390.93	101.33	34.22	257.55	69.00	2.78	5.67
	U_o	0.42	0.44	0.50	0.41	0.42	0.89	0.10
	U_d	0.24	0.25	0.31	0.23	0.26	0.85	0.04
	p	0.000	0.000	0.000	0.000	0.000	0.001	0.04
B2	Minimum	33.23	8.83	3.70	20.70	6.26	2.01	0.22
	Average	91.14	24.48	9.38	57.29	17.31	2.34	1.70
	Maximum	278.72	71.61	25.65	181.46	54.90	2.61	4.33
	U_o	0.36	0.36	0.39	0.36	0.36	0.86	0.13
	U_d	0.12	0.12	0.14	0.11	0.11	0.77	0.05
	p	0.000	0.000	0.000	0.000	0.000	0.175	0.05

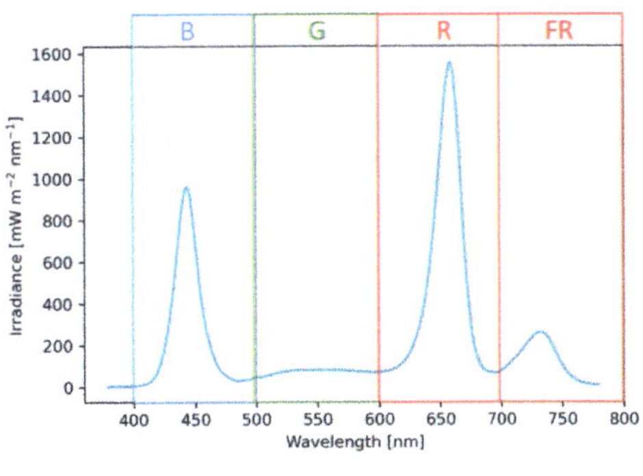

Figure 4. Representative spectral distribution of irradiance measured on tray B1, position row 1, column 1. In this spectrum, the R/B and R/FR ratios were 2.5 and 3.6, respectively. In the experiment, there was only a minor change in the relative intensity of the peaks in the B, G, R, and FR wavebands; only the absolute irradiance changed across the illuminated plane.

The power ratios of the color channels were held constant throughout the experiment; therefore, the quality of light, i.e., the shape of the spectrum, was expected to be the same at any point of the illuminated work plane. The quantitative measure of light, however, varied according to the power of LED luminaires and the position of the light intensity measurement.

The minimum, average, and maximum values measured on trays A, B1, and B2 are listed in Table 1 for the photosynthetic photon flux density (PPFD) as well as the photon irradiances in the B, G, R, and FR wavebands. The R/B photon irradiance ratio quantifying the spectral property of light was calculated for each tray, and the related statistical parameters, including the overall uniformity (U_o), are also listed in Table 1. According to expectations, the difference between the average PPFDs of A and B1 was small, only 7% in absolute value. The range of data points, however, was much broader in the case of B1 compared with A. This difference between the two settings is reflected by the uniformity parameters: $U_o = 0.94$ in A, indicating a highly uniform PPFD distribution, whereas $U_o = 0.42$ represents the low uniformity case. The difference between the maximum and minimum B2 PPFD values was even broader than in the case of B1, but the average value was 91 μmol m^{-2} s^{-1}, 40% lower than the average of B1. Similar trends can be seen in the B, G, R, and FR wavebands, indicating the stability of the irradiance spectrum across the cultivation trays.

The R/B ratio is a frequently used measure of the quality of light in horticulture. There was only a minor difference between the means of A and B1. B2 had a slightly lower mean R/B value, indicating a less than 10% shift in the spectral peak ratios at low irradiance, but this difference is negligible considering the broad light response sensitivity range of plants [34].

The two-dimensional photosynthetic photon irradiance distributions are visualized in Figure 5a–c. The color scale of the contour plots is the same in all the subfigures, ranging from 0 to 400 μmol m^{-2} s^{-1}. The colored patterns in the contour plots exhibit a reflectional symmetry relative to the axis of symmetry in the middle of the illuminated area. On tray B2 in Figure 5a the highest intensities were measured in the upper and bottom left corners, resulting from the edge effect between the high-power LED B-2 and low-power LED B-3 luminaire pairs, as shown in the side view of Figure 1. The low-intensity region is in the center, extending towards the right-hand side. In Figure 5c, there are only two adjacent colors, indicating highly uniform PPFD distribution across tray A, with a minor increase from low to high row numbers. In Figure 5b, warm colors on the top and bottom edges indicate high PPFD regions, whereas cool colors in the middle represent low PPFD values. The line symmetry of the irradiance in B1 is reflected by the contour lines running parallel to the x-axis along the shelves.

The histograms in Figure 6a–c provide a more quantitative description of the PPFD distributions on trays B2, B1, and A. In Figure 6a, the histogram is skewed towards the left in accordance with the high proportion of low irradiance values on tray B2. The PPFD values on B1 in Figure 6b are evenly distributed between the minimum and maximum values. The histogram of A is characterized by a high, narrow peak in Figure 6c. Neither of the PPFD histograms can be described with a normal distribution, as indicated by the p-values in Table 1.

The boxplot in Figure 7a compares the averages and the spread of the PPFD distributions of B2, B1, and A. The horizontal red lines in the boxes show the medians; the crosses stand for the mean values.

Beyond the graphical representation of the distributions in Figure 7, non-parametric hypothesis tests were used to check the equality of the medians (Kruskal–Wallis test) and variances (Levene's test) [35]. The p-values of the pairwise comparisons are shown in Table 2. The $p < 0.05$ values indicate statistically significant differences between the tested parameters. The p-value of the Kruskal–Wallis test was 0.786; consequently, the null hypothesis that the median PPFD values of A and B1 are equal cannot be rejected. The median PPFD of B2 proved to be significantly lower than that of B1 and A. The p-values of Levene's test were all zero, indicating significantly different PPFD variances on trays B2, B1, and A.

Table 2. The p-values of the pairwise non-parametric hypothesis tests for the medians (Kruskal-Wallis test) and variances (Levene's test). The boldface numbers indicate p-value < 0.05 corresponding to statistically significant differences between the medians or variances.

		Kruskal–Wallis Test (Medians)			Levene's Test (Variances)		
Parameter	**Tray**	**A**	**B1**	**B2**	**A**	**B1**	**B2**
PPFD	A	1.000	-	-	1.000	-	-
	B1	0.107	1.000	-	**0.000**	1.000	-
	B2	**0.000**	**0.000**	1.000	**0.000**	**0.000**	1.000
R/B	A	1.000	-	-	1.000	-	-
	B1	0.335	1.000	-	**0.000**	1.000	-
	B2	**0.000**	**0.000**	1.000	**0.000**	0.050	1.000
FW	A	1.000	-	-	1.000	-	-
	B1	0.786	1.000	-	0.101	1.000	-
	B2	**0.000**	**0.000**	1.000	**0.015**	**0.000**	1.000

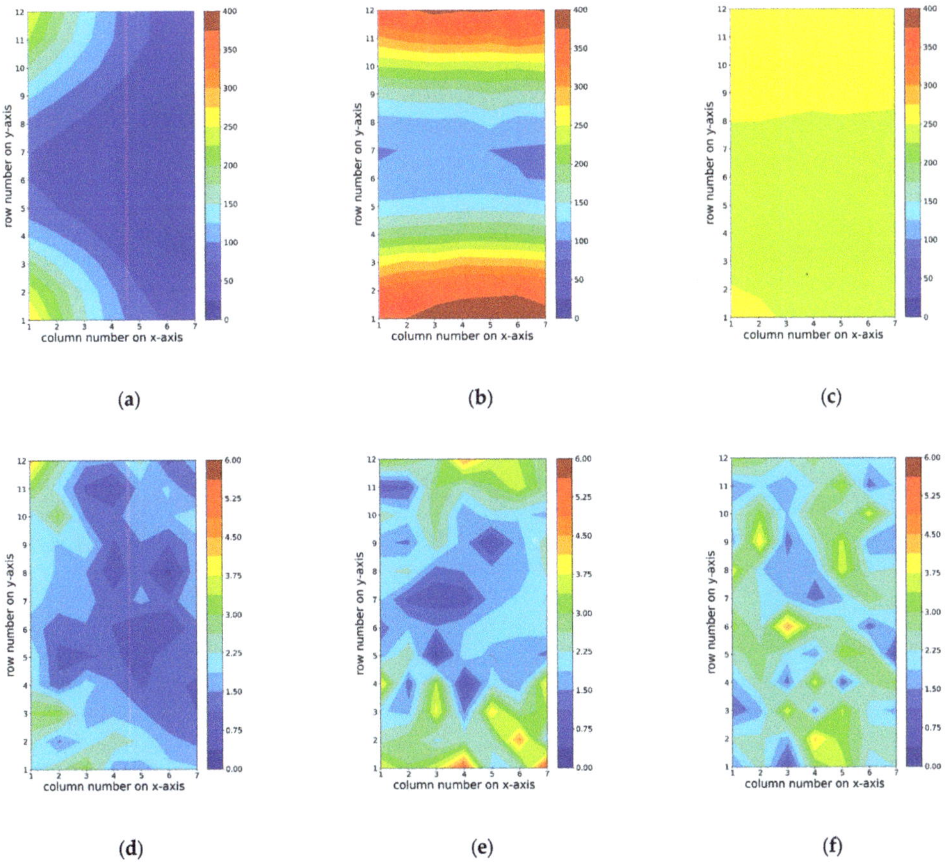

Figure 5. Horizontal PPFD distributions measured on trays (**a**) B2, mean = 91 μmol m^{-2} s^{-1}; (**b**) B1; mean = 232 μmol m^{-2} s^{-1} and (**c**) A, mean = 249 μmol m^{-2} s^{-1}. The color scale ranges from 0 to 400 μmol m^{-2} s^{-1} in (**a–c**). Fresh weight distribution of individual seedlings on trays: (**d**) B2, (**e**) B1, and (**c**) A. The color scale ranges from 0 to 6 g in (**d–f**).

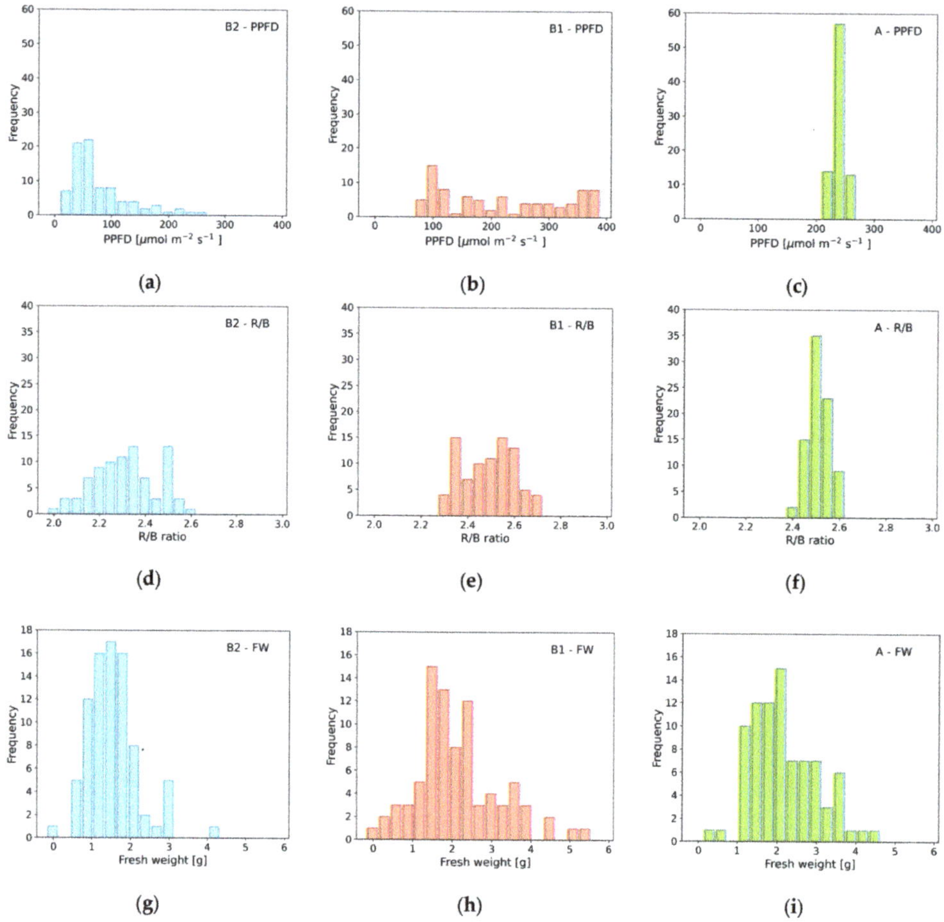

Figure 6. Histogram of PPFD distributions: (**a**) B2; (**b**) B1; (**c**) A; R/B ratio: (**d**) B2; (**e**) B1; (**f**) A and: fresh weight (**g**) B2; (**h**) B1; (**i**) A.

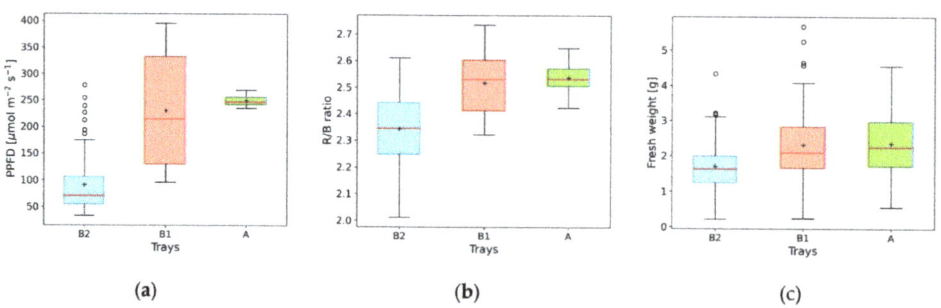

Figure 7. Comparison of spatial distributions on trays B2, B1, and A: (**a**) PPFD; (**b**) R/B ratio; (**c**) Shoot fresh weight. Pairwise comparisons in Table 2 indicate that the medians of B2 are significantly lower than those of B1 and A, whereas the differences between B1 and A are not statistically different.

Continuing the analysis with the R/B ratios in Figure 7b, the medians of B1 and A are not statistically different according to $p = 0.335$ of the Kruskal–Wallis test. The median R/B ratio of B2, however, is significantly different from the group of B1 and A. The variance of B2 is significantly higher than the variance of A. The comparison of B2 and B1 variances resulted in $p = 0.050$, indicating a borderline case.

All these data indicate that the lighting conditions in the three environments were in line with our objectives. The mean values of A and B1 were statistically not different, but the range of data values was 8.5 times broader in the case of B1 relative to A. The mean value of the B2 was 37% of the A average. The relative spectral distribution of the irradiance was constant in all three cultivation trays.

3.2. Fresh Weight Analysis

The statistical parameters determined for the shoot fresh weight measurements are summarized in the rightmost column of Table 1. There was a minor, 1.7% difference between the mean values of A (2.35 g) and B1 (2.31 g), whereas the average fresh weight of B2 was only 1.7 g. Comparing the minima, maxima, and the U_o and U_d values, all distributions exhibit high spread and a low level of uniformity.

Noise dominates the fresh weight contour plots in Figure 5d–f, which vaguely reflect the symmetry of the spatial PPFD distributions in Figure 5a–c. The fresh weight histograms of B1 and A in Figure 6h,i are close to each other both in position and spread in sharp contrast with the broad PPFD distribution in Figure 6b and narrow PPFD distribution in Figure 6c. Comparing the boxplots in Figure 7a,c one can conclude that the range or uniformity of PPFD distributions had little effect on the fresh weight distributions. Inferential statistics confirmed that neither the medians nor the variances of the shoot fresh weights on B1 and A are statistically different. In Table 2, $p = 0.786$ and $p = 0.101$ for the Kruskal–Wallis and Levene's tests, respectively. B2, however, can be regarded as an outlier from the group of A and B1 both in terms of median and variance values.

In Figure 8, the fresh shoot weights of individual plants are plotted against the local PPFD values. The three different markers—triangles, squares, and circles—represent data points measured on trays A, B1, and B2, respectively. Although there is a large variation in the fresh weight, the dotted trendline indicates a linear relationship between the light intensity and biomass accumulated in the individual plants. The correlation is statistically significant, with an F-test value of 111.0 and a significance of $p = 0.000$. From the value of the coefficient of determination, $R^2 = 0.31$, one can conclude that 31% of the variation in the fresh weight can be attributed to the PPFD changes; the rest is due to other factors.

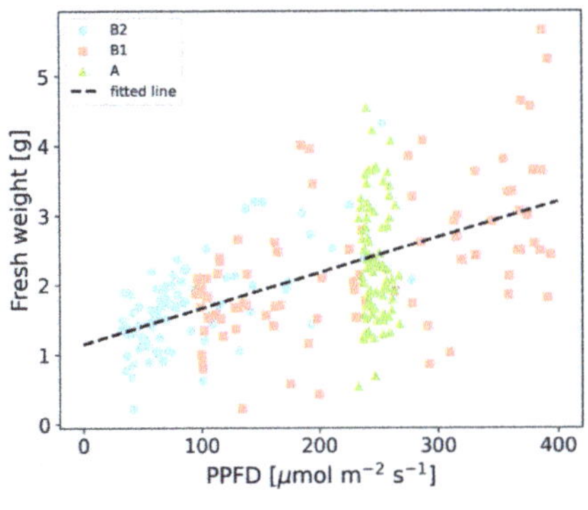

Figure 8. Correlation between individual fresh weight and PPFD data in the pea growth experiment. Markers differentiate data points related to the cultivation trays: (○) B2; (≤) B1; and (△) A. The dotted line represents the linear fit to all data points. Estimated parameters: slope = 0.005 ± 0.001, intercept = 1.145 ± 0.21. $R^2 = 0.31$. 31% of the variations can be attributed to the local PPFD changes.

The slope and the intercept of the least squares regression line were 0.0051 ($\pm$0.00096) and 1.14 ($\pm$0.21). The values in brackets indicate the 95% confidence interval about the estimated parameter means.

Looking up the parameters of trays A and B1 in Table 1 and comparing Figure 7a with Figure 7c, one can conclude that the uniformity of the PPFD distribution had no measurable effect on the mean fresh weight (i.e., crop yield) in our experiment. As long as the PPFD uniformities corresponded to the extremely high and low cases with U_o values of 0.94 (A) and 0.42 (B1), the mean fresh weights of 2.35 g (A), and 2.31 (B1) did not differ significantly. The coefficient of determination, $R^2 = 0.31$, indicated that the fresh weight variance is driven by the different behaviors of individual seeds, and only 31% of the fresh weight variance can be attributed to photon irradiance changes. On the other hand, the growth test was carried out in the linear regime of the light response curve, far below the light saturation point of pea seedlings [36]. Seedlings exposed to higher than the mean PPFD value grew faster, while plants below the PPFD mean developed slower compared with the average; therefore, the effect of photon irradiances on the mean is expected to cancel out. The range of the fresh weight values increased: the minimum was 0.55 in A and 0.23 in B1, whereas the maximum fresh weight values were 4.55 in A and 5.67 in B1.

This assumption is supported by the analysis of measurement data grouped by rows. The linear array of LED luminaire pairs created a symmetrical light intensity distribution with an axis of symmetry running in the center line of the trays, as indicated in Figure 1. On trays A and B1, seedlings in a row were exposed to the same microenvironmental conditions independently of the column number. By averaging measurement data in one row, we can reduce variations due to the differences between individual seeds. The dotted line box in Figure 1 shows an example of creating the group of rows #3.

In Figure 9a, the row averages of PPFD distributions are plotted as a function of the row numbers. The triangles representing A data indicate a minor upward trend from row 1 up to 12 in accordance with Figure 5c. Similarly, squares representing B1 values show high row averages on the lower and upper parts of the trays, and the minimum values can be found at the position of the axis of symmetry (c.f. Figure 5b). Row averages in Figure 9b reveal trends in B1 and A fresh weight data, which were hidden for visual inspection in Figure 5e,f. Triangles representing row averages of A are randomly scattered about the grand mean of 2.35 g in Figure 9b. The trend shown by the dotted green line is not statistically significant. The fresh weight row averages of B1, represented by the red squares, however, can be approximated by a second-order polynomial. The polynomial regression shows a statistically significant trend with $R^2 = 0.91$. Tray B2 was left out of the grouped average calculations since positions on one row were not equivalent from the photon irradiance perspective, as is obvious in Figure 5a.

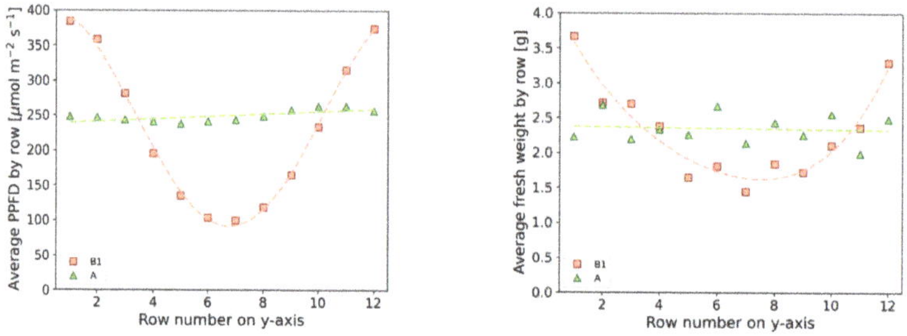

(**a**) (**b**)

Figure 9. Row average of measurement data as defined in Figure 1. Dotted lines represent polynomial fit to the data points: (**a**) average PPFD by row number on trays: (≤) B1: $R^2 = 0.998$; (△) A: $R^2 = 0.51$. (**b**) Average shoot weight by row number on trays. (□) B1: $R^2 = 0.92$; and (△) A: $R^2 = 0.006$.

The correlation between the row averages of shoot fresh weight and PPFD values is shown in Figure 10. Data points A and B1 are scattered about a straight line, indicating a linear relationship between the fresh weight and PPFD. The slope = 0.0056 ± 0.0014 and intercept = 0.97 ± 0.36 fall within the confidence intervals of the linear relationship determined for individual data. The $R^2 = 0.75$ indicates a significantly reduced variance in fresh weight relative to the regression of individual data.

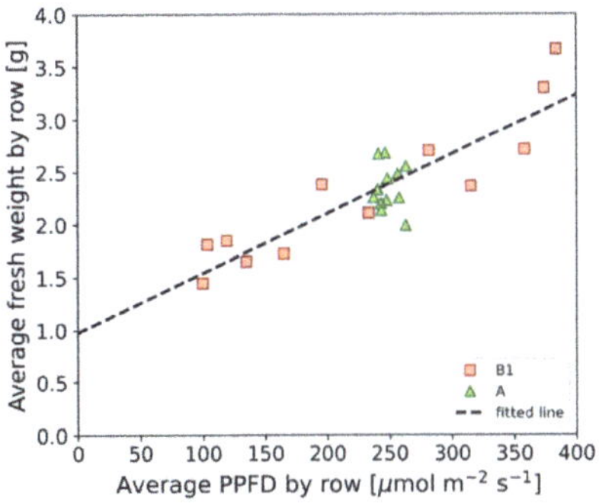

Figure 10. The correlation between the row average fresh weight and the row average PPFD in the pea growth experiment was measured on cultivation trays B1 (□) and A (△). Individual measurement points were grouped by rows. The dotted line represents the linear fit to all plotted data points. Estimated parameters: slope = 0.0056 ± 0.0014, intercept = 0.97 ± 0.36. $R^2 = 0.75$.

4. Discussion

The objective of this study was to analyze the differences in seedling growth traits cultivated under uniform (A) and non-uniform (B1, B2) lighting conditions. Although there is no standardized threshold for PPFD uniformity in horticultural lighting, in commercial cultivation facilities, the criteria $U_o > 0.8$ or $U_d > 0.7$ are applied [17]. From this perspective, the $U_o = 0.94$, and $U_d = 0.87$ of environment A can be regarded as highly uniform, whereas the $U_o = 0.42$ and $U_d = 0.24$ for B1 is a low uniformity case. The overall uniformity and diversity for B2 were even lower compared with B1.

We observed that the mean PPFD determined the average weight of individual plants, and PPFD uniformity had no statistically significant effect on the crop yield. This finding has important implications for the lighting design of vertical farms. Reducing the separation distance between the LED luminaires and the crop canopy is an opportunity to improve the space utilization and energy efficiency of vertical farms [8]. The close-canopy approach increases the photon capture efficiency or utilization factor of the horticultural lighting, defined as the quotient of useful photon flux incident on the crop and the total photon flux emitted by the lighting equipment [37]. Maintaining high PPFD uniformity at reduced mounting heights of the luminaires is not a trivial task. It requires additional investment in sophisticated lighting equipment to enable high photon capture efficiency at high photon irradiance uniformity. Our results demonstrated that at moderate photon irradiances, far away from the light saturation point, the PPFD uniformity criteria can be relaxed assuming the crop is sold in bulk by mass.

The reduction of the separation distance between the plant canopy and the LED luminaires raises two questions:

1. How does the photon irradiance at the top canopy level change as the seedlings grow close to the luminaire?
2. Has the upper leaves' temperature increased due to thermal radiation from the luminaires?

It is beyond the scope of this work to provide a detailed analysis of the three-dimensional evolution of environmental conditions during plant growth. Nevertheless, we carried out two additional control tests to determine the scale of changes with the increasing height of the seedlings. First, we measured the PPFD distributions across horizontal planes corresponding to 6 cm and 12 cm plant heights in trays A and B1. Data presented in Figures S3 and S4 indicate that photon irradiance values increase underneath the luminaires and decrease in between the space of the luminaire pairs as the height of seedlings rises, pointing out the limitations of our experiments. In our analysis, we used only the horizontal photon irradiance at the seed level, which reflects the lighting conditions during the early stages of plant growth. As plants grow taller, the light intensity at the top of the canopy will be different from that at the lower levels. Additionally, plants create shade, influencing the growth of neighboring plants. The light interception changes both horizontally and vertically within the canopy. To provide a more accurate description of the lighting environment and predict crop yield, we recommend using three-dimensional canopy models [27–29].

To test the heating effect of the luminaires, we measured the surface temperature at the uppermost leaf of a seedling at 17 cm above the tray (B1) surface and at the lowest leaf of the same seedling at 3 cm height. The distance between the upper leaf and the cover glass of the luminaire was 4 cm. The average surface temperature was 20.8 °C at the upper leaf and 20.2 °C at the bottom leaf. The sample size was 16 in both cases. The temperature difference is significant at $p = 0.003$, but nevertheless, the absolute value of the difference shown in Figure S5 is in the range of leaf surface temperature variations at various positions of the vertical farm (Figure S6).

We plan further experiments to reveal the physiological response of the plant leaves to the local environmental conditions at various horizontal locations and heights of the canopy.

An important assumption of this study was that the quality of light is constant in all cultivation trays. Although the R/B ratio of B2 was statistically different from the groups A and B1, the difference was minor compared with the spectral sensitivity of plant growth traits measured by various research groups for different species [22,24,34]. The statistically significant differences in R/B ratios may be attributed to a slight shift in the emission spectrum of the luminaires at low power or to the increased portion of wall reflections in the low end of the R/B distribution, especially on the extreme dark zone of tray B2, as indicated by the dark blue colors in Figure 5a. Adjusting the power of the red and blue channels of the luminaires above B2 (cf. LED-B3 luminaire pair in Figure 3) might have eliminated the difference in the R/B means between B2 and the groups of B1 and A. We did not make any further attempt to fine-tune the photon irradiance distributions because the 7% difference in the mean R/B ratios of B2 and A was expected to have a negligible effect on plant growth compared with the 64% difference in PPFD averages.

Individual shoot fresh weight data exhibited high variability both in uniformly and non-uniformly illuminated environments. Despite the large variations, a statistically significant relationship was found between the local PPFD values and the shoot fresh weight. The low value of the coefficient of determination of the linear regression ($R^2 = 0.31$) in Figure 8 is in line with the observations of other researchers measuring the growth of cotyledons and initial true leaves of individual lettuce seedlings [15]. A large amount of the variability in seedling growth rate is possibly due to the genetic differences of the seeds.

The opportunity to aggregate individual observations into groups has several practical implications. Instead of weighing individual seedlings one by one, it is possible to make measurements in rows, reducing measurement time. The procedure described can be extended to other species and implemented in commercial vertical farms, enabling the light response of any crops to be retrieved under the conditions of commercial production.

This paper highlights the advantages of establishing a light intensity gradient within the crop canopy to obtain accurate light response functions. Additionally, we measured the shoot height and root weight of B2 seedlings, which showed a correlation with the shoot fresh weight data (Figure S2). Given the higher margin of error associated with the shoot height and root weight measurements, analyzing other growth traits is unlikely to yield further insights.

5. Conclusions

Understanding the variance causes is the prerequisite for the optimization of vertical farm settings. Our experiments have effectively quantified the impact of horizontal PPFD variation on microgreen crop yield in a vertical farm. We utilized high-spatial-resolution measurements to uncover the correlation between the shoot fresh weight of pea seedlings and light intensity. Our experiment presents a methodology to separate and quantify light intensity-related variations from other microenvironmental and genetic factors, enabling data-driven decisions in the lighting design process. This methodology can be applied to determine a crop's light response under production conditions on vertical farms.

Supplementary Materials: The following supporting information can be downloaded at: https://www.mdpi.com/article/10.3390/horticulturae9111187/s1, Figure S1: Photon intensity distribution of the luminaire with 80° beam angle used in the experiments; Figure S2: Individual shoot height (a) and root weight (b) as a function of local PPFD for tray B2; Figure S3: PPFD distribution at the canopy level on tray A; Figure S4: PPFD distribution at the canopy level on tray B1; Figure S5: Leaf surface temperature on the top and bottom of a seedling with a height of 17 cm on tray B1; Figure S6: Variation of leaf surface temperature on various trays of the vertical farm.

Author Contributions: Conceptualization, L.B., Á.T. and Z.K.; methodology, L.B. and Z.K; software, L.B.; validation, L.B., P.P. and Z.K.; investigation, L.B., P.P. and Z.K.; resources, G.P.K. and C.G.; writing—original draft preparation, L.B.; writing—review and editing, L.B., Á.T., Z.K. and G.P.K.; visualization, L.B.; project administration, G.P.K.; funding acquisition, C.G. All authors have read and agreed to the published version of the manuscript.

Funding: This research received no external funding.

Data Availability Statement: Not applicable.

Acknowledgments: The authors acknowledge the support from bedrock.farm Ltd., Budapest, Hungary, by providing access to their commercial vertical farm to carry out preliminary experiments of this study.

Conflicts of Interest: The authors declare no conflict of interest.

References

1. Research, S. Microgreens Market Analysis, Growth, Forecast to 2030. Available online: https://straitsresearch.com/report/microgreens-market (accessed on 27 August 2023).
2. Teng, Z.; Luo, Y.; Pearlstein, D.J.; Wheeler, R.M.; Johnson, C.M.; Wang, Q.; Fonseca, J.M. Microgreens for Home, Commercial, and Space Farming: A Comprehensive Update of the Most Recent Developments. *Annu. Rev. Food Sci. Technol.* **2023**, *14*, 539–562. [CrossRef]
3. Kyriacou, M.C.; Rouphael, Y.; Di Gioia, F.; Kyratzis, A.; Serio, F.; Renna, M.; De Pascale, S.; Santamaria, P. Micro-Scale Vegetable Production and the Rise of Microgreens. *Trends Food Sci. Technol.* **2016**, *57*, 103–115. [CrossRef]
4. Wojdyło, A.; Nowicka, P.; Tkacz, K.; Turkiewicz, I.P. Sprouts vs. Microgreens as Novel Functional Foods: Variation of Nutritional and Phytochemical Profiles and Their In Vitro Bioactive Properties. *Molecules* **2020**, *25*, 4648. [CrossRef]
5. Xiao, Z.; Lester, G.E.; Luo, Y.; Wang, Q. Assessment of Vitamin and Carotenoid Concentrations of Emerging Food Products: Edible Microgreens. *J. Agric. Food Chem.* **2012**, *60*, 7644–7651. [CrossRef]
6. Xiao, Z. Nutrition, Sensory, Quality and Safety Evaluation of a New Specialty Produce: Microgreens. Doctoral Thesis, Faculty of the Graduate School, the University of Maryland, College Park, MD, USA, 2013.
7. Al-Kodmany, K. The Vertical Farm: A Review of Developments and Implications for the Vertical City. *Buildings* **2018**, *8*, 24. [CrossRef]
8. Sheibani, F.; Bourget, M.; Morrow, R.C.; Mitchell, C.A. Close-Canopy Lighting, an Effective Energy-Saving Strategy for Overhead Sole-Source LED Lighting in Indoor Farming. *Front. Plant Sci.* **2023**, *14*, 1215919. [CrossRef]
9. Kozai, T.; Lu, N.; Hasegawa, R.; Nunomura, O.; Nozaki, T.; Amagai, Y.; Hayashi, E. Plant Cohort Research and Its Application. In *Smart Plant Factory: The Next Generation Indoor Vertical Farms*; Kozai, T., Ed.; Springer: Singapore, 2018; pp. 413–431. ISBN 9789811310652.
10. Balázs, L.; Dombi, Z.; Csambalik, L.; Sipos, L. Characterizing the Spatial Uniformity of Light Intensity and Spectrum for Indoor Crop Production. *Horticulturae* **2022**, *8*, 644. [CrossRef]

11. Jones-Baumgardt, C.; Llewellyn, D.; Ying, Q.; Zheng, Y. Intensity of Sole-Source Light-Emitting Diodes Affects Growth, Yield, and Quality of Brassicaceae Microgreens. *HortScience* **2019**, *54*, 1168–1174. [CrossRef]
12. Zheng, W.; Wang, P.; Zhang, H.-X.; Zhou, D. Photosynthetic Characteristics of the Cotyledon and First True Leaf of Castor (*Ricinus communis* L.). *Aust. J. Crop Sci.* **2011**, *5*, 702–708.
13. Santos, H.P.; Buckeridge, M.S. The Role of the Storage Carbon of Cotyledons in the Establishment of Seedlings of Hymenaea Courbaril Under Different Light Conditions. *Ann. Bot.* **2004**, *94*, 819–830. [CrossRef]
14. Walter, A.; Scharr, H.; Gilmer, F.; Zierer, R.; Nagel, K.A.; Ernst, M.; Wiese, A.; Virnich, O.; Christ, M.M.; Uhlig, B.; et al. Dynamics of Seedling Growth Acclimation towards Altered Light Conditions Can Be Quantified via GROWSCREEN: A Setup and Procedure Designed for Rapid Optical Phenotyping of Different Plant Species. *New Phytol.* **2007**, *174*, 447–455. [CrossRef]
15. Hayashi, E.; Amagai, Y.; Kozai, T.; Maruo, T.; Tsukagoshi, S.; Nakano, A.; Johkan, M. Variations in the Growth of Cotyledons and Initial True Leaves as Affected by Photosynthetic Photon Flux Density at Individual Seedlings and Nutrients. *Agronomy* **2022**, *12*, 194. [CrossRef]
16. Both, A.J.; Benjamin, L.; Franklin, J.; Holroyd, G.; Incoll, L.D.; Lefsrud, M.G.; Pitkin, G. Guidelines for Measuring and Reporting Environmental Parameters for Experiments in Greenhouses. *Plant Methods* **2015**, *11*, 43. [CrossRef]
17. Both, A.J.; Ciolkosz, D.E.; Albright, L.D. Evaluation of light uniformity underneath supplemental lighting systems. *Acta Hortic.* **2002**, *580*, 183–190. [CrossRef]
18. Zhen, S.; Bugbee, B. Far-Red Photons Have Equivalent Efficiency to Traditional Photosynthetic Photons: Implications for Redefining Photosynthetically Active Radiation. *Plant Cell Environ.* **2020**, *43*, 1259–1272. [CrossRef]
19. Zhen, S.; van Iersel, M.; Bugbee, B. Why Far-Red Photons Should Be Included in the Definition of Photosynthetic Photons and the Measurement of Horticultural Fixture Efficacy. *Front. Plant Sci.* **2021**, *12*, 693445. [CrossRef]
20. DesignLights Consortium. Available online: https://www.designlights.org/wp-content/uploads/2023/03/DLC_HORT-Technical-Requirements-V3-0_03312023.pdf (accessed on 1 September 2023).
21. Trouwborst, G.; Hogewoning, S.W.; van Kooten, O.; Harbinson, J.; van Ieperen, W. Plasticity of Photosynthesis after the 'Red Light Syndrome' in Cucumber. *Environ. Exp. Bot.* **2016**, *121*, 75–82. [CrossRef]
22. Lin, K.-H.; Huang, M.-Y.; Hsu, M.-H. Morphological and Physiological Response in Green and Purple Basil Plants (*Ocimum basilicum*) under Different Proportions of Red, Green, and Blue LED Lightings. *Sci. Hortic.* **2021**, *275*, 109677. [CrossRef]
23. Paradiso, R.; Proietti, S. Light-Quality Manipulation to Control Plant Growth and Photomorphogenesis in Greenhouse Horticulture: The State of the Art and the Opportunities of Modern LED Systems. *J. Plant Growth Regul.* **2022**, *41*, 742–780. [CrossRef]
24. Pennisi, G.; Blasioli, S.; Cellini, A.; Maia, L.; Crepaldi, A.; Braschi, I.; Spinelli, F.; Nicola, S.; Fernandez, J.A.; Stanghellini, C.; et al. Unraveling the Role of Red:Blue LED Lights on Resource Use Efficiency and Nutritional Properties of Indoor Grown Sweet Basil. *Front. Plant Sci.* **2019**, *10*, 305. [CrossRef]
25. Pennisi, G.; Orsini, F.; Blasioli, S.; Cellini, A.; Crepaldi, A.; Braschi, I.; Spinelli, F.; Nicola, S.; Fernandez, J.A.; Stanghellini, C.; et al. Resource Use Efficiency of Indoor Lettuce (*Lactuca sativa* L.) Cultivation as Affected by Red:Blue Ratio Provided by LED Lighting. *Sci. Rep.* **2019**, *9*, 14127. [CrossRef]
26. Jin, W.; Urbina, J.L.; Heuvelink, E.; Marcelis, L.F.M. Adding Far-Red to Red-Blue Light-Emitting Diode Light Promotes Yield of Lettuce at Different Planting Densities. *Front. Plant Sci.* **2021**, *11*, 609977. [CrossRef]
27. Sarlikioti, V.; de Visser, P.H.B.; Marcelis, L.F.M. Exploring the Spatial Distribution of Light Interception and Photosynthesis of Canopies by Means of a Functional–Structural Plant Model. *Ann. Bot.* **2011**, *107*, 875–883. [CrossRef]
28. Kim, J.; Kang, W.H.; Son, J.E. Interpretation and Evaluation of Electrical Lighting in Plant Factories with Ray-Tracing Simulation and 3D Plant Modeling. *Agronomy* **2020**, *10*, 1545. [CrossRef]
29. Gu, S.; Wen, W.; Xu, T.; Lu, X.; Yu, Z.; Guo, X.; Zhao, C. Use of 3D Modeling to Refine Predictions of Canopy Light Utilization: A Comparative Study on Canopy Photosynthesis Models with Different Dimensions. *Front. Plant Sci.* **2022**, *13*, 735981. [CrossRef]
30. Cabrera-Bosquet, L.; Fournier, C.; Brichet, N.; Welcker, C.; Suard, B.; Tardieu, F. High-Throughput Estimation of Incident Light, Light Interception and Radiation-Use Efficiency of Thousands of Plants in a Phenotyping Platform. *New Phytol.* **2016**, *212*, 269–281. [CrossRef]
31. Thrash, T.; Lee, H.; Baker, R.L. A Low-Cost High-Throughput Phenotyping System for Automatically Quantifying Foliar Area and Greenness. *Appl. Plant Sci.* **2022**, *10*, e11502. [CrossRef]
32. Virtanen, P.; Gommers, R.; Oliphant, T.E.; Haberland, M.; Reddy, T.; Cournapeau, D.; Burovski, E.; Peterson, P.; Weckesser, W.; Bright, J. SciPy 1.0: Fundamental Algorithms for Scientific Computing in Python. *Nat. Methods* **2020**, *17*, 261–272. [CrossRef]
33. D'Agostino, R.; Pearson, E.S. Tests for Departure from Normality. *Biometrika* **1973**, *60*, 613–622. [CrossRef]
34. Piovene, C.; Orsini, F.; Bosi, S.; Sanoubar, R.; Bregola, V.; Dinelli, G.; Gianquinto, G. Optimal Red:Blue Ratio in Led Lighting for Nutraceutical Indoor Horticulture. *Sci. Hortic.* **2015**, *193*, 202–208. [CrossRef]
35. Statistical Functions (Scipy.Stats)—SciPy v1.11.2 Manual. Available online: https://docs.scipy.org/doc/scipy/reference/stats.html (accessed on 14 September 2023).

36. Ariz, I.; Esteban, R.; García-Plazaola, J.I.; Becerril, J.M.; Aparicio-Tejo, P.M.; Moran, J.F. High Irradiance Induces Photoprotective Mechanisms and a Positive Effect on NH4+ Stress in *Pisum sativum* L. *J. Plant Physiol.* **2010**, *167*, 1038–1045. [CrossRef]
37. Balázs, L.; Dombi, Z.; Csalambik, L.; Sipos, L. Characterization and Optimization of Photon Irradiance Distribution in Vertical Farms. *Acta Hortic.* **2023**, *1369*, 149–156. [CrossRef]

horticulturae

MDPI

Article

Mesh Crop Cover Optimizes the Microenvironment in a Tropical Region and Modifies the Physiology and Metabolome in Tomato

Victoria A. Delgado-Vargas [1], Gloria I. Hernández-Bolio [2], Emanuel Hernández-Núñez [2], Hélène Gautier [3], Oscar J. Ayala-Garay [1,*] and René Garruña [4,*]

[1] Colegio de Postgraduados, Postgrado de RGP-Fisiología Vegetal, Carretera México-Texcoco Km 36.5, Montecillo, Texcoco 56230, México, Mexico; delgado.victoria@colpos.mx

[2] Departamento de Recursos del Mar, Centro de Investigación y de Estudios Avanzados del IPN, Unidad Mérida, Antigua Carretera a Progreso Km 6, Cordemex, Loma Bonita Xcumpich, Mérida 97310, Yucatán, Mexico; gloria.hernandez@cinvestav.mx (G.I.H.-B.); emanuel.hernandez@cinvestav.mx (E.H.-N.)

[3] INRAE, UR1115, Plantes et Systèmes de Culture Horticoles, Domaine St Paul, Site Agroparc, 84914 Avignon, France; helene.gautier@inrae.fr

[4] CONACYT-Instituto Tecnológico de Conkal. Avenida Tecnológico s/n. Conkal 97345, Yucatán, Mexico

* Correspondence: oayala@colpos.mx (O.J.A.-G.); rene.garruna@itconkal.edu.mx (R.G.); Tel.: +52-595-116-4504 (O.J.A.-G.); +52-999-222-3553 (R.G.)

Abstract: In tropical regions, high light levels can lead to increased photooxidative damage in plants. Thus, reducing solar radiation could have a substantial impact on crop performance. This study aimed to evaluate the physiological responses and metabolic profile of two tomato varieties grown in microenvironments modified with cover meshes under a high light level and a warm climate. The experiment was achieved under high solar irradiance and an unfavorably high temperature. The varieties "Moneymaker" (MM) and "Campeche 40" (C40) were grown from 45 to 130 days after sowing at four solar irradiance levels: 100% (T1), 80% (T2), 75% (T3), and 50% (T4). In both varieties, the plants grown under the lowest irradiances (T3 and T4) were the tallest, with larger leaf areas, and accumulated more aerial and root biomass. Under moderate shading (T2), plants took better advantage of the light and had the highest photochemical quenching coefficient (qP) (C40 = 0.60 and MM = 0.48) and the highest electron transport rate (ETR). However, T3 and T4 plants had the highest net assimilation rate (23.6 and 23.9 μmol m^{-2} s^{-1} in C40, and 22.7 and 22.6 μmol m^{-2} s^{-1} in MM, respectively) and the highest A/Ci coefficients. Although both tomato varieties accumulate similar metabolites, MM leaves accumulate more glucose and C40 leaves accumulate more proline and valine. Furthermore, MM leaves accumulate more glycine and GABA under high radiation, and C40 leaves accumulate more proline and valine than leaves under 50% shade (T4). We conclude that using meshes in areas with high irradiance could be an alternative to reduce abiotic stress factors in plants.

Keywords: temperature; tomato quality; nutrients; phenolic compounds; carotenoids; minerals

1. Introduction

Tomato (*Solanum lycopersicum* L.) is the world's second most economically important vegetable [1]. These plants are typically cultivated in warm, tropical, and subtropical climates, under both open field and protected agriculture conditions. However, high solar radiation and temperature can affect production in tropical regions since the optimum crop temperature is 25/15 °C (day/night) [2]. Indeed, Boote et al. [3] consider tomato to be a species sensitive to high temperatures.

Xu et al. [4] reported that increased maximum temperatures negatively affect reproductive development and crop physiology. In some cases, it has been observed that temperatures above 34 °C and high environmental humidity reduced flower pollen and ovule development and increased flower malformation [5]. One of the problems in tropical regions is that their environmental characteristics can cause morpho-anatomical, physiological, and biochemical changes in tomatoes; for example, between 35 and 40 °C, the Rubisco enzyme undergoes reversible inhibition, but at higher temperatures, the inhibition is irreversible [6]. In addition to photosynthesis, water relations and hormonal balance can also be affected [7], negatively impacting both fresh and dry mass of fruits due to changes in primary or secondary metabolism [8]. A study in tomato plants showed that photosynthesis and growth parameters were enhanced when solar radiation increased [9], and the temperature was close to optimal for tomato. However, in a tropical region with unfavorably high temperatures, both solar radiation and temperature can influence the growth of tomato plants [10].

Thus, projections of rising global temperatures pose a challenge to agricultural production worldwide [11], especially in tropical areas where the excess electromagnetic radiation from the sun will increase air temperature above the thermal optimum for crops [12], especially tomato. In this sense, employing crop cover meshes could modify the microenvironment by reducing the radiation reaching the plants and generating a near-optimum microenvironment [13]. The meshes most commonly used in agriculture are black, as they provide shade equally throughout the entire band of the electromagnetic spectrum; their main objective is to reduce irradiance without modifying the quality of light [14]. While numerous types of plastic mesh are currently used to promote optimal crop growth and development [13], more information needs to be available regarding the effects of meshes on plant physiology, growth, and, in particular, leaf composition. For example, a metabolomic study was conducted to distinguish between mature green and red ripe tomato fruits, enabling the authors to create a list of amino acids and secondary components that define each of the tomato ripening stages [15]. In this sense, detecting the presence and measuring the concentration of specific metabolites are essential to understanding the functioning of a biological system. A previous study demonstrated that nuclear magnetic resonance (NMR) could be utilized for metabolomics since it yields substantial qualitative and quantitative information about plant metabolites [16]. Additional advantages of 1H-NMR metabolomics include its non-destructive nature, the possibility of detecting signals from diverse polarity metabolites, and simple spectra processing [17]. The interpretation of spectroscopic data is now easier since they can be compared with available databases [18].

Therefore, this research aimed to evaluate the physiological responses and metabolic profile of two tomato varieties grown in microenvironments modified with cover meshes in a tropical region with a warm climate.

2. Materials and Methods

2.1. Experimental Site and Plant Material

The experiment was carried out in the experimental area of the Instituto Tecnológico de Conkal in Yucatán, Mexico (21.07° NL; 89.52° WL and 8 m.a.s.l.). Two tomato varieties were planted: (1) "Moneymaker" (MM), which is a commercial temperate-climate variety with indeterminate growth, produces ball-type fruits, and is considered a world reference in studies of the species [19]; and (2) "Campeche 40 " (C40), which is a landrace variety of the state of Campeche in Mexico, where the climate is warm–subhumid, has indeterminate growth, and produces kidney-type fruits [20].

2.2. Crop Establishment and Management

Seed sowing was performed in 200-cavity polystyrene trays, with Canadian moss (Sunshine, Springfield, OH, USA) used as substrate. Fertilization began 15 days after sowing (das) with the appearance of the first pair of leaflets; the fertilizer 19-19-19 (N:P:K) + 1% M.E. (Poly-Feed, Haifa, Mexico) was applied in the irrigation water three times a week at a concentration of 1 g L^{-1}.

At 45 das, the plants were transplanted into 40 × 50 cm black polyethylene bags; the substrate used was a mixture of soil and vermicompost at a 70:30 (v/v) ratio, previously disinfected by the vaporization method. The population density was 3.5 plants m^{-2}. Agronomic management was performed according to Guzmán et al. [21]. Conventional tutoring was performed throughout the crop, and the leaves below the first fruit cluster and the shoots in axillary buds were pruned every 15 days. Steiner's [22] solution (electrical conductivity of 3.5 dS m^{-1} and a pH of 5.5 to 6) was applied for fertilization at the time of transplanting.

2.3. Treatments and Characterization of Microenvironments

The experiment consisted of eight treatments: two varieties (MM and C40) and four solar irradiance intensities: T1 = open field, 100% irradiance; T2 = white anti-aphid mesh tunnel, 80% irradiance; T3 = gray anti-aphid mesh tunnel, 75% irradiance; T4 = tunnel with white anti-aphid mesh plus black shade mesh, 50% irradiance.

In each treatment, the microenvironment was characterized by a weather station (Onset HOBO U30, Bourne, MA, USA). Sensors were placed inside the tunnel at canopy height, and the station was programmed to record data every 30 s and average them every 10 min. The meteorological variables evaluated included solar radiation (R), photosynthetic photon flux density (PPFD), air temperature (AT), and relative humidity (RH). Diurnal curves (6 am to 7 pm) were constructed using the data.

2.4. Morphological Variables and Biomass Distribution

Destructive sampling was carried out at 130 das during the fruit-filling stage. In each sampling, four plants were used for each treatment and were evaluated for height, the total number of leaves, and leaf area. An area integrator (LICOR LI-3100, Lincoln, NE, USA) was used to measure the leaf area. The plants were separated by organs and dried in a forced air oven at 70 °C until constant weight mass (~72 h).

2.5. Leaf Photochemistry and Gas Exchange

The quantum efficiency of photosystem II (PS_{II}) was evaluated with a pulse-amplitude-modulated fluorometer (PAM Walz, Effeltrich, Germany). The following variables were measured as proposed by Samaniego-Gámez et al. [23]: maximum photochemical quantum yield of PSII (Fv/Fm) and potential activity of PSII (Fv/F0) (where Fv is variable fluorescence (Fm-F0), F0 is initial fluorescence, and Fm is maximum fluorescence), photochemical quenching coefficient (qP) and non-photochemical quenching coefficient (NPQ), electron transport rate (ETR), and effective quantum yield of PS_{II} (Φ_{PSII}). The saturating light was an 8000 μmol m^{-2} s^{-1} pulse of actinic light. Nine light pulses (from 0 to 1500 μmol m^{-2} s^{-1}) were used for the ETR and Φ_{PSII} curves. Measurements on the third fully developed leaf from the apex were taken at noon on 120 das.

Gas exchange variables were measured at 115 das (at the reproductive stage of the third cluster). An infrared gas analyzer (LICOR LI-6400, Lincoln, NE, USA) was used to evaluate the net CO_2 assimilation rate (A_N), stomatal conductance (g_s), intercellular carbon (C_i), transpiration (E), and water-use efficiency (WUE). Five plants per treatment and three leaves per plant were evaluated; the leaves were from the upper part of the canopy and were fully expanded. Measurements were made from 6 am to 6 pm to record physiological responses during the diurnal course [24].

2.6. Metabolic Profile by NMR

From each plant, 6 g of leaves was collected and dried for 12 h at room temperature (25 °C) and then at 50 °C for 24 h in a convection oven. The dried leaves were ground, and 48 mg was stored in amber glass jars until extraction of the metabolites [25]. For extraction, the samples were transferred to a 2 mL Eppendorf tube, to which 750 μL of phosphate buffer in deuterated water with 0.05% trimethylsilylpropanoic acid sodium salt (D_2O/TSP) and 750 μL of deuterated methanol (MeOD) were added. Extraction was carried out by sonication for 20 min, and, subsequently, the samples were centrifuged at 13,000 rpm for 10 min, after which 800 μL of the supernatant was taken and transferred to a 5 mm NMR tube for recording.

The ^{1}H-RMN spectra were recorded on a Varian 600 MHz AR Premium Compact spectrometer (Agilent, Santa Clara, CA, USA). Each ^{1}H-RMN spectrum was recorded at 25 °C with 128 scans (nt) under the following parameters: acquisition time (at) of 3.2 s, pulse width (pw) of 30 °C, and a relaxation time (d1) of 1.5 s, requiring about 10 min to record each sample. A presaturation sequence (PRESAT) was used to suppress the residual water signal. The raw spectra obtained from the ^{1}H-NMR were processed with MNOVA software and spectral intensities were normalized with a value of 100 with respect to the TSP signal. A manual TSP reference was performed by placing it at δ 0.0 ppm, and it was apodized with a Gaussian basis function (LB = 0.3 Hz). Spectra were reduced to 0.04 ppm integrated regions (bins) from δ −0.5 to 10. The residual water signal (δ 4.75–4.90 region) and methanol signal (δ 3.29–3.32 region) were excluded from the matrix. The data matrix obtained from the ^{1}H-NMR analysis of the metabolic profile of both varieties contained the intensities of 247 bins (integrated regions) for each of the 48 samples.

Metabolite identification was performed using a representative ^{1}H-RMN spectrum (600 MHz) (nt = 1024), from which the chemical shifts of metabolites characteristic of the species were obtained using ^{1}H-NMR spectra libraries. Assignment of the metabolites corresponding to the selected signals with respect to the VIP (variable importance in projection) statistics and the loading plot was performed by comparison of the chemical shifts and coupling patterns of the detected signals with those reported in previous studies [25,26], the Chenomx NMR Mixture Analysis database, "https://www.chenomx.com/ (accessed on 15 January 2023)", and the Human Metabolome Database (HMDB), "https://hmdb.ca/ (accessed on 15 January 2023)".

2.7. Experimental Design and Statistical Analysis

A completely randomized split-plot experimental design was used; the main plots were the irradiance levels (T1, T2, T3, and T4) and the secondary plots were the varieties (MM and C40). Each plot had 20 replicates. An analysis of variance (ANOVA) was applied, and means were compared using Tukey's test ($p < 0.05$). The results were analyzed using InfoSat Ver 2013 and Sigmaplot Ver 2004 statistical software. Principal component analysis (PCA) was performed using the matrix of spectral intensities with Pareto scaling in MetaboAnalyst software "https://www.metaboanalyst.ca/MetaboAnalyst/home.xhtml (accessed on 15 January 2023)". Subsequently, a partial least squares discriminant analysis (PLS-DA) was performed to obtain the model's most important and influential variables. The model's quality was determined by the R^2 value (variation percentage of the set explained by the Y-predicted components) and Q2 (variation percentage of the set predicted by the model according to cross-validation). Important variables in the PLS-DA model were detected using the VIP plot.

3. Results and Discussion

3.1. Characterization of the Microenvironments

In both solar radiation (R) and photosynthetic photon flux density (PPFD), the open field treatment (T1) had the highest values throughout the day; at 3 pm, the maximum values were reached (1032 W m^{-2} and 1844 μmol m^{-2} s^{-1}, respectively) (Figure 1A,B), while the maximum values of R and PPFD in the mesh treatments were 764 W m^{-2} and 1304 μmol m^{-2} s^{-1} for T2 (80%), 635 W m^{-2} and 1118 μmol m^{-2} s^{-1} for T3 (75%), and 529 W m^{-2} and 854 μmol m^{-2} s^{-1} for T4 (50%). The percentages in each treatment indicate the proportion of incident solar radiation on the plants.

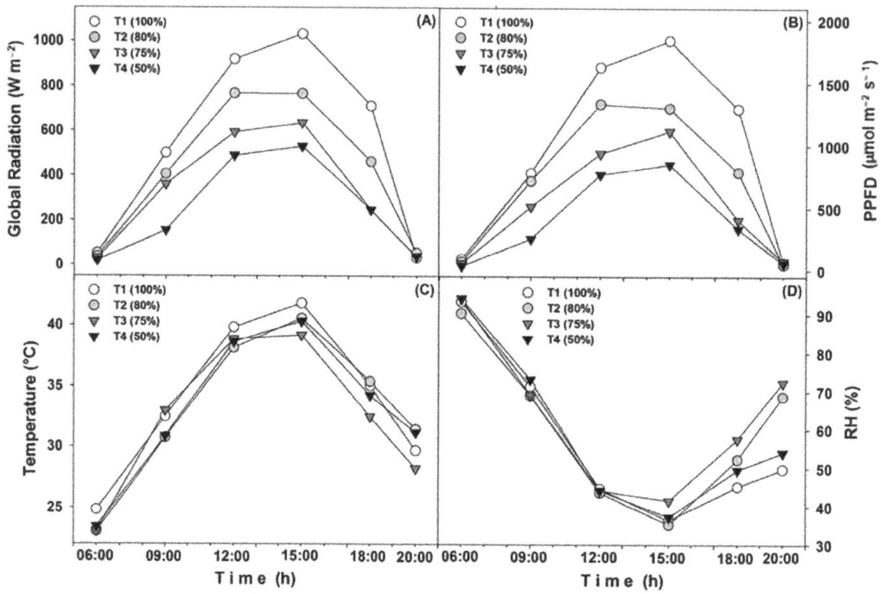

Figure 1. (**A**) Solar radiation (R), (**B**) photosynthetic photon flux density (PPFD), (**C**) air temperature (AT), and (**D**) relative humidity (RH) of four environments generated by meshes that allowed light to pass through at 80% (T2), 75% (T3), and 50% (T4), and a control (100% = T1). Measurements were made on a clear sunny day 120 das.

Furthermore, the maximum air temperature (AT) recorded at 3 pm (41.8 °C) in T1 was 16.7 °C higher than the optimum temperature (25 °C) reported for tomato cultivation [27]; it was at this time of day when the maximum heat point was reached in T1, with the meshes used in T2, T3, and T4 barely decreasing AT by 1.2, 2.6, and 1.5 °C, respectively (Figure 1C). According to Peet [28], daytime temperatures above 35 °C drastically reduce fruit production and seed formation; in this experiment, all treatments were above the optimum temperature for the crop. However, this is an inherent condition in tropical areas. The average night temperature for all treatments was 27.3 °C.

Relative humidity (RH) was maximal at the beginning of the day (above 90%) and minimal at 3 pm in all treatments (T1 = 36%, T2 = 35%, T3 = 41%, and T4 = 37%), and then it began to increase slightly towards dusk (Figure 1D).

According to the characterization of the microenvironments, despite having excellent PPFD conditions, the large amount of R that affects the site causes stressful situations in the functioning of the plants due to the excessive increase in AT and very low RH.

3.2. Morphological Variables and Biomass Distribution

According to the analysis of variance, all morphometric variables (height, number of leaves, and leaf area) were statistically affected by the shading treatments. Under the highest irradiances, the plants were shorter; in T1, the height was 86.5 cm and 74.3 cm in C40 and MM, respectively, while in T2, it was 93 cm (C40) and 90.6 cm (MM). The plants of T3 (135 and 158 cm in C40 and MM, respectively) and T4 (154.8 and 161.3 cm in C40 and MM, respectively) were the tallest (Table 1). A similar trend was observed in the leaf area, with C40 and MM presenting reduced leaf area in the treatments with more light. However, in the number of leaves, significant differences were only observed in T1-MM compared to T3 or T4. This result indicates that the higher the irradiance, the shorter the plants, the lower the leaf area, and the lower the biomass.

Table 1. Plant height, number of leaves, and plant leaf area of two tomato varieties (C40 and MM) grown at four different solar irradiance levels (T1: 100%, T2: 80%, T3: 75%, and T4: 50%). Measurements were made 130 das.

Treatments	Variety	Plant Height (cm)	Number of Leaves	Leaf Area (cm^2)
T1	C40	86.5 ± 1.66 ^{b}	13.2 ± 1.65 ^{ab}	882.8 ± 97.75 ^{cd}
	MM	74.3 ± 3.26 ^{b}	6.5 ± 0.87 ^{b}	358.1 ± 29.75 ^{d}
T2	C40	93.0 ± 2.64 ^{b}	12.7 ± 1.31 ^{ab}	716.5 ± 127.97 ^{cd}
	MM	90.6 ± 2.59 ^{b}	14.3 ± 1.49 ^{a}	879.6 ± 43.90 ^{cd}
T3	C40	135.0 ± 3.94 ^{a}	19.8 ± 0.75 ^{a}	2075.4 ± 202.61 ^{ab}
	MM	158.0 ± 7.40 ^{a}	18.0 ± 1.58 ^{a}	1459.3 ± 61.18 ^{bc}
T4	C40	154.8 ± 1.80 ^{a}	19.8 ± 0.75 ^{a}	3100.8 ± 264.90 ^{a}
	MM	161.3 ± 8.96 ^{a}	14 ± 2.48 ^{a}	1481.7 ± 209.40 ^{bc}
LSD		27.15	7.09	1051.4

Data are means ± standard error; n = 9. Different letters in the same column indicate significant statistical differences (Tukey, $p \leq 0.05$). LSD = least significant difference.

We also observed that the treatments with lower light intensity promoted plant growth, possibly in search of the resource [29]. By comparison, treatments with more light decreased leaf size, which is a typical response in plants growing in environments with high radiation and temperature, as they reduce their boundary layer to avoid water loss [30]. In this regard, Ayala-Tafoya et al. [29] observed that using mesh to reduce total radiation (30 and 50%) increased leaf size and, consequently, leaf area with photosynthetically more efficient tomato leaves. In some cases, leaves exposed to low light intensities may have a higher photosynthetic efficiency than leaves with higher exposure; this is because they make the most of the resource to be able to maintain themselves, while exposed leaves may present a photosynthetic acclimation that limits their maximum light saturation rate [31].

According to the analysis of variance, biomass accumulation was statistically different in all organs depending on the treatment. In the lower irradiance treatments (T3 and T4), the C40 and MM plants accumulated greater biomass in the roots (T3: 5.52 and 5.21 g plant^{-1}, T4: 6.34 and 5.33 g plant^{-1}, respectively), stems (T3: 10.96 and 10.23 g plant^{-1}, T4: 13.33 and 12.16 g plant^{-1}, respectively) and leaves (T3: 14.52 and 15.24 g plant^{-1}, T4: 18.64 and 14.81 g plant^{-1}, respectively) compared to the T1 and T2 plants (Figure 2). On the other hand, only flower biomass was found to be significantly different due to the effect of the variety, with C40 in T4 (1.63 g plant^{-1}) exhibiting significantly increased flower biomass compared to the plants in T1 (0.42 and 0.21 g in C40 and MM, respectively) and T2 (0.39 and 0.28 g plant^{-1} in C40 and MM, respectively). As shown in Figure 2, the highest fruit dry mass accumulation was observed in T4 (C40, 8.85 g plant^{-1} and MM, 8.17 g plant^{-1}). In this sense, Garruña-Hernández et al. [24] noted that in tropical climates the biomass distribution of some vegetables is an indicator of the effect generated by temperature and irradiance on the accumulation of photoassimilates in each plant organ. When heat stress is constant, it can induce morpho-anatomical, physiological, and biochemical changes [11]; it is likely that the MM variety, being of temperate origin, had lower biomass values than C40, which is of tropical origin. Some studies found that high irradiance and temperature affect the development of tomato plants, causing burning and abscission of leaves, branches, and stems, premature leaf senescence, attenuated root growth, floral abortions, and fruit drop. Several studies reported that the latter is due to these environmental conditions inducing flower malformation caused by deficient fertilization processes that damage reproductive structures, resulting in deficient fruit setting and reduced yields [7,12].

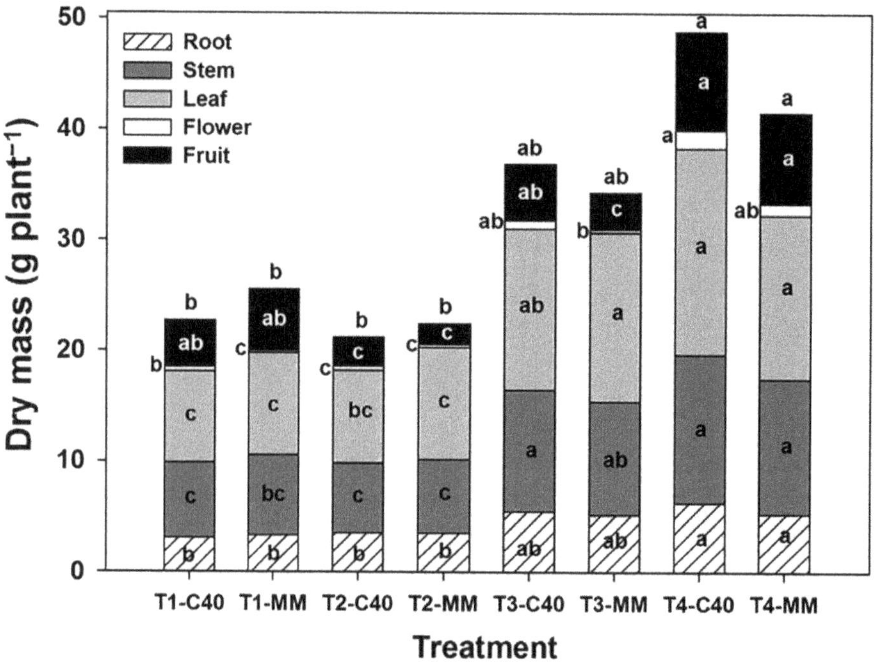

Figure 2. Biomass per organ of the C40 and MM tomato varieties grown at four different solar irradiance levels (T1: 100%, T2: 80%, T3: 75%, and T4: 50%). Data are means; $n = 9$. Different letters in the same organ indicate significant statistical differences (Tukey, $p \leq 0.05$).

3.3. Leaf Photochemistry

According to the analysis of variance, there were no statistical differences ($p \leq 0.05$) in the maximum quantum yield of photosystem II (Fv/Fm = chlorophyll fluorescence) among treatments. However, there were significant differences in qP, such that it decreased in extreme environments (T1 and T4 with 100 and 50% irradiance, respectively). In contrast, the highest qP values were observed in T2 (C40 = 0.60 and MM = 0.48), indicating that there is not necessarily a linear trend between the amount of light reaching the plant and the amount of energy allocated to photochemical processes in tomato (Figure 3A). Furthermore, in the non-photochemical quenching coefficient (NPQ), the C40 variety in T2 (with PPFD of 1300 μmol m^{-2} s^{-1}) had the lowest values (0.1) (Figure 3B). This observation was reflected in the electron transport rate, where the plants of the two varieties grown with 80% solar irradiance (T2) statistically outperformed the rest of the treatments, followed by the plants that received 75% irradiance (T3) (Figure 3C,D, respectively). The excess and lack of light (T1 = 100 and T4 = 50% solar irradiance) affected leaf photochemistry in this case. Alternatively, a moderate decrease in irradiance decreased the amount of energy going to non-photochemical processes and caused more energy to be channeled to photochemical processes, which favored the electron transport rate of photosystem II in tomato plants regardless of the variety. In this case, the amount of radiation received at the site is likely excessive, and the rate of D1 protein regeneration of tomato plants is not adequate for the site's environmental conditions, to the extent of saturating the photosystems [32]. In places with high radiation levels, utilizing meshes could be an alternative to reduce the quantity of light or modify its quality. However, an increase in leaf photochemistry does not always result in greater carbon assimilation or increased biomass, as other factors could limit these processes [31].

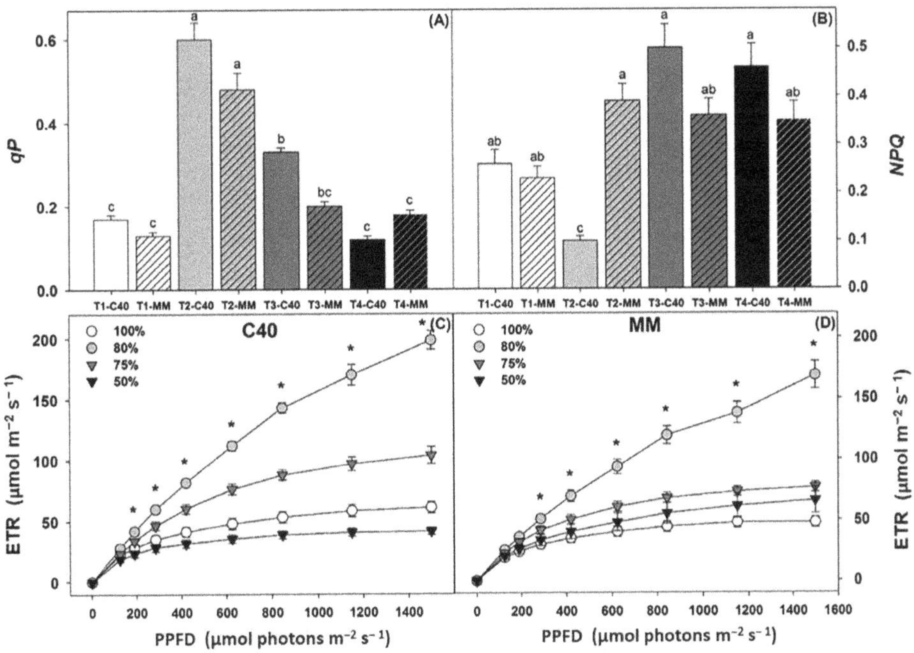

Figure 3. (**A**) Photochemical quenching coefficient (qP), (**B**) non-photochemical quenching coefficient (NPQ) and electron transport rate of PSII of tomato varieties C40, (**C**), and MM (**D**), grown at four different solar radiation levels (100%, 80%, 75%, and 50%). Data are means ± standard error; $n = 9$. Different letters in the same column and * between PPDF levels indicate significant statistical differences (Tukey, $p \leq 0.05$).

3.4. Gas Exchange

At noon, all treatments reached the highest net CO_2 assimilation rate (NA), which decreased as the sun set. However, the treatments with the lowest irradiance (with 75% and 50% solar radiation) produced the highest NA values throughout the day, reaching up to 23.6 and 23.9 µmol m^{-2} s^{-1} in C40 and 22.7 and 22.6 µmol m^{-2} s^{-1} in MM, respectively (Figure 4A,B). This result suggests that a higher incidence of PPFD will not necessarily be reflected in a higher carbon assimilation rate. In this sense, there is likely some biochemical limitation in the photosynthetic mechanism caused by excess light or high temperature [12]. There are species that, when faced with excess light energy, suffer damage to their photosystems and do not recover adequately [33]. The damage can increase when the growth temperature rises above the optimum for the crop [34], as was the case in this experiment. In this case, PPFD values between 800 and 1200 µmol m^{-2} s^{-1} were sufficient to reach the highest NA values without inflicting photodamage. The latter coincides with the results obtained in the light saturation curves (A/PPFD), where it was observed that, except for MM in T4 (50% irradiance), the treatments had photosynthetic acclimation above 1200 µmol m^{-2} s^{-1} of PPFD (Figure 4C,D). The response of MM in the 50% irradiance treatment is likely due to the ability of that genotype to increase carbon assimilation in response to light increases. However, in the CO_2 saturation curves (A/C_i), photosynthetic acclimation was not observed in any treatment. Instead, the trend was similar to that observed in the diurnal courses, with the highest photosynthetic values detected in the treatments with the lowest irradiance (Figure 4E,F). Only in the MM genotype at 50% irradiance did we observe a clear difference from 200 µmol^{-1} mol^{-1} of CO_2, where it obtained its compensation point, to 1500 µmol^{-1} mol^{-1} of atmospheric CO_2.

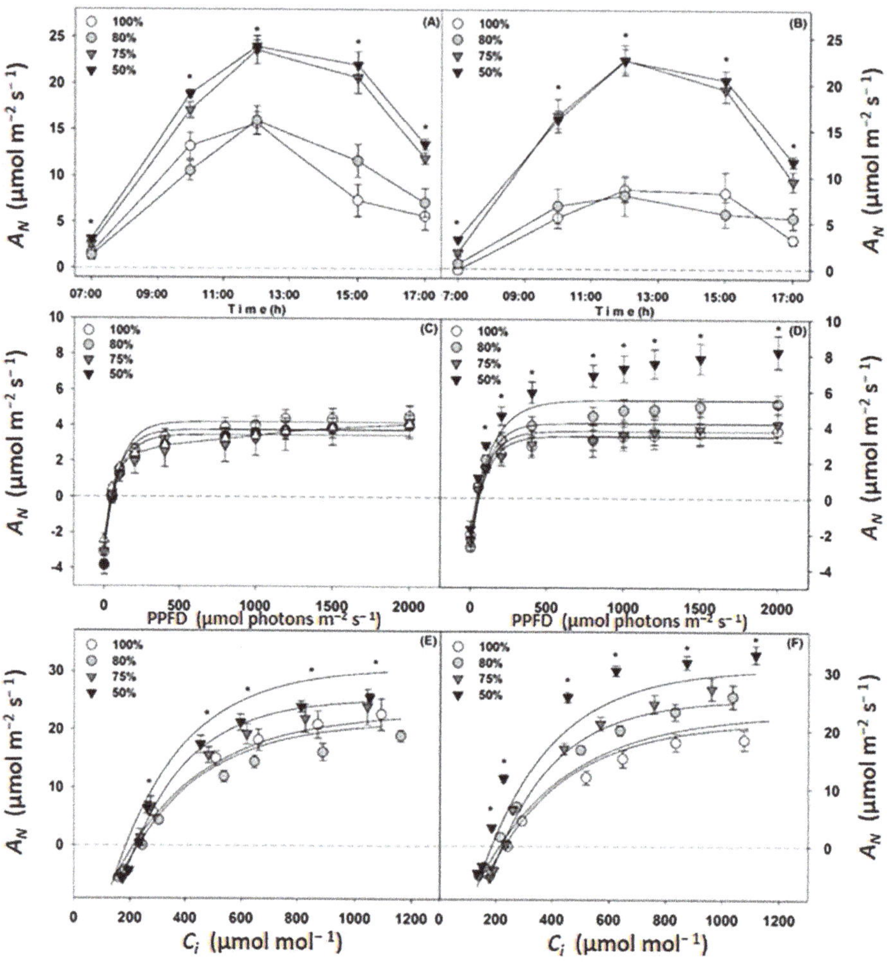

Figure 4. Photosynthesis throughout the day, A/PPFD, and A/C_i response curves of tomato varieties C40 (**A**, **C**, and **E**, respectively) and MM (**B**, **D**, and **F**, respectively), grown at four different solar irradiance levels (100%, 80%, 75%, and 50%). Data are means ± standard error; $n = 9$. * = significant statistical differences (ANOVA, $p \leq 0.05$).

3.5. Principal Component Analysis

In the principal component analysis (PCA) score plot of the leaf extracts (PC1 vs. PC2, 72.5% explained variance), the four solar irradiance treatments and the two tomato varieties were clustered in the central part of the plot, suggesting that all samples have a similar metabolic profile and that the differences among treatments consist primarily of variations in the abundance of the metabolites present (Figures 5 and S1).

PLS-DA was used to analyze the differences between tomato varieties. Thus, a clear separation between the two varieties was observed (Figure 6A) in the VIP, and loading plots were analyzed to identify the signals responsible for this separation and the differences between the varieties. It was found that the signals with the most significant influence on the separation of the samples are those with chemical shifts in the range of δ 2–4. Amino acid and carbohydrate resonances commonly occur in this spectral region, so it can be inferred that these compounds mark a difference between the two varieties, particularly the signals at δ 3.25, 2.09, and 2.01.

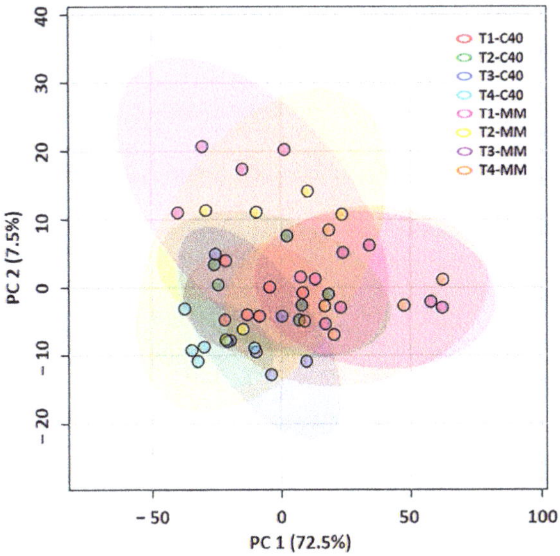

Figure 5. PCA score plot (PC1 vs. PC2) of leaf metabolic profiles of two tomato varieties (C40 and MM) grown at four different solar irradiance levels (T1 = 100%, T2 = 80%, T3 = 75%, and T4 = 50%).

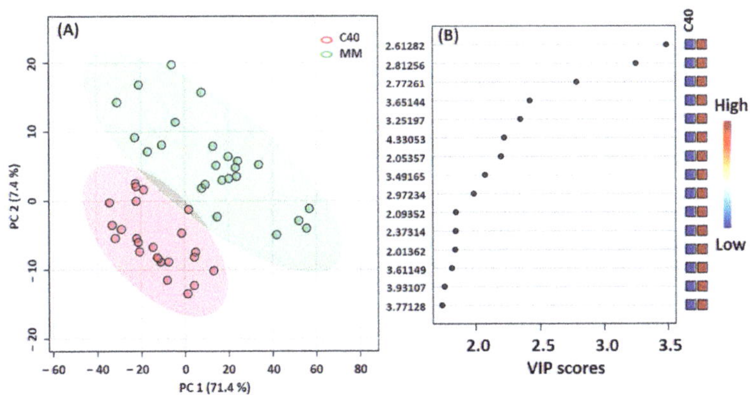

Figure 6. (**A**) Partial least squares discriminant analysis (PLS-DA, PC1 vs. PC2) and (**B**) main VIPs with the most significant influence on PC1 variation of two tomato varieties (C40 and MM).

Of the 15 VIPs with the most significant influence on principal component 1, different abundances were identified among the metabolites with the values of *δ* 3.25, 2.61, 2.09, and 2.01; these metabolites corresponded to proline (*δ* 4.12, 3.41, 3.32, 2.34, 2.09, 2.01), glucose (*δ* 3.25, 3.40, 3.46, 3.52, 3.728, 3.82, 3.89, 4.63, 5.22), and aspartate (*δ* 7.92, 4.41, 2.72, 2.51, 2.03) (Table S1) [35]; therefore, the commercial variety MM presented a higher abundance of these metabolites in its metabolic profile compared to the wild variety (C40), with amino acids being the predominant metabolites. The separation of the tomato varieties was due to a high concentration of amino acids, such as proline, glycine, and aspartate, and sugars such as glucose. In general, the main difference between the C40 and MM varieties was due to the abundance of metabolites, mainly amino acids.

The maximum and minimum solar radiation treatments (T1 = 100 and T4 = 50%) were analyzed to identify differences due to the environment. The PLS-DA-generated model produced an R^2 of 0.61547 and 0.56906 and a predictive capacity (Q2) of 0.4592 and 0.43004 for C40 and MM, respectively (Figure 7A,B). These results indicate that there is a noticeable effect on the abundance of metabolites present in tomato leaves. Therefore, considering the 15 VIPs with the greatest influence on principal component 1, in the commercial tomato variety (MM), metabolites with chemical shifts of 3.25 (glucose) were identified, while in the wild variety (C40) only chemical shifts corresponding to amino acids (proline δ 1.09 and valine δ 1.01) were identified. The metabolites identified are part of the projection that best discriminates between the treatment conditions (100% and 50% solar irradiance) (Figure 7C,D). At higher light availability, an increase in the intensity of the signals was observed, while in less light, the abundance of metabolites was lower. In MM, the values with the most significant influence were amino acids such as glycine and γ-amino butyric acid (GABA), which correspond to a chemical shift of 3.57 and 1.89, respectively. On the other hand, in C40, other metabolites were identified, with a chemical shift of 1.09 and 1.01 corresponding to proline and valine, respectively (Table S1).

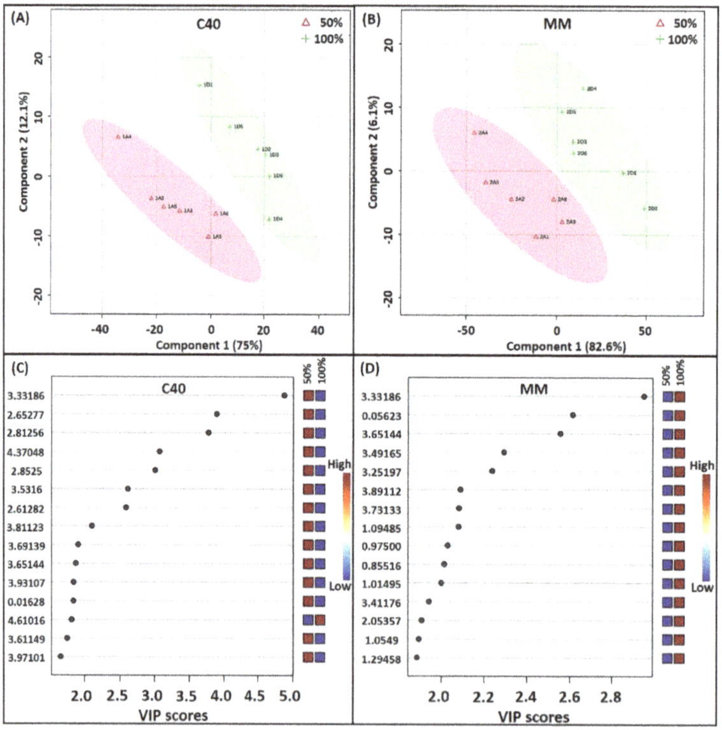

Figure 7. (**A**) Partial least squares discriminant analysis (PLS-DA, PC1 vs. PC2) and (**B**) main VIPs with the most significant influence on PC1 variation of treatments with minimum (50%) and maximum (100%) solar irradiance on tomato varieties C40 ((**A**) and (**C**), respectively) and MM ((**B**) and (**D**), respectively).

In general, the separation of metabolic profiles in tomato plants was observed when the incidence of solar radiation was reduced to 50%. Likewise, there was a greater abundance of metabolites in the commercial tomato variety (MM). In this sense, it is known that abiotic factors induce the production of secondary metabolites in plants [7], and if these factors cause abiotic stress, they can generate the accumulation of proline, GABA, and a variety of carbohydrates [36]. In this regard, Hüther et al. [37] note that an increase in metabolites in tomato plants may be related to light utilization by the photosynthetic electron transport chain. However, in this experiment, the highest ETR and qP were observed in T2 (80% irradiance). That is, even though at higher irradiance, there were more metabolites; the same trend was not observed in the utilization of the light by photosystem II since the plants in the treatment with the most irradiance (T1 = 100%) were the least efficient in using light energy, possibly generating a level of abiotic stress due to excess irradiance. In this regard, Baracaldo et al. [38] state that stress due to light intensity reduces biomass accumulation in tomato plants, with a detrimental effect on photosynthesis, which coincides with the results of this research where the plants in the treatments with the highest light intensity had the lowest photosynthetic rate and consequently the lowest values of total biomass. In this sense, Carrari et al. [39] indicate that light-related stress can decrease fruit size and modify sugar content.

The metabolic profile of MM had a higher abundance of α-glucose, which is generally associated with plant resistance against infections caused by biotic agents such as *Meloidogyne incognita* [26]. Amino acids play an essential role in plants, whether to overcome stress or disease. Previously, Chaves-Barrantes and Gutiérrez-Soto [40] applied abiotic stress by temperature and observed an increase in the accumulation of soluble sugars, sugar alcohols (mannitol, sorbitol, and glycerol), proline, glycine, betaine, and ternary sulfur compounds. Some authors [41–43] noted that the amount of proline in plants rises in response to abiotic stress (drought, high temperature, luminosity, ultraviolet radiation, salinity, and heavy metals in the soil). Regarding the abundance of proline in tomato plants, Schwacke et al. [44] observed an evident increase in response to abiotic stresses such as water stress. Furthermore, Hare et al. [45] observed that proline in plants could play vital roles in different tissues or conditions, while glycine protects the plant against pests [46]. Another important osmolyte is GABA, a non-protein amino acid synthesized from glutamic acid, through a reaction catalyzed by glutamate decarboxylase, for which studies by Wahid et al. [7] indicate that it confers heat tolerance to plants.

Furthermore, although tomato plants can grow in a wide range of climatic conditions, their vegetative and reproductive growth can be seriously affected in conditions of high temperature and irradiance [12].

4. Conclusions

The use of meshes modified the microenvironment. In the most critical hours of the day, solar radiation decreased by 26% to 49% and photosynthetic photon flux density by 29 to 54%, which caused the temperature to fall by 1.2 to 2.6 °C. The treatments with the least light availability (T3 and T4) had the plants with the most remarkable growth in height and leaf area and the most significant accumulation of total biomass. In the photochemical parameters, although in the maximum quantum yield of photosystem II (Fv/Fm), there were no statistical differences among treatments, and the plants grown at 80% irradiance (T2) allocated more light energy to the electron transport chain (ETR) and the photochemical quenching of photosystem II (qP). However, plants exposed to the least irradiance (T3 and T4) displayed the highest photosynthetic rates in the diurnal gas exchange and A/Ci response curves. On the other hand, changes in the abundance of leaf metabolites were observed in the tomato varieties (MM and C40), with the MM variety having a higher abundance of metabolites than the C40 variety. The abiotic irradiance factor directly influenced the leaf metabolome of the two tomato varieties by modifying the abundance of metabolites such as sugars and amino acids. The MM leaves contained more sugars and amino acids at higher irradiance, which reflected their metabolic change under abiotic stress conditions. This study confirmed the potential of using shading meshes to limit irradiance in tropical climates to maintain tomato leaf photosynthetic activity, plant growth, and biomass accumulation.

Supplementary Materials: The following supporting information can be downloaded at: https://www.mdpi.com/article/10.3390/horticulturae9060636/s1, Figure S1: Overlay of ^{1}H-NMR spectra of *Solanum lycopersicum* leaf extracts from 1 h; Table S1: Metabolites identified from *Solanum lycopersicum* leaf extracts using ^{1}H-NMR (D2O/MeOH [1:1], 600 MHz).

Author Contributions: Conceptualization, O.J.A.-G. and R.G.; methodology, V.A.D.-V., O.J.A.-G., R.G., G.I.H.-B. and E.H.-N.; validation, V.A.D.-V., O.J.A.-G., H.G. and R.G.; formal analysis, V.A.D.-V., O.J.A.-G., G.I.H.-B., E.H.-N. and R.G.; investigation, V.A.D.-V., O.J.A.-G. and R.G.; data curation, V.A.D.-V. and R.G.; writing—original draft preparation, V.A.D.-V., O.J.A.-G. and R.G.; writing—review and editing, V.A.D.-V., O.J.A.-G. and R.G.; visualization, V.A.D.-V., O.J.A.-G. and R.G.; supervision, V.A.D.-V., O.J.A.-G., H.G. and R.G.; project administration, V.A.D.-V. and R.G. All authors have read and agreed to the published version of the manuscript.

Funding: This research was funded by National Council of Science and Technology (CONACyT, Mexico), project CB 286756 for R.G. and CB 254321 for E.H.-N.

Data Availability Statement: The data presented in this study are available in article.

Acknowledgments: To the National Council of Science and Technology (CONACyT, Mexico) for the postgraduate scholarship granted to the first author V.A.D.-V.

Conflicts of Interest: The funders had no role in the design of the study; in the collection, analyses, or interpretation of data; in the writing of the manuscript; or in the decision to publish the results.

References

1. FAOSTAT. Food and Agriculture Organization of the United Nations Statistics Division. Available online: http://www.fao.org/faostat/en/#data/ (accessed on 10 May 2023).
2. Adams, S.R.; Cockshull, K.E.; Cave, C.R.J. Effect of temperature on the growth and development of tomato fruits. *Ann. Bot.* **2001**, *88*, 869–877. [CrossRef]
3. Boote, J.K.; Rybak, M.R.; Scolberg, J.M.; Jones, J.W. Improving the cropgrow-tomato model for predicting growth yield response to temperature. *HortScience* **2012**, *47*, 1038–1049. [CrossRef]
4. Xu, J.; Wolters-Arts, M.; Mariani, C.; Huber, H.; Rieu, I. Heat stress affects vegetative and reproductive performance and trait correlations in tomato (*Solanum lycopersicum*). *Euphytica* **2017**, *213*, 156. [CrossRef]
5. Orozco, A.J.; Ayala, C.C.; Tatis, H.A. Efecto del cambio climático sobre la fisiología de las plantas cultivadas: Una revisión. *Rev. UDCA Actual. Divulg. Científica* **2012**, *15*, 63–76. [CrossRef]
6. Kubien, D.; Von Caemmerer, S.; Furbank, R.; Sage, R. C4 photosynthesis at low temperature. A study using transgenic plants with reduced amounts of Rubisco. *Plant Physiol.* **2003**, *132*, 1577–1585. [CrossRef]
7. Wahid, A.; Gelani, S.; Ashraf, M.; Foolad, M.R. Heat tolerance in plants: An overview. *Environ. Exp. Bot.* **2007**, *61*, 199–223. [CrossRef]
8. Ruíz-Nieves, J.M.; Ayala-Garay, O.J.; Serra, V.; Dumont, D.; Vercambre, G.; Génard, M.; Gautier, H. The effects of diurnal temperature rise on tomato fruit quality. Can the management of the greenhouse climate mitigate such effects. *Sci. Hortic.* **2021**, *278*, 109836. [CrossRef]
9. Kläring, H.P.; Krumbein, A. The effect of constraining the intensity of solar radiation on the photosynthesis, growth, yield and product quality of tomato. *J. Agron. Crop. Sci.* **2013**, *199*, 351–359. [CrossRef]
10. Perin, L.; Nogueira Peil, R.M.; Trentin, R.; Anibele Streck, E.; Bergmann da Rosa, D.S.; Hohn, D.; Silveira Schaun, W. Solar radiation threshold and growth of mini tomato plants in mild autumn/winter condition. *Sci. Hortic.* **2018**, *239*, 156–162. [CrossRef]
11. Karipcin, M.Z.; Dinç, S.; Kara, M.; Kahraman, S.D.; Alp, I.E.; Cicekci, H. High temperature-tolerant tomato lines: Bioactive compounds. *J. Verbr. Leb.* **2016**, *11*, 117–125. [CrossRef]
12. Florido-Bacallao, M.; Alvarez-Gil, M. Aspectos relacionados con el estrés de calor en tomate (*Solanum lycopersicum* L.). *Cultiv. Trop.* **2015**, *36*, 77–95.
13. Geilfus, C.M. Controlled environment horticulture. In *Improving Quality of Vegetables and Medicinal Plants*; Springer: Cham, Switzerland, 2019; pp. 1–233.
14. Valera, D.; Molina, F.; Gil, J. Las mallas como técnica de control climático en invernaderos. *Vida Rural* **2001**, *8*, 50–52.
15. Meza, S.L.; Egea, I.; Massaretto, I.L.; Morales, B.; Purgatto, E.; Egea-Fernández, J.M.; Bolarin, M.C.; Flores, F.B. Traditional tomato varieties improve fruit quality without affecting fruit yield under moderate salt stress. *Front. Plant Sci.* **2020**, *11*, 587754. [CrossRef]
16. Kumar, D. Nuclear magnetic resonance (NMR) spectroscopy: Metabolic profiling of medicinal plants and their products. *Crit. Rev. Anal. Chem.* **2016**, *46*, 400–412. [CrossRef]
17. Kim, H.K.; Choi, Y.H.; Verpoorte, R. NMR-based metabolomic analysis of plants. *Nat. Protoc.* **2010**, *5*, 536–549. [CrossRef] [PubMed]

18. Verpoorte, R.; Choi, Y.; Kim, H. Metabolomics: What's new? *Flavour Fragr. J.* **2010**, *25*, 128–131. [CrossRef]
19. Ruíz-Nieves, J.M.; Magdaleno-Villar, J.J.; Sánchez-Alonso, M.G.; Delgado-Vargas, V.A.; Gautier, H.; Ayala-Garay, O.J. Parameters of physical and physiological quality in tomato seeds produced under high temperature condition during different periods of development. *Agroproductividad* **2021**, *14*, 45–50. [CrossRef]
20. Delgado-Vargas, V.A.; Magdaleno-Villar, J.J.; Ayala-Garay, Ó.J.; Garfias-Sánchez, D. Calidad de semillas de tres variedades nativas y una comercial de tomate producidas bajo temperaturas altas. *Rev. Chapingo Ser. Hortic.* **2018**, *24*, 215–227. [CrossRef]
21. Guzmán, A.; Corradini, F.; Martínez, P.; Allende, M.; Abarca, P.; Ferlmer, S. *Manual de Cultivo Del Tomate al Aire Libre*; Instituto de Investigaciones Agropecuarias (INIA): Santiago de Chile, Chile, 2017.
22. Steiner, A.A. The universal nutrient solution. In Proceedings of the 6th International Congress on Soilless Culture, Lunteren, The Netherlands, 29 April 1984.
23. Samaniego-Gámez, B.Y.; Garruña, R.; Tun-Suárez, J.M.; Kantun-Can, J.; Reyes-Ramírez, A.; Cervantes-Díaz, L. Bacillus spp. inoculation improves photosystem II efficiency and enhances photosynthesis in pepper plants. *Chil. J. Agric. Res.* **2016**, *76*, 409–416. [CrossRef]
24. Garruña-Hernández, R.; Orellana, R.; Larque-Saavedra, A.; Canto, A. Understanding the physiological responses of a tropical crop (*Capsicum chinense* Jacq.) at high temperature. *PLoS ONE* **2014**, *9*, e111402. [CrossRef]
25. Gall, G.; Colquhoun, I.J.; Davis, A.L.; Collins, G.J.; Verhoeyen, M.E. Metabolite profiling of tomato (*Lycopersicon esculentum*) using ^{1}H NMR spectroscopy as a tool to detect potential unintended effects following a genetic modification. *J. Agric. Food Chem.* **2003**, *51*, 2447–2456. [CrossRef] [PubMed]
26. Afifah, E.N.; Murti, R.H.; Nuringtyas, T.R. Metabolomics approach for the analysis of resistance of four tomato genotypes (*Solanum lycopersicum* L.) to root-knot nematodes (*Meloidogyne incognita*). *Open Life Sci.* **2019**, *14*, 141–149. [CrossRef] [PubMed]
27. Lorenzo, P.; Sánchez-Guerrero, M.C.; Medrano, E.; García, M.L.; Caparrós, I.; Giménez, M. External greenhouse mobile shading: Effect on microclimate, water use efficiency and yield of a tomato crop grown under different salinity levels of the nutrient solution. *Acta Hortic.* **2003**, *609*, 181–186. [CrossRef]
28. Peet, M.M. Physiological disorders in tomato fruit development. *Acta Hortic.* **2009**, *821*, 151–160. [CrossRef]
29. Ayala-Tafoya, F.; Zatarain-López, D.M.; Valenzuela-López, M.; Partida-Ruvalcaba, L.; Velázquez-Alcaraz, T.D.J.; Díaz-Valdés, T.; Osuna-Sánchez, J.A. Crecimiento y rendimiento de tomate en respuesta a radiación solar transmitida por mallas sombra. *Terra Latinoam.* **2011**, *29*, 403–410.
30. Hang, T.; Lu, N.; Takagaki, M.; Mao, H. Leaf area model based on thermal effectiveness and photosynthetically active radiation in lettuce grown in mini-plant factories under different light cycles. *Sci. Hortic.* **2019**, *252*, 113–120. [CrossRef]
31. Arenas-Corraliza, M.G.; Rolo, V.; López-Díaz, M.L.; Moreno, G. Wheat and barley can increase grain yield in shade through acclimation of physiological and morphological traits in Mediterranean conditions. *Sci. Rep.* **2019**, *9*, 9547. [CrossRef]
32. Li, L.; Aro, E.M.; Millar, H. Mechanisms of photodamage and protein turnover in photoinhibition. *Trends Plant Sci.* **2018**, *23*, 667–676. [CrossRef]
33. Trojak, M.; Skowron, E. Light Quality-Dependent Regulation of Non-Photochemical Quenching in Tomato Plants. *Biology* **2021**, *10*, 721. [CrossRef] [PubMed]
34. Yepes, A.; Buckeridge, M.S. Respuestas de las plantas ante los factores ambientales del cambio climático global: Revisión. *Colomb. For.* **2011**, *14*, 213–232. [CrossRef]
35. Wishart, D.S.; Knox, C.; Guo, A.C.; Eisner, R.; Young, N.; Gautam, B.; Hau, D.D.; Psychogios, N.; Dong, E.; Bouatra, S.; et al. HMDB: A knowledgebase for the human metabolome. *Nucleic Acids Res.* **2009**, *37*, D603–D610. [CrossRef]
36. Serrano, R. Salt tolerance in plants and microorganisms: Toxicity targets and defense responses. *Int. Rev. Cytol.* **1996**, *165*, 1–52. [CrossRef] [PubMed]
37. Hüther, C.M.; Martinazzo, E.G.; Rombaldi, C.V.; Bacarin, M.A. Effects of flooding stress in 'Micro-Tom' tomato plants transformed with different levels of mitochondrial sHSP23.6. *Braz. J. Biol.* **2016**, *77*, 43–51. [CrossRef]
38. Baracaldo, A.; Carvajal, R.; Romero, A.P.; Prieto, A.M.; García, F.J.; Fischer, G.; Miranda, D. El anegamiento afecta el crecimiento y producción de biomasa en tomate chonto (*Solanum lycopersicum* L.), cultivado bajo sombrío. *Rev. Colomb. De Cienc. Hortícolas* **2014**, *8*, 92–102. [CrossRef]
39. Carrari, F.; Asis, R.; Fernie, A.R. The metabolic shifts underlying tomato fruit development. *Plant Biotechnol.* **2007**, *24*, 45–55. [CrossRef]
40. Chaves-Barrantes, N.F.; Gutiérrez-Soto, M.V. Respuestas al estrés por calor en los cultivos. I. Aspectos moleculares, bioquímicos y fisiológicos. *Agron. Mesoam.* **2017**, *28*, 237–253. [CrossRef]
41. Kavi-Kishor, P.; Sangam, S.; Amrutha, R.N.; Laxmi, P.S.; Naidu, K.R.; Rao, K.S.; Reddy, K.J.; Theriappan, P.; Sreenivasulu, N. Regulation of proline biosynthesis, degradation, uptake and transport in higher plants: Its implications in plant growth and abiotic stress tolerance. *Curr. Sci.* **2005**, *88*, 424–438.
42. Verbruggen, N.; Hermans, C. Proline accumulation in plants: A review. *Amino Acids* **2008**, *35*, 753–759. [CrossRef] [PubMed]
43. Szabados, L.; Savouré, A. Proline: A multifunctional amino acid. *Trends Plant Sci.* **2010**, *15*, 89–97. [CrossRef] [PubMed]
44. Schwacke, R.; Grallath, S.; Breitkreuz, K.E.; Stransky, E.; Stransky, H.; Frommer, W.B.; Rentsch, D. LeProT1, a transporter for proline, glycine betaine, and γ-amino butyric acid in tomato pollen. *Plant Cell* **1999**, *11*, 377–391. [CrossRef]

45. Hare, P.D.; Cress, W.A.; Van Staden, J. Dissecting the roles of osmolyte accumulation during stress. *Plant Cell Environ.* **1998**, *21*, 535–553. [CrossRef]
46. Rodríguez, A.; Campo-Costa, A.; Batista-Ricardo, E.; Morales-Miranda, A.; Camejo-Gálvez, A.I. Influencia del Biobras 16 y Fitomas-E contra el tizón temprano y el geminivirus (TYLCV) en cultivo de tomate (*Solanum licopersicum*). *ICIDCA. Sobre Los Deriv. De La Caña De Azúcar* **2017**, *51*, 3–7.

horticulturae

Article

Effects of Light Intensity on Endogenous Hormones and Key Enzyme Activities of Anthocyanin Synthesis in Blueberry Leaves

Xiaoli An [1], Tianyu Tan [2], Xinyu Zhang [1], Xiaolan Guo [3], Yunzheng Zhu [1], Zejun Song [1] and Delu Wang [1,*]

[1] College of Forestry, Guizhou University, Huaxi, Guiyang 550025, China; anxiaoli61@163.com (X.A.); zhangxinyu_x@163.com (X.Z.); zhuyunzheng04@163.com (Y.Z.); zejun202212@163.com (Z.S.)
[2] Forestry Bureau of Kaili, Kaili 556000, China; 18786671054@163.com
[3] College of Life Sciences, Huizhou University, Huizhou 516007, China; gxl2022@hzu.edu.cn
* Correspondence: dlwang@gzu.edu.cn

Abstract: Plant anthocyanin is a secondary metabolite widely distributed in the roots, stems, leaves, flowers and fruits of plants, and its synthesis is significantly affected by light intensity. To reveal the physiological response mechanism of anthocyanin synthesis in blueberry leaves at different light intensities, four light intensities (100% (CK), 75%, 50% and 25%) were set for the 'O'Neal' southern highbush blueberry as the experimental material in our study. The relationship between endogenous hormone contents, key enzyme activities, and variations in the anthocyanin content in blueberry leaves under various light intensities during the white fruit stage (S1), purple fruit stage (S2) and blue fruit stage (S3) of fruit development were studied. The results showed that the anthocyanin content of blueberry leaves increased first and then decreased, and decreased first and then increased with the increase in light intensity and development stage, respectively. The appropriate light intensity could significantly promote the synthesis of anthocyanin, and the anthocyanin content in leaves treated with 75% light intensity was 1.09~4.08 times that of other light intensity treatments. The content or activities of gibberellin (GA_3), indoleacetic acid (IAA), jasmonic acid (JA), abscisic acid (ABA), ethylene (ETH), phenylalanine ammonia lyase (PAL), chalcone isomerase (CHI), dihydroflavonol reductase (DFR) and UDP-glucose: flavonoid 3-glucosyltransferase (UFGT) were significantly or extremely significantly correlated with the content of anthocyanin in leaves. This indicated that light intensity significantly promoted anthocyanin synthesis in blueberry leaves by affecting endogenous hormone contents and key enzyme activities in the anthocyanin synthesis pathway. This study lays a foundation for further research on the molecular mechanism of light intensity regulating anthocyanin synthesis in blueberry leaves.

Keywords: light intensity; blueberry; anthocyanin; endogenous hormones; key enzyme activities

1. Introduction

Blueberry (Ericaceae, *Vaccinium*) leaves are rich in various nutrients such as anthocyanins, flavonoids and polyphenols, and they can be used as a sustainable and low-cost plant material to extract anthocyanins. Thus far, over 600 anthocyanins have been identified in nature [1], and anthocyanins in leaves mainly exist in vacuoles of leaf epidermal cells [2,3] or glandular hairs [4], which can be used as antioxidants to protect plants from damage caused by UV radiation [5], freezing and drought stress [6]. Light is one of the key environmental factors affecting anthocyanin synthesis in many plants, among which light intensity is the most significant [7,8].

Phytohormones have been shown to play an important role in regulating plant responses to environmental stress [9], and can participate in the regulation of anthocyanin synthesis [10]. For example, Luo et al. [11] found that genes related to IAA, ABA, ETH, JA and GA in rapeseed seedlings responded to high light stress. Aux (Auxin), ABA, JA and GA (gibberellic acid) regulate the function and expression of transcriptions factors of the MYB-bHLH-WD40 complex and flavonoid biosynthesis pathway genes involved in the anthocyanin branch [12]. MeJA (methyl jasmonate) and SA (salicylic acid) were both found to stimulate anthocyanin production in the callus cultures of *Daucus carota* [13]. Accumulation of anthocyanin was suppressed by shading in grape berry skins [14]. In Arabidopsis thaliana, strong light can regulate anthocyanin content by stimulating JA content [15].

The biosynthetic pathway of anthocyanins is mainly divided into three stages. Firstly, 4-coumaroyl-CoA is synthesized by the precursor phenylalanine via PAL, C4H (cinnamate-4-hydroxylase) and 4CL (4-Coumarate: CoA ligase) [16]. Secondly, 4-coumaroyl-CoA and malonyl CoA are catalyzed by CHS (Chalcone synthase) to synthesize tetrahydroxychalcone, which is isomerized by CHI to form the colorless compound trihydroxyflavanone, which is then further catalyzed by F3H (flavanone-3-hydroxylase) to synthesize flavanones and dihydroflavonols [17]. Finally, flavanones and dihydroflavonols are catalyzed by DFR to reduce the 4-position of C ring to produce different colorless anthocyanins. These colorless anthocyanins are catalyzed by ANS (anthocyanidin synthase) to produce colored anthocyanins, and UFGT catalyzes the combination of colored anthocyanins with glycosides to transform those into colored anthocyanins [18]. Studies have found that light intensity regulates plant anthocyanin synthesis by inducing the expression of associated genes and enzyme activities in metabolic pathways [19]. Increasing light intensity promoted the expression levels of *MYB*, *CHS* and *F3H* genes of anthocyanin in coleus, thus inducing anthocyanin synthesis and increasing anthocyanin content [20]. Zhu et al. [21] clarified that under low light stress, the activities of CHI, CHS and F3H involved in the anthocyanin biosynthesis pathway of purple cabbage decreased, resulting in a decrease in anthocyanin content.

The molecular mechanism of anthocyanin biosynthesis induced by light intensity has been reported in blueberry [8,22]. However, it is not clear how the light intensity affects the physiological mechanism of anthocyanin content by affecting the endogenous hormone contents and the key enzyme activities in the anthocyanin synthesis pathway. Therefore, this study analyzed the changes in anthocyanin content in blueberry leaves under different light intensities at stage S1, S2 and S3, and its correlation with the content of endogenous hormones (GA_3, JA, IAA, ABA and ETH) and the activities of key enzymes (PAL, CHI, DFR and UFGT). We sought to explore the correlation between anthocyanin content, endogenous hormone contents and key activities under different light intensities, and to analyze the synergistic regulation of anthocyanin biosynthesis by light intensity, hormones and key enzymes from the physiological level, to provide a scientific basis for the control of light intensity in blueberry production.

2. Materials and Methods

2.1. Overview of the Experimental Site

The experimental site is located in the Experimental Nursery of the College of Forestry, South Campus of Guizhou University, Huaxi District, Guiyang, with an altitude of 1159 m, 104°34′ east longitude, and 26°34′ north latitude. It is a subtropical humid and moderate climate. The maximum temperature is 39.5 °C, the minimum temperature is −9.5 °C and the average annual temperature is 15.8 °C. The yearly effective accumulated temperature above 10 °C is 4637.5 °C, the annual precipitation is 1229 mm, the annual average relative humidity is 79% and the total integrated solar radiation is 3567 MJ/m^2.

2.2. Experimental Materials

Four-year-old southern highbush blueberry variety 'O'Neal' with the same maturity and growth was used as the experimental material, and the test seedlings were transplanted into plastic flower pots (inner diameter 26.5 cm, bottom diameter 17.5 cm, height 19.7 cm). One seedling per pot was cultured with pine forest humus as the substrate. The nutrient content of the substrate is high, with a pH of about 4.8, which can satisfy the normal growth of blueberries, and weeding and irrigation were carried out regularly.

2.3. Experimental Design

As shown in Table 1, the four light intensities were 100% (CK group, natural light), 75% (light shading), 50% (moderate shading) and 25% (severe shading) full light intensity, which were controlled by an illuminance meter (Shenzhen Jumaoyuan Technology Co., Ltd., Shenzhen, China) and black sunshade nets of different densities with 2, 3, 4, 6 and 8 needles. Three replicates were set for each treatment, with 10 plants in each group. The light intensity was measured at three random locations under the sunshade nets. The trial began after the blueberries had bloomed (1 April 2020).

Table 1. Actual light intensity corresponding to relative light intensity.

Light Intensity	S1/μmol·m^{-2}s^{-1}	S2/μmol·m^{-2}s^{-1}	S3/μmol·m^{-2}s^{-1}
25%	372 ± 34.06 Ad	369 ± 29.44 Ad	379 ± 28.29 Ad
50%	750 ± 31.18 Ac	699 ± 24.83 Ac	778 ± 30.02 Ac
75%	1123 ± 40.99 Ab	1094 ± 48.50 Ab	1143 ± 36.37 Ab
CK	1498 ± 39.26 Aa	1456 ± 44.46 Aa	1587 ± 37.53 Aa

Note: The above table shows the light intensity at 10 a.m. as measured with a photometer. S1: white fruit stage, S2: purple fruit stage, S3: blue fruit stage. In the table, different uppercase letters indicate significant differences in the same light intensity during different stages, and different lowercase letters indicate significant differences in different light intensity treatments at the same stage ($p < 0.05$); values represent mean ± standard error.

2.4. Sample Collection

After one month of treatment, according to the test scheme, blueberry plants with consistent growth were selected for sample collection. The healthy leaves with the same size were randomly sampled in the group at three fruit development stages of 28 days (white fruit stage, S1), 35 days (purple fruit stage, S2) and 42 days (blue fruit stage, S3) after full bloom. We sampled 10 g from each biological replicate in each stage, and a total of 3 biological replicates were used for experimental research. The samples were placed in a screw-tip-bottom centrifuge tube wrapped with tin foil paper, stored in liquid nitrogen and returned to the laboratory for storage in an ultra-low temperature refrigerator at −80 °C.

2.5. Method of Index Determination

2.5.1. Methods for Determination of Endogenous Hormone Contents and Enzyme Activities

The contents of endogenous hormones and the activities of key enzymes in the anthocyanin synthesis pathway were determined by double antibody sandwich enzyme-linked immunosorbent assay (ELISA) [23]. The collected blueberry leaves were tested using an ELISA kit produced by Guizhou Wela Technology Limited Liability Company. Sample treatment: The tissue was rinsed with pre-cooled PBS (0.01 M, pH = 7.4), and the weighed 0.1 g leaf and the corresponding volume of PBS (according to the weight to volume ratio of 1:9) were added to the homogenizer for grinding. To further lyse the tissue cells, the homogenate was broken by ultrasound. Finally, the homogenate was centrifuged at 10,000 rpm for 5 min, and the supernatant was taken for detection. The content of gibberellin 3 (GA_3), jasmonic acid (JA), indoleacetic acid (IAA), abscisic acid (ABA) and ethylene (ETH), and the activities of phenylalanine ammonia lyase (PAL), chalcone isomerase (CHI), dihydroflavonol reductase (DFR) and UDP-glucose: flavonoid 3-glucosyltransferase (UFGT) were detected.

2.5.2. Method for Determination of Anthocyanin Content

The anthocyanin content of blueberry leaves was determined using a Solarbio biochemical kit. (1) We weighed and ground 0.1g of blueberry leaf samples with a low-temperature grinding machine. In order to further lyse tissue cells, appropriate ultrasonic fragmentation was performed. We added 1 mL of the extract, and it was transferred to the EP tube after being fully homogenized. The extract was diluted to 1 mL, covered and extracted at 60 °C for 40 min, during which time it was high-speed shocked 8 times. We then centrifuged at 12,000 rpm, held at 4 °C for 15 min and took the supernatant for testing. (2) The microplate reader was preheated for 30 min, recalibrated and the wavelength was adjusted. (3) We added 40 µL of samples to the determination tubes (1.5 mL EP tube) 1 and 2, and then we added 160 µL of reagent 1 and reagent 2, respectively. (4) After mixing, we centrifuged at 10,000 rpm for 15 min, placed 150 µL of supernatant in 96-well plates and detected its absorbance. The absorbance values of tube 1 at 530 nm and 700 nm were recorded as A1 and A1′, respectively, and the absorbance values of tube 2 at 530 nm and 700 nm were recorded as A2 and A2′, respectively.

The anthocyanin content was calculated according to Formulas (1) and (2).

$$\Delta A = (A1 - A1') - (A2 - A2') \quad (1)$$

Anthocyanin content (µmol/gFW):

$$[\Delta A \div (\varepsilon \times d) \times 103 \times F] \times V \div W = 0.31 \times \Delta A \div W \quad (2)$$

where ε: molar extinction coefficient, 2.69×10^4 mL/mmol/cm; d: 96-hole plate optical path length, 0.6 cm; 103: 1 mmol = 103 µmol; F: dilution multiple, 5; V: total volume of extract, 1 mL; W: fresh weight of sample, g.

2.6. Data Analysis

Excel 2019 and Origin 2022 were used for sorting, calculating, mapping data and correlation analysis. One-way ANOVA and Tukey's test were performed using SPSS 19.0. Statistical differences were marked by sequential letter labeling.

3. Results

3.1. Effects of Light Intensity on Endogenous Hormone Contents in Blueberry Leaves

As shown in Figure 1, the content of five endogenous hormones in blueberry leaves was significantly affected by light intensity and development stage. Except for IAA at S3, the content of GA_3 and IAA in other treatments decreased gradually with the increase in light intensity and leaf development, while the content of JA, ABA and ETH increased gradually with the increase in light intensity and leaf development.

The effects of light intensity and development stage on the content of GA_3 and IAA in blueberry leaves are shown in Figure 1a,b. Under the same light intensity treatment, the contents of GA_3 and IAA at S1 were significantly higher than those at S2 and S3. The contents of GA_3 and IAA at 25% light intensity treatment were significantly higher than those under other light intensity treatments during the same development stage, while the content of IAA at S3 was the opposite. Under 25% light intensity treatment, the GA_3 content in leaves of S1 was 13.94%, 11.30% and 6.68% higher than that of CK, 75% and 50% light intensity treatments at the same stage, while S2 was 3.04%, 3.27% and 1.76% higher, and S3 was 17.07%, 8.62% and 2.29% higher, respectively. From S1 to S3, the content of IAA in leaves treated with 25% light intensity was 1.04~1.07 times, 1.03~1.48 times and 0.54~0.91 times that of other light intensity treatments during the same stage, respectively.

The effects of light intensity and developmental stage on the content of JA, ABA and ETH in blueberry leaves are shown in Figure 1c–e. Under the same light intensity, the contents of JA, ABA and ETH in leaves at S3 were significantly higher than those at S1 and S2. From S1 to S3, the contents of JA, ABA and ETH in leaves treated with 25% light intensity were significantly lower than those treated with other light intensities at the same stage. Compared with CK, the JA content of leaves during S1 decreased by 6.78%~52.85%, with an average decrease of 28.00%; the decrease rate at S2 was 4.70%~28.12%, with the average decrease rate was 14.83%; and the decrease at S3 was 2.62%~15.15%, with an average decrease of 7.56%. The ABA content under CK treatment at S3 was as high as 127.10 ng/g, which was 1.06~1.2 times that of other light intensity treatments at the same stage, and 1.78 and 1.16 times that of the same light intensity treatment at S1 and S2, respectively. The content of ETH in leaves was less, but it also had a similar change rule with the content of JA and ABA. Among them, the content of ETH under CK treatment at S3 was the highest, which was 46.71 μg/g, that is, 1.02~1.18 times that of other light intensity treatments during the same stage. The results showed that full light and late development were more conducive to the synthesis of JA, ABA and ETH content in leaves, but the increase rate of these hormone contents gradually decreased with the continuous development of leaves.

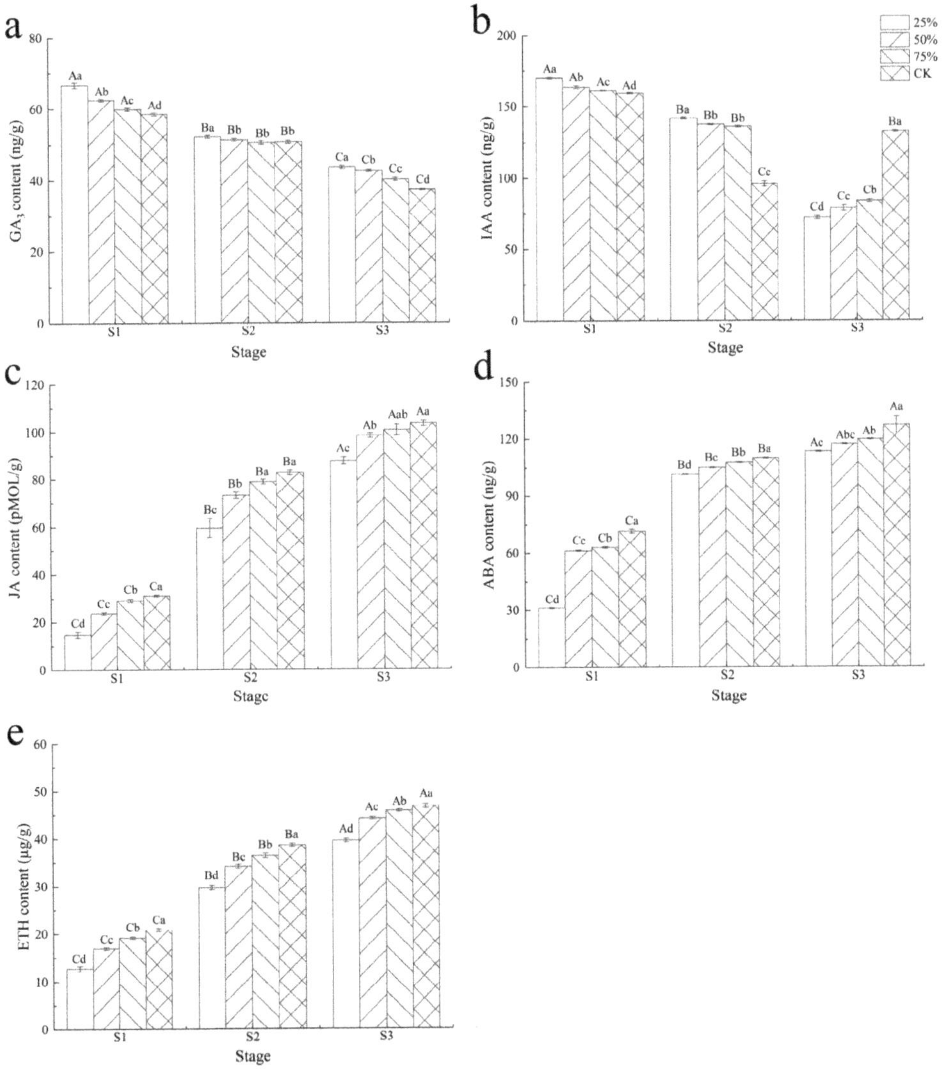

Figure 1. Effects of light intensity on endogenous hormone (GA_3 (**a**), IAA (**b**), JA (**c**), ABA (**d**) and ETH (**e**)) contents in blueberry leaves. Note: In the figure, different uppercase letters indicate significant differences at the same light intensity during different stages, and different lowercase letters indicate significant differences under different light intensity treatments during the same stage ($p < 0.05$). Error bars represent SD.

3.2. Effects of Light Intensity on Key Enzyme Activities in the Anthocyanin Synthesis Pathway of Blueberry Leaves

The effects of different light intensities and development stages on the activities of key enzymes in the anthocyanin synthesis pathway in blueberry leaves are shown in Figure 2. The activities of PAL, CHI, DFR and UFGT in leaves were significantly affected by light intensity and development stage. These four enzyme activities gradually increased with the increase in light intensity and leaf development, but the extent of the increase gradually decreased.

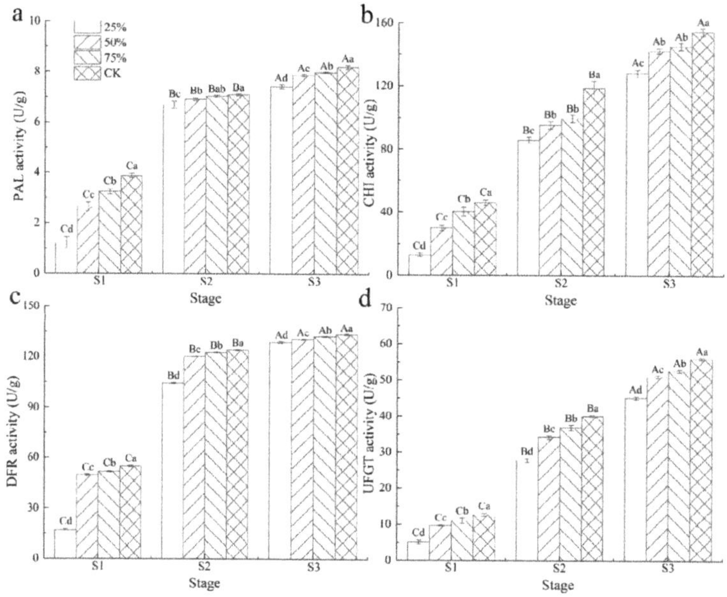

Figure 2. Effects of light intensity on key enzyme activities (PAL (**a**), CHI (**b**), DFR (**c**) and UFGT (**d**)) in the anthocyanin synthesis pathway of blueberry leaves. Note: In the figure, different uppercase letters indicate significant differences at the same light intensity during different stages, and different lowercase letters indicate significant differences under different light intensity treatments during the same stage ($p < 0.05$). Error bars represent SD.

Under the same light intensity, the activities of four enzymes in leaves at S3 were significantly higher than those of S1 and S2, and the four enzyme activities of CK treatment at each stage were significantly higher than those of shading treatment at the same stage. Compared with the other three key enzymes, the PAL enzyme activity in blueberry leaves was the lowest, which was 1.22 U/g under 25% light intensity at S1, and 8.20 U/g under CK treatment at S3. Compared with CK, the CHI activity of leaves at S1 decreased by 12.22%~72.03%, with an average decrease of 39.60%. At S2, it decreased by 16.28%~27.83%, with an average decrease of 21.32%, and at S3, it decreased by 6.01%~17.03%, with an average decrease of 10.31%. The activities of DFR and UFGT in leaves under different light intensities also had the same variation pattern at different developmental stages. It can be seen that the four enzyme activities in the anthocyanin synthesis pathway of blueberry leaves are positively correlated with light intensity and development stage.

3.3. Effect of Light Intensity on Anthocyanin Content in Blueberry Leaves

As shown in Figure 3, light intensity and development stage had significant effects on the anthocyanin content in blueberry leaves, and the anthocyanin content at S3 was significantly higher than that at S1 and S2 under the same light intensity treatment. At the same development stage, the anthocyanin content in leaves increased first and then decreased with the increase in light intensity, and reached the peak at 75% light intensity. Under the same light intensity treatment, the anthocyanin content in leaves treated with 25% light intensity gradually increased with leaf development, while the anthocyanin content in leaves treated with the other three light intensities decreased first and then increased with leaf development, and the decrease was smaller than the increase. The anthocyanin content of leaves treated with CK and 75% light intensity at S1 and S3 was significantly higher than that of the other two treatments, while the anthocyanin content of leaves treated with 75% light intensity at S2 was significantly higher than that of the other three treatments. The anthocyanin content of leaves under 75% light intensity at S3 was as high as 1.122 μmol/gFW, which was 1.19 times, 1.76 times and 2.45 times those of CK, 50% and 25% light intensity treatments at the same stage, and 1.42 times and 1.47 times those of 75% light intensity treatment at S1 and S2, respectively. This indicated that too low or too high light intensity was not conducive to the synthesis of anthocyanin in blueberry leaves, and 75% light intensity was more conducive to the synthesis of anthocyanin in leaves.

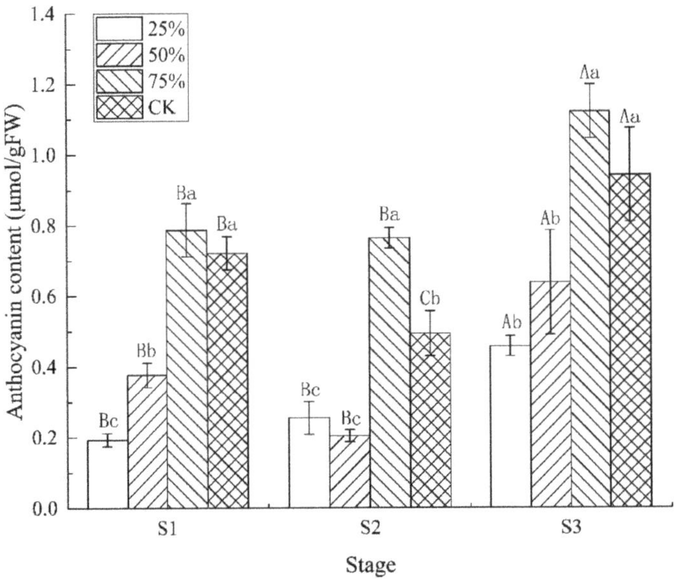

Figure 3. Effect of light intensity on anthocyanin content in blueberry leaves. Note: In the figure, different uppercase letters indicate significant differences at the same light intensity during different stages, and different lowercase letters indicate significant differences under different light intensity treatments during the same stage ($p < 0.05$). Error bars represent SD.

3.4. Correlation Analysis

The correlation analysis of anthocyanin content with light intensity, endogenous hormones (GA_3, JA, IAA, ABA, ETH) and the activities of key enzymes in the anthocyanin synthesis pathway (PAL, CHI, DFR, UFGT) in blueberry leaves at different developmental stages under different light intensity conditions is shown in Figure 4. From S1 to S2, the content of GA_3 and IAA in leaves had an extremely significant negative correlation with light intensity, but at S3, the former had an extremely significant negative correlation and the latter had an extremely significant positive correlation. The anthocyanin content, other three hormones and four enzyme activities at the three stages were extremely significantly or significantly positively correlated with light intensity. This was consistent with the trend of changes in anthocyanin content, endogenous hormone contents and key enzyme activities in the anthocyanin synthesis pathway with light intensity, as mentioned above.

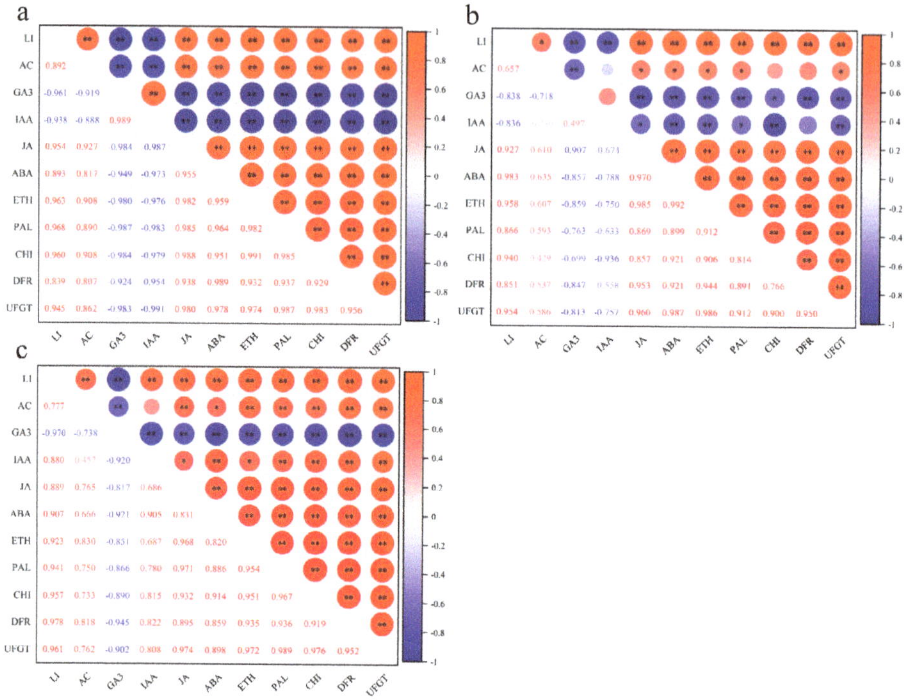

Figure 4. Correlation analysis of anthocyanin content in blueberry leaves with light intensity, endogenous hormones and enzyme activities. Note: LI: light intensity, AC: anthocyanin content; (**a**): white fruit stage (S1), (**b**): purple fruit stage (S2), (**c**): blue fruit stage (S3); * $p \leq 0.05$, significant correlation; ** $p \leq 0.01$, extremely significant correlation.

From S1 to S3, the anthocyanin content in leaves treated with different light intensities showed an extremely significant negative correlation with GA_3 content, and an extremely significant negative correlation with IAA content at first, then a negative correlation and then a positive correlation. The anthocyanin content of leaves treated with different light intensities had an extremely significant or significant positive correlation with the other three hormone contents and four enzyme activities at S1 and S2, and had a significant positive correlation with the content of JA, ABA and ETH, and the activities of PAL and UFGT at S2. At the same time, from S1 to S3, the correlation coefficients between the contents of five hormones, the activities of four enzymes and anthocyanin content decreased first and then increased, which was similar to the trend of anthocyanin content changing with light intensity.

In addition, light intensity, anthocyanin content, five hormone contents and four enzyme activities were significantly correlated with any two factors at S1. There was no significant correlation between anthocyanin content and the IAA content, CHI activity and DFR activity, and between IAA content and the GA_3 content and DFR activity, at S2; there was no significant correlation between anthocyanin content and the IAA content at S3; and there was significant correlation between any other two factors. From S1 to S3, the content of JA, ABA and ETH and the activities of PAL, CHI, DFR and UFGT in leaves had a significant or extremely significant negative correlation with GA_3 content, and showed an extremely significant negative correlation with IAA content at first, though that then turned to a negative correlation, and then to a significant positive correlation. However, the IAA content was first significantly positively correlated with GA_3 content, then positively correlated, and then significantly negatively correlated. The results showed that the light intensity's regulation of anthocyanin synthesis in blueberry leaves was closely related to five endogenous hormones and four key enzyme activities in the anthocyanin synthesis pathway, and there was also a certain relationship between hormones and enzyme activities.

It follows that the anthocyanin content, endogenous hormone contents and key enzyme activities in the anthocyanin synthesis pathway in blueberry leaves are significantly correlated with light intensity and development stage. However, the correlations of some factors at different development stages and at different light intensities are quite different.

4. Discussion

4.1. Light Intensity Promotes Anthocyanin Synthesis by Regulating the Content of Endogenous Hormones

Hormones affect all aspects of plant development and growth physiology, including the biosynthesis of anthocyanin [24], and the prerequisite for inducing anthocyanin synthesis in vegetative tissues is light [2]. In the leaves of oilseed peony, light, moderate and severe shadings decreased the ABA concentration by 8.8%, 14.4% and 22.7% but increased the IAA concentration by 38.1%, 45.5% and 49.0% and the GA_3 concentration by 6.3%, 7.6% and 11.7%, respectively [25]. The GA_3 content in *Carpinus betulus* L. seedlings increased with the decrease in light intensity, and the IAA content decreased with the decrease in light intensity [26]. In the two studies cited, the IAA contents showed different trends with light intensity, but they were consistent with the IAA content trend at S1, S2 and S3 in this study. At the same time, they were consistent with the results in this study in that the IAA content in blueberry leaves and light intensity showed a highly significant negative correlation at S1 and S2, and a highly significant positive correlation at S3. It has been reported that high light intensity triggers the biosynthesis of ABA, which, in turn, promotes the expression of anthocyanin biosynthetic genes and enhances anthocyanin biosynthesis [27]. Exogenous ethylene treatment markedly improved the expression levels of the anthocyanin-biosynthesis-related genes (*PsPAL, PsDRF, PsANS, PsUFGT*, etc.) in plum, thus accelerating anthocyanin accumulation [28]. JA promoted anthocyanin biosynthesis in leaves of rapeseed seedlings [11] and apple [29]. Exogenous ABA treatment enhanced anthocyanin accumulation in grape berry skins [14]. Plant growth regulators ABA and ethephon promoted anthocyanin synthesis in chicory (*Cichorium intybus* L.), while GA_3 inhibited its anthocyanin synthesis [30]. In summary, the literature indicates that light intensity can promote the synthesis of anthocyanins by regulating the biosynthesis of hormones.

During the whole process of blueberry fruit development, the anthocyanin content was highly significantly positively correlated with the ABA and ETH content, and highly significantly negatively correlated with the IAA content [31]. The GA_4 content was strongly negatively correlated with the anthocyanin content in sweet cherry [32]. There was a strong positive correlation between the ABA content and anthocyanin content in *Lycium* fruit [33] and purple-leaved cultivars of tea [34]. The IAA content was positively correlated with the anthocyanin content during bicolor leaf development [35]. In our study, the anthocyanin content was significantly or extremely significantly positively correlated with the ETH and ABA contents, and extremely significantly negatively correlated with the GA_3 content, which supported the above views. However, from S1 to S3, the IAA content and anthocyanin content were extremely significantly negatively correlated, negatively correlated and positively correlated, respectively. The differences in findings between studies may have been caused by differences in plant species, sampling organs and leaf growth and development. Nonetheless, the contents of JA and ETH were strongly positively correlated with the anthocyanin content in our study, which was confirmed in *Saxifraga longifolia* leaves [36] and plum fruits [28]. This indicates that the synergistic effect of different hormones promotes the biosynthesis of anthocyanin.

4.2. Light Intensity Regulates Anthocyanin Synthesis by Inducing the Expression of Enzyme Activities

The activities of four enzymes (PAL, CHI, DFR, UFGT) in this study showed an upward trend with the increase in light intensity and leaf development, and the enzyme activities at three development stages were significantly positively correlated with light intensity. It has been reported that light intensity can induce the expression of CHS, CHI, F3′5′H and DFR genes in the peel of sand pear [37] and grape [38], thus significantly inducing anthocyanin accumulation. That supports the proposal that the expression of enzyme genes in the anthocyanin synthesis pathway can regulate the level of enzyme activities to a certain extent, which regulates the content of anthocyanin, a proposal that is consistent with the results of this study. In *Perilla frutescens* var. *crispa*, the PAL activity under low light intensity was lower than that under normal light intensity [39]. DFR activity in 'Fuji' apple peel increased with the increase in light intensity, and anthocyanin synthesis was regulated by DFR activity [40]. The lack of anthocyanin accumulation in Matthiola line white flowers [41] and ivy [42] is due to a lack of DFR activity, while the lack of anthocyanin in white grapes [43] is due to a lack of UFGT activity. The results indicate that light intensity can affect anthocyanin synthesis by regulating the activity of key enzymes in the anthocyanin biosynthesis pathway through the signal transduction pathway.

The anthocyanin content in eggplant was significantly positively correlated with the expression levels of *SmCHI* and *SmDFR* [44], while the DFR activity was significantly correlated with anthocyanin accumulation in apple [45]. Meanwhile, the PAL activity was strongly positively correlated with anthocyanin biosynthesis in eggplant peel and fruit [46], which was consistent with the results of this study. At the same time, there was no significant relationship between CHI and DFR activities and the anthocyanin content at S2 in this study, which may be related to the decrease in anthocyanin biosynthesis and the increase in anthocyanin degradation at S2. Combined with the study of enzyme activities under different light intensity conditions, the comparative analysis of its variation law confirmed that the light intensity will affect the biosynthesis of anthocyanin by inducing key enzyme activities.

4.3. Regulation of Light Intensity in Anthocyanin Biosynthesis

Generally speaking, inducing the synthesis of anthocyanin requires high light intensity, and the anthocyanin content in plant leaves is related to light levels [2]. Manetas [47] pointed out that anthocyanins are ubiquitous in green leaves, but their content is not well covered by the color of chlorophyll, meaning it is not easy to detect. Later, the absorption value of anthocyanin was instead detected in green leaves, which confirmed the presence of anthocyanin in green leaves [48]. The blueberry leaf samples collected in this study were all green leaves with low anthocyanin content, but they were significantly affected by light intensity and leaf development. Research has found that the concentration of anthocyanin in grapes gradually decreases with a decrease in light transmittance [49]. Meanwhile, strong light (100% light transmittance) inhibited anthocyanin synthesis in *Petunia* corollas [50]. Elsewhere, the anthocyanin content in leaves of four subtropical dominant tree species gradually decreased with their growth and development, and shading (30% light transmittance) inhibited the accumulation of anthocyanin in leaves [51]. These results are consistent with the results of this study, indicating that anthocyanin synthesis of blueberry leaves is closely related to light intensity and leaf development. In addition, the anthocyanin content in the leaves of the three stages reached its highest under 75% light intensity treatment, indicating that excessive light intensity may reduce anthocyanin biosynthesis or increase anthocyanin degradation, leading to a decrease in anthocyanin content in the leaves.

5. Conclusions

Light is a fundamental requirement for plant growth and development, but excessive light intensity can cause irreversible damage to chloroplasts and cell metabolism. Correlation analysis showed that light intensity, endogenous hormones and key enzyme activities had significant effects on anthocyanin content in blueberry leaves. Among them, the content of GA_3 was negatively correlated with the content of anthocyanin in all three stages. This study showed that light intensity affected anthocyanin synthesis by regulating the content of endogenous hormones in blueberry leaves and the activities of key enzymes in the anthocyanin synthesis pathway, and 75% light intensity was the most conducive to anthocyanin biosynthesis in blueberry leaves. Blueberry leaves are byproducts with potential economic value, and these findings help us understand the potential mechanisms by which light intensity regulates anthocyanin synthesis and accumulation in blueberry leaves.

Author Contributions: Data curation, visualization and writing—original draft, X.A.; investigation, T.T.; data curation, X.Z.; formal analysis, X.G., Y.Z. and Z.S.; resources, writing—review and editing, funding acquisition and supervision, D.W. All authors have read and agreed to the published version of the manuscript.

Funding: This work was supported by the National Natural Science Foundation of China (grant number 31760205).

Institutional Review Board Statement: Not applicable.

Informed Consent Statement: Not applicable.

Data Availability Statement: Not applicable.

Conflicts of Interest: The authors declare no conflict of interest.

References

1. Smeriglio, A.; Barreca, D.; Bellocco, E.; Trombetta, D. Chemistry, Pharmacology and Health Benefits of Anthocyanins. *Phytother. Res.* **2016**, *30*, 1265–1286. [CrossRef]
2. Steyn, W.J.; Wand, S.J.E.; Holcroft, D.M.; Jacobs, G. Anthocyanins in vegetative tissues: A proposed unified function in photoprotection. *New Phytol.* **2002**, *155*, 349–361. [CrossRef] [PubMed]
3. Merzlyak, M.N.; Chivkunova, O.B.; Solovchenko, A.E.; Naqvi, K.R. Light absorption by anthocyanins in juvenile, stressed, and senescing leaves. *J. Exp. Bot.* **2008**, *59*, 3903–3911. [CrossRef]
4. Ntefidou, M.; Manetas, Y. Optical Properties of Hairs During the Early Stages of Leaf Development in *Platanus orientalis*. *Funct. Plant Biol.* **1996**, *23*, 535–538. [CrossRef]
5. Gould, K.S. Nature's Swiss Army Knife: The Diverse Protective Roles of Anthocyanins in Leaves. *J. Biomed. Biotechnol.* **2004**, *2004*, 314–320. [CrossRef]
6. Chalker-Scott, L. Environmental Significance of Anthocyanins in Plant Stress Responses. *Photochem. Photobiol.* **1999**, *70*, 1–9. [CrossRef]
7. Zhang, H.-N.; Li, W.-C.; Wang, H.-C.; Shi, S.-Y.; Shu, B.; Liu, L.-Q.; Wei, Y.-Z.; Xie, J.-H. Transcriptome Profiling of Light-Regulated Anthocyanin Biosynthesis in the Pericarp of Litchi. *Front. Plant Sci.* **2016**, *7*, 963. [CrossRef]
8. Guo, X.; Shakeel, M.; Wang, D.; Qu, C.; Yang, S.; Ahmad, S.; Song, Z. Metabolome and transcriptome profiling unveil the mechanisms of light-induced anthocyanin synthesis in rabbiteye blueberry (*vaccinium ashei*: Reade). *BMC Plant Biol.* **2022**, *22*, 223. [CrossRef]
9. Verma, V.; Ravindran, P.; Kumar, P.P. Plant hormone-mediated regulation of stress responses. *BMC Plant Biol.* **2016**, *16*, 86. [CrossRef]
10. Shan, X.; Zhang, Y.; Peng, W.; Wang, Z.; Xie, D. Molecular mechanism for jasmonate-induction of anthocyanin accumulation in *Arabidopsis*. *J. Exp. Bot.* **2009**, *60*, 3849–3860. [CrossRef]
11. Luo, Y.; Teng, S.; Yin, H.; Zhang, S.; Tuo, X.; Tran, L.-S.P. Transcriptome Analysis Reveals Roles of Anthocyanin- and Jasmonic Acid-Biosynthetic Pathways in Rapeseed in Response to High Light Stress. *Int. J. Mol. Sci.* **2021**, *22*, 13027. [CrossRef] [PubMed]
12. Araguirang, G.E.; Richter, A.S. Activation of anthocyanin biosynthesis in high light—What is the initial signal? *New Phytol.* **2022**, *236*, 2037–2043. [CrossRef] [PubMed]
13. Sudha, G.; Ravishankar, G.A. Elicitation of anthocyanin production in callus cultures of *Daucus carota* and the involvement of methyl jasmonate and salicylic acid. *Acta Physiol. Plant.* **2003**, *25*, 249–256. [CrossRef]
14. Jeong, S.; Goto-Yamamoto, N.; Kobayashi, S.; Esaka, M. Effects of plant hormones and shading on the accumulation of anthocyanins and the expression of anthocyanin biosynthetic genes in grape berry skins. *Plant Sci.* **2004**, *167*, 247–252. [CrossRef]
15. Qi, T.; Song, S.; Ren, Q.; Wu, D.; Huang, H.; Chen, Y.; Fan, M.; Peng, W.; Ren, C.; Xie, D. The Jasmonate-ZIM-Domain Proteins Interact with the WD-Repeat/bHLH/MYB Complexes to Regulate Jasmonate-Mediated Anthocyanin Accumulation and Trichome Initiation in *Arabidopsis thaliana*. *Plant Cell* **2011**, *23*, 1795–1814. [CrossRef]

16. Zhang, Y.; Chu, G.; Hu, Z.; Gao, Q.; Cui, B.; Tian, S.; Wang, B.; Chen, G. Genetically engineered anthocyanin pathway for high health-promoting pigment production in eggplant. *Mol. Breed.* **2016**, *36*, 54. [CrossRef]
17. Zorenc, Z.; Veberic, R.; Koron, D.; Miosic, S.; Hutabarat, O.S.; Halbwirth, H.; Mikulic-Petkovsek, M. Polyphenol metabolism in differently colored cultivars of red currant (*Ribes rubrum* L.) through fruit ripening. *Planta* **2017**, *246*, 217–226. [CrossRef]
18. Hichri, I.; Barrieu, F.; Bogs, J.; Kappel, C.; Delrot, S.; Lauvergeat, V. Recent advances in the transcriptional regulation of the flavonoid biosynthetic pathway. *J. Exp. Bot.* **2011**, *62*, 2465–2483. [CrossRef]
19. Albert, N.; Lewis, D.H.; Zhang, H.; Irving, L.; Jameson, P.; Davies, K.M. Light-induced vegetative anthocyanin pigmentation in *Petunia*. *J. Exp. Bot.* **2009**, *60*, 2191–2202. [CrossRef]
20. Nguyen, P.; Cin, V.D. The role of light on foliage colour development in coleus (*Solenostemon scutellarioides* (L.) Codd). *Plant Physiol. Biochem.* **2009**, *47*, 934–945. [CrossRef]
21. Zhu, H.; Li, X.; Zhai, W.; Liu, Y.; Gao, Q.; Liu, J.; Ren, L.; Chen, H.; Zhu, Y. Effects of low light on photosynthetic properties, antioxidant enzyme activity, and anthocyanin accumulation in purple pak-choi (*Brassica campestris* ssp. Chinensis Makino). *PLoS ONE* **2017**, *12*, e0179305. [CrossRef]
22. Guo, X.; Wang, D.; Shakeel, M. Transcriptome analysis reveals light-induced anthocyanin synthesis candidate genes in rabbiteye blueberry (*Vaccinium ashei*: Reade). *Biotechnol. Biotechnol. Equip.* **2021**, *35*, 746–757. [CrossRef]
23. Schraer, S.M.; Shaw, D.R.; Boyette, M.; Coupe, R.H.; Thurman, E.M. Comparison of Enzyme-Linked Immunosorbent Assay and Gas Chromatography Procedures for the Detection of Cyanazine and Metolachlor in Surface Water Samples. *J. Agric. Food Chem.* **2000**, *48*, 5881–5886. [CrossRef]
24. LaFountain, A.M.; Yuan, Y. Repressors of anthocyanin biosynthesis. *New Phytol.* **2021**, *231*, 933–949. [CrossRef] [PubMed]
25. Han, C.-J.; Wang, Q.; Zhang, H.-B.; Wang, S.-H.; Song, H.-D.; Hao, J.-M.; Dong, H.-Z. Light shading improves the yield and quality of seed in oil-seed peony (*Paeonia ostii* Feng Dan). *J. Integr. Agric.* **2018**, *17*, 1631–1640. [CrossRef]
26. Zhou, Q.; Zhao, F.; Zhang, H.; Zhu, Z. Responses of the growth, photosynthetic characteristics, endogenous hormones and antioxidant activity of *Carpinus betulus* L. seedlings to different light intensities. *Front. Plant Sci.* **2022**, *13*, 1055984. [CrossRef]
27. Zhang, M.-Y.; Cai, X.; Wan, Y.-T.; Fu, Y.-F.; Yang, X.-Y.; Zhang, Z.-W.; Yuan, S. Relatively Low Light Intensity Promotes Phosphorus Absorption and Enhances the Ethylene Signaling Component EIN3 in Maize, Wheat, and Oilseed Rape. *Agronomy* **2022**, *12*, 427. [CrossRef]
28. Li, X.; Cheng, Y.; Wang, Y.; Yang, X.; Wei, C.; Guan, J. Ethylene Signal Is Involved in the Regulation of Anthocyanin Accumulation in Flesh of Postharvest Plums (*Prunus salicina* Lindl.). *Plants* **2023**, *12*, 893. [CrossRef]
29. An, J.; Xu, R.; Liu, X.; Zhang, J.; Wang, X.; You, C.; Hao, Y. Jasmonate induces biosynthesis of anthocyanin and proanthocyanidin in apple by mediating the JAZ1–TRB1–MYB9 complex. *Plant J.* **2021**, *106*, 1414–1430. [CrossRef]
30. Boo, H.O.; Chon, S.U.; Lee, S.Y. Effects of temperature and plant growth regulators on anthocyanin synthesis and phenylalanine ammonia-lyase activity in chicory (*Cichorium intybus* L.). *J. Hortic. Sci. Biotechnol.* **2006**, *81*, 478–482. [CrossRef]
31. Li, Y.; Nie, P.; Zhang, H.; Wang, L.; Wang, H.; Zhang, L. Dynamic changes of anthocyanin accumulation and endogenous hormone contents in blueberry. *J. Beijing For. Univ.* **2017**, *39*, 64–71. [CrossRef]
32. Teribia, N.; Tijero, V.; Munné-Bosch, S. Linking hormonal profiles with variations in sugar and anthocyanin contents during the natural development and ripening of sweet cherries. *New Biotechnol.* **2016**, *33*, 824–833. [CrossRef] [PubMed]
33. Li, G.; Zhao, J.; Qin, B.; Yin, Y.; An, W.; Mu, Z.; Cao, Y. ABA mediates development-dependent anthocyanin biosynthesis and fruit coloration in *Lycium* plants. *BMC Plant Biol.* **2019**, *19*, 317. [CrossRef]
34. Gao, C.; Sun, Y.; Li, J.; Zhou, Z.; Deng, X.; Wang, Z.; Wu, S.; Lin, L.; Huang, Y.; Zeng, W.; et al. High Light Intensity Triggered Abscisic Acid Biosynthesis Mediates Anthocyanin Accumulation in Young Leaves of Tea Plant (*Camellia sinensis*). *Antioxidants* **2023**, *12*, 392. [CrossRef]
35. Ren, J.; Liu, Z.; Chen, W.; Xu, H.; Feng, H. Anthocyanin Degrading and Chlorophyll Accumulation Lead to the Formation of Bicolor Leaf in Ornamental Kale. *Int. J. Mol. Sci.* **2019**, *20*, 603. [CrossRef]
36. Cotado, A.; Müller, M.; Morales, M.; Munné-Bosch, S. Linking jasmonates with pigment accumulation and photoprotection in a high-mountain endemic plant, *Saxifraga longifolia*. *Environ. Exp. Bot.* **2018**, *154*, 56–65. [CrossRef]
37. Bai, S.; Sun, Y.; Qian, M.; Yang, F.; Ni, J.; Tao, R.; Li, L.; Shu, Q.; Zhang, D.; Teng, Y. Transcriptome analysis of bagging-treated red Chinese sand pear peels reveals light-responsive pathway functions in anthocyanin accumulation. *Sci. Rep.* **2017**, *7*, 63. [CrossRef]
38. Azuma, A.; Yakushiji, H.; Koshita, Y.; Kobayashi, S. Flavonoid biosynthesis-related genes in grape skin are differentially regulated by temperature and light conditions. *Planta* **2012**, *236*, 1067–1080. [CrossRef] [PubMed]
39. Miki, S.; Wada, K.C.; Takeno, K. A possible role of an anthocyanin filter in low-intensity light stress-induced flowering in *Perilla frutescens* var. *crispa*. *J. Plant Physiol.* **2015**, *175*, 157–162. [CrossRef]
40. Chen, W.; Zhang, M.; Zhang, G.; Li, P.; Ma, F. Differential Regulation of Anthocyanin Synthesis in Apple Peel under Different Sunlight Intensities. *Int. J. Mol. Sci.* **2019**, *20*, 6060. [CrossRef]
41. Heller, W.; Forkmann, G.; Britsch, L.; Grisebach, H. Enzymatic reduction of (+)-dihydroflavonols to flavan-3,4-cis-diols with flower extracts from *Matthiola incana* and its role in anthocyanin biosynthesis. *Planta* **1985**, *165*, 284–287. [CrossRef] [PubMed]
42. Murray, J.R.; Hackett, W.P. Dihydroflavonol Reductase Activity in Relation to Differential Anthocyanin Accumulation in Juvenile and Mature Phase *Hedera helix* L. *Plant Physiol.* **1991**, *97*, 343–351. [CrossRef] [PubMed]
43. Boss, P.K.; Davies, C.; Robinson, S.P. Expression of anthocyanin biosynthesis pathway genes in red and white grapes. *Plant Mol. Biol.* **1996**, *32*, 565–569. [CrossRef] [PubMed]

44. Jiang, M.; Liu, Y.; Ren, L.; Lian, H.; Chen, H. Molecular cloning and characterization of anthocyanin biosynthesis genes in eggplant (*Solanum melongena* L.). *Acta Physiol. Plant.* **2016**, *38*, 163. [CrossRef]
45. Bizjak, J.; Weber, N.; Mikulic-Petkovsek, M.; Slatnar, A.; Stampar, F.; Alam, Z.; Stich, K.; Halbwirth, H.; Veberic, R. Influence of Phostrade Ca on Color Development and Anthocyanin Content of 'Braeburn' Apple (*Malus domestica* Borkh.). *Hortscience* **2013**, *48*, 193–199. [CrossRef]
46. Sharma, H.; Chawla, N.; Dhatt, A.S. Role of phenylalanine/tyrosine ammonia lyase and anthocyanidin synthase enzymes for anthocyanin biosynthesis in developing *Solanum melongena* L. genotypes. *Physiol. Plant.* **2022**, *174*, e13756. [CrossRef]
47. Manetas, Y. Why some leaves are anthocyanic and why most anthocyanic leaves are red? *Flora Morphol. Distrib. Funct. Ecol. Plants* **2005**, *201*, 163–177. [CrossRef]
48. Ibañez, S.; Rosa, M.; Hilal, M.; González, J.A.; Prado, F.E. Leaves of *Citrus aurantifolia* exhibit a different sensibility to solar UV-B radiation according to development stage in relation to photosynthetic pigments and UV-B absorbing compounds production. *J. Photochem. Photobiol. B Biol.* **2008**, *90*, 163–169. [CrossRef]
49. Ma, Z.-H.; Li, W.-F.; Mao, J.; Li, W.; Zuo, C.-W.; Zhao, X.; Dawuda, M.M.; Shi, X.-Y.; Chen, B.-H. Synthesis of light-inducible and light-independent anthocyanins regulated by specific genes in grape 'Marselan' (*V. vinifera* L.). *PeerJ* **2019**, *7*, e6521. [CrossRef]
50. Weiss, D.; Halevy, A.H. The role of light reactions in the regulation of anthocyanin synthesis in *Petunia* corollas. *Physiol. Plant.* **1991**, *81*, 127–133. [CrossRef]
51. Yu, Z.; Zhang, P.; Lin, W.; Zheng, X.; Cai, M.; Peng, C. Sequencing of anthocyanin synthesis-related enzyme genes and screening of reference genes in leaves of four dominant subtropical forest tree species. *Gene* **2019**, *716*, 144024. [CrossRef] [PubMed]

horticulturae

MDPI

Article

Effects of LED Red and Blue Spectra Irradiance Levels and Nutrient Solution EC on the Growth, Yield, and Phenolic Content of Lemon Basil (*Ocimum citriodurum* Vis.)

Zurafni Mat Daud [1], Mohd Firdaus Ismail [1] and Mansor Hakiman [1,2,*]

1 Department of Crop Science, Faculty of Agriculture, Universiti Putra Malaysia, Serdang 43400 UPM, Selangor, Malaysia
2 Laboratory of Sustainable Resources Management, Institute of Tropical Forestry and Forest Products, Universiti Putra Malaysia, Serdang 43400 UPM, Selangor, Malaysia
* Correspondence: mhakiman@upm.edu.my

Abstract: This research was conducted to study the effects of LED red and blue spectra irradiance levels and nutrient solution (electrical conductivity) and their interaction on the plant growth, yield, and phytochemical contents of lemon basil (*Ocimum citriodorum* Vis.) in a controlled environment. The controlled environment was equipped with red and blue spectra at a 4:1 ratio with irradiance levels of 80 and 160 μmol m^{-2} s^{-1} and irrigated with four different nutrient solution ECs at 1.0, 1.8, 2.6, and 3.4 mS cm^{-1}, cultivated on a vertical structure. The temperature and relative humidity of the controlled environment and the pH of the nutrient solution were maintained at 26 and 18 °C day and night, 65 ± 5%, and pH 6, respectively. It was observed that plant height, canopy diameter, and the number of leaves of lemon basil had significantly increased under the irradiance levels of 160 μmol m^{-2} s^{-1} in combination with a nutrient solution EC of 2.6 mS cm^{-1}. In addition, there was an interaction observed between the LED irradiance levels and the nutrient solution EC on the fresh weight of the stem and the dry weight of all the plant parts (leaves, stem, and roots). Lemon basil cultivated at 160 μmol m^{-2} s^{-1} and irrigated with 2.6 mS cm^{-1} was significantly higher in fresh stem weight and dry leaf, stem, and root weight at 17.36, 1.79, 1.82, and 0.22 g, respectively. The ascorbic acid of lemon basil was significantly higher under a treatment of 160 μmol m^{-2} s^{-1} irradiance level and an EC of 2.6 mS cm^{-1}, and no interaction was observed. At the same time, there was an interaction observed between the LED irradiance level and the nutrient solution EC on the total phenolic content (TPC), total flavonoid content (TFC), and caftaric acid concentration of lemon basil. Lemon basil cultivated at 160 μmol m^{-2} s^{-1} and irrigated with 2.6 mS cm^{-1} was significantly higher in TPC, TFC, and caftaric acid concentration, with 1440.62 mg gallic acid equivalent to 100 g^{-1} DW, 1148.79 mg quercetin equivalent to 100 g^{-1} DW, and 2812.50 mg 100 g^{-1} DW, respectively. This result indicates that the irradiance levels of red and blue LED spectra at 160 μmol m^{-2} s^{-1} and irrigated with a nutrient solution EC of 2.6 mS cm^{-1} enhances the growth, yield production, and phenolic content of lemon basil in a controlled environment facility.

Keywords: light-emitting diodes; electrical conductivity; caftaric acid; rosmarinic acid; controlled environment

1. Introduction

Lemon basil (*Ocimum citriodorum* Vis.) is an ornamental, culinary, and medicinal herb that is planted widely and has flourished under a variety of planting conditions [1]. The species belong to the family Lamiaceae and contains an abundant source of phenolic compounds. Phenolic compounds, such as rosmarinic, chicoric, caffeic, and caftaric acids, have been reported to be obtained (in vast concentrations) from various basil cultivars [2–4], which have been documented to be a rich source of antioxidants [2]. The leaves have been used, either fresh or dried, as spices. Meanwhile, the essential oil extracted from basil can be used as an aromatic additive in food, pharmaceuticals, and cosmetics [4]. Usually, basil is cultivated in open fields, and the United States is known to be the largest producer and importer of basil in the world [5]. However, the yield and quality in terms of phytochemical contents are hard to control, and it varies with season, cultivar, and cultivation location [6–8].

Indoor vertical farming, known as "plant factory", is a highly controlled environmental system for plant production that uses multiple-layer culture shelves with artificial lighting [9,10]. Light is one of the most important environmental factors that affect plant development and regulate plant behavior, depending on the light quality, quantity, direction, and duration [11–14]. Light quantity is known as the irradiance level of light. The irradiance levels of red and blue LEDs significantly affect lemon basil growth and development. Red light seems to be more effective in improving photosynthesis compared to blue or green light [12]. In contrast, blue light has been shown to lead to an increase in ascorbic acid, phenolic contents, and chlorophyll in various species [15]. Darko et al. [16] stated that a combination of red and blue light is more efficient than monochromatic light. In indoor-grown basil and lettuce, a range of optimal intensities is used, ranging from 50–150 [17]. In addition, lettuce can be grown under light intensities ranging from 40–200, leading to increases in ascorbic acid, phenolic, carotenoid, tocopherol, flavonoids, glucosinolate, and anthocyanin content and reduced postharvest decay [18–22].

In addition, irrigation systems are one of the most important parts of controlled environment systems that use hydroponic systems. The nutrient solution (electrical conductivity) supplied to the plants plays a crucial role as it significantly affects plant growth performance, such as stem height and dry weight, and can also influence plant appearance, nutritional values, and the shelf life of plants [23,24]. Supplementing with a high level of electrical conductivity in the nutrient solution has been said to stimulate ion toxicity, osmotic stress, and nutrient disparity, while insufficient electrical conductivity, in general, leads to nutrient scarcity [25]. According to Vendrame et al. [26] and Poorter and Nagel [27], nutrient uptake is generally affected by irradiance levels (photosynthetic photon-flux density). A previous study by Lu et al. [28] showed that the growth parameters and anthocyanin concentration of red and blue perilla were higher under high light intensity (PPFD) and EC. However, the rosmarinic acid concentration was higher under the lowest EC with high light intensity. Samarakoon et al. [29] stated that plants supplemented with nutrient solution at an electrical conductivity of 2.0 to 3.0 mS cm^{-1} provide the optimum rate for better plant growth performance. However, farmers tended to over or undersupply the nutrient solution, which impacted plant growth performance, yield production, and quality.

The red and blue spectra irradiance levels of LEDs significantly affected lemon basil growth and development. However, there is scarce information on the interaction between the irradiance levels of LEDs with red and blue spectra and the EC of the nutrient solution in regulating herb production and the accumulation of phytochemical contents. Therefore, the current study was implemented to ascertain the changes in growth, yield production, and quality in terms of the phytochemical contents of lemon basil plants in response to different combinations of irradiance levels of LEDs of red and blue spectra and the EC of the nutrient solution. The study envisages the possibility of providing valuable insights into the regulation of the irradiance levels of LEDs with red and blue spectra and the EC of the nutrient solution in attempts to improve the growth performance, yield production, and phytochemical contents of lemon basil grown under vertical structures in a hydroponic system in a controlled environment facility.

2. Materials and Methods

2.1. Plant Materials and Treatments

The research was conducted in a controlled environment growth room at the Faculty of Agriculture, Universiti Putra Malaysia. Seeds of lemon basil (*Ocimum citriodorum* Vis.) were germinated and raised in peat moss. On day 14, the seedlings were transplanted into a pot (3.5 × 5.0 × 5.7 cm) and placed on the vertical structure in a growth room. The seedlings were grown in a closed-circulating water culture under two different irradiance levels of LEDs (brand Philips) and supplied by Elite Scientific Instruments Sdn. Bhd. (80 and 160 μmol m^{-2} s^{-1}) with red and blue spectra in a ratio of 4:1. The essential nutrient solution in each tank was set at different electrical conductivity (EC) values: 1.0, 1.8, 2.6, and 3.4 mS cm^{-1}, which were regularly checked using an EC meter (DIST 4 EC Meter by Hanna Instruments) (Table 1) (FERTITRADE, Petaling Jaya, Malaysia). The growth chamber's relative humidity and day/night temperature during the study were 65 ± 5% and 26/18 °C, respectively. The photoperiod was set up for 14 h (06.00 a.m. to 08.00 p.m.). The pH value of the nutrient solution was amended to 6.0 and maintained throughout the experiment.

Table 1. Mineral composition of nutrient solution.

Electrical Conductivity (EC)	Nutrient Concentrations (mg L^{-1})
1.0	N = 92.80, P = 26.80, K = 95.60, Ca = 40.00, Mg = 12.00, S = 32.00, Fe = 1.20, Mn = 0.248, B = 0.176, Cu = 0.008, Zn = 0.044, Mo = 0.019
1.8	N = 232.00, P = 67.00, K = 239.00, Ca = 100.00, Mg = 30.00, S = 80.00, Fe = 3.00, Mn = 0.62, B = 0.44, Cu = 0.02, Zn = 0.11, Mo = 0.048
2.6	N = 278.00, P = 80.40, K = 286.80, Ca = 120.00, Mg = 36.00, S = 96.00, Fe = 3.60, Mn = 0.744, B = 0.528, Cu = 0.024, Zn = 0.132, Mo = 0.058
3.4	N = 324.80, P = 93.80, K = 334.60, Ca = 140.00, Mg = 42.00, S = 112.00, Fe = 4.20, Mn = 0.868, B = 0.616, Cu = 0.028, Zn = 0.154, Mo = 0.067

Note: N-nitrogen, P-phosphorus, K-potassium, Ca-calcium, Mg-magnesium, S-sulfur, Mn-manganese, Fe-iron, Cu-copper, B-boron, Zn-zinc, and Mo-molybdenum.

2.2. *Plant Growth Measurement*

Data on lemon basil height, canopy diameter, and the number of leaves were collected at three-day intervals for 30 days after transplanting (DAT). Lemon basil height was evaluated using a ruler from the sponge surface to the shoot tip. Each plant was viewed from all sides for the plant canopy to determine the side where the canopy was broadest. The distance between the two opposite sites was recorded as the canopy width (cm) and was measured with a ruler, and the number of leaves was counted manually.

2.3. *Yield Parameters*

Three plants from each experimental unit were harvested and separated into three different parts: leaf, stem, and root, and the fresh weight was taken. The leaf area of detached leaves was measured (before dry mass measurement) using a leaf area meter (LI-300 LI-COR, USA) and expressed as cm^2 $plant^{-1}$. Leaf, stem, and root were oven-dried at 65 °C for three days, and the dry mass of leaves, stem, and roots was recorded.

2.4. *Phytochemical Contents*

2.4.1. Ascorbic Acid

Approximately 0.2 g of a fresh lemon basil leaf was weighed and put in mortar. Then, 2 mL of 10% (*v*/*v*) trichloroacetic acid was added and homogenized under dull light and on ice for cold conditions using a pestle. The prepared sample was then centrifuged for 10 min at 4 °C at 5000 rpm. Subsequently, 0.3 mL of supernatant was added with 0.2 mL of 10% (*v*/*v*) Folin-Ciocalteu reagent and 1.7 mL distilled water. The absorbance was measured at 760 nm using a UV-Vis spectrophotometer. A standard curve was arranged using several concentrations of ascorbic acid from 0 to 60 μg ml^{-1} [30]. The determination of ascorbic acid was carried out in triplicate.

2.4.2. Sample Extraction for Total Phenolic Content and Total Flavonoid Content

About 0.5 g of dried lemon basil leaves was extracted in 10 mL of 80% (*v*/*v*) methanol by shaking for 4 h at room temperature. Then the samples were centrifuged for 30 min at 13,200 rpm. The methanolic extract was stored at −20 °C for further analysis of total phenolic contents and total flavonoid contents [31].

2.4.3. Total Phenolic Content

Total phenolic contents of lemon basil were established using a modified Folin–Ciocalteu colorimetric assay [32]. A combination of 1 mL of methanolic extract, 0.525 mL Folin–Ciocalteu reagent, 0.945 mL of distilled water, and 2.625 mL of 2.1% (*w*/*v*) aqueous sodium carbonate were prepared and incubated for 20 min in the dark at room temperature. The absorbance of the samples was measured using a UV-Vis spectrophotometer at 735 nm against a blank solution containing 0.21 mL of 80% (*v*/*v*) methanol, 1.89 mL of distilled water, and 0.525 mL of 2.1% (*w*/*v*) aqueous sodium carbonate. Total phenolic content was computed by comparing the sample absorbance to a calibration curve of gallic acid. The value of total phenolic content was expressed as gallic acid equivalents (GAE) mg 100 g^{-1} dry weight. The equation of calibration curve was expressed as y = 1.8185x + 0.1359 (R^2 = 0.9958).

2.4.4. Total Flavonoid Content

For total flavonoid content analysis, about 1 mL of methanolic extract was added to 0.3 mL of 5% (*w*/*v*) sodium nitrite and incubated at room temperature for 5 min. Afterward, 0.3 mL of 10% (*w*/*v*) aluminum chloride and 2 mL of 1 N sodium hydroxide were added, and the total volume was made up to 5 mL with distilled water. The samples were measured using a UV-Vis spectrophotometer at an absorbance of 510 nm. The calibration curve was prepared using quercetin, and the total flavonoid contents were then calculated using this calibration curve. Then the total flavonoid content was expressed as mg quercetin equivalents 100 g^{-1} dry weight [33]. The equation of calibration curve was expressed as y = 0.0783x + 0.0676 (R^2 = 0.9845).

2.5. Individual Phenolic Compounds

Methanolic lemon basil extracts were sieved using a 0.22 μm Whatman nylon filter and then analyzed for rosmarinic, chicoric, caftaric, and gentisic acids using a prescribed method by Flanigan and Niemeyer [34] on a Waters dual-pump high-performance liquid chromatography (HPLC) using a Waters C-18 Symmetry column (Milford, MA). The detection wavelength was 330 nm. Eluent A was 3% (*v*/*v*) methanol and 1% (*v*/*v*) formic acid in waters, and eluent B was 0.1% (*v*/*v*) formic acid in acetonitrile. The linear gradient was used with a mobile phase flow rate of 0.1 mL/minute: hold at 95% A, 0–2 min; 95–75% A, 2–12 min; hold at 75% A, 12–17 min; 75–10% A, 17–18 min; hold at 10% A, 18–23 min; 10–95% A, 23–24 min; equilibrate at 95% A, 24–34 min. The phenolic compound was identified in the aqueous methanolic lemon basil extracts by comparison of chromatographic retention times against analytical standards. The phenolic compound was quantified by comparing integrated peak areas to standard calibration curves: chicoric and caftaric acids = 0.26–50.0 mg/L, rosmarinic acid = 0.11–75.0 mg/L, and gentisic acid= 0.87–65.0 mg/L.

2.6. Experimental Design and Statistical Analysis

The treatments were arranged as a randomized complete block design (RCBD) with four replications with 2 factorials (irradiance levels of the LEDs and the EC of the nutrient solution). All data collected were analyzed using a statistical analysis system [35]. Analysis of variance (ANOVA) was conducted, and significant differences among the treatments were determined using Duncan multiple range test (DMRT) at $p \leq 0.05$.

3. Results

3.1. Growth Parameters

3.1.1. Plant Height

The performance of the lemon basil cultivated under different red and blue spectra irradiance levels of LEDs and different electrical conductivities of the nutrient solution are shown in Figure 1, which indicates that the growth pattern is well-fitted to the growth function of $y = A/(1 + be^{-cx})$. Plants grown under all treatments had no significant differences in plant height at Day 3 until Day 18 after transplantation.

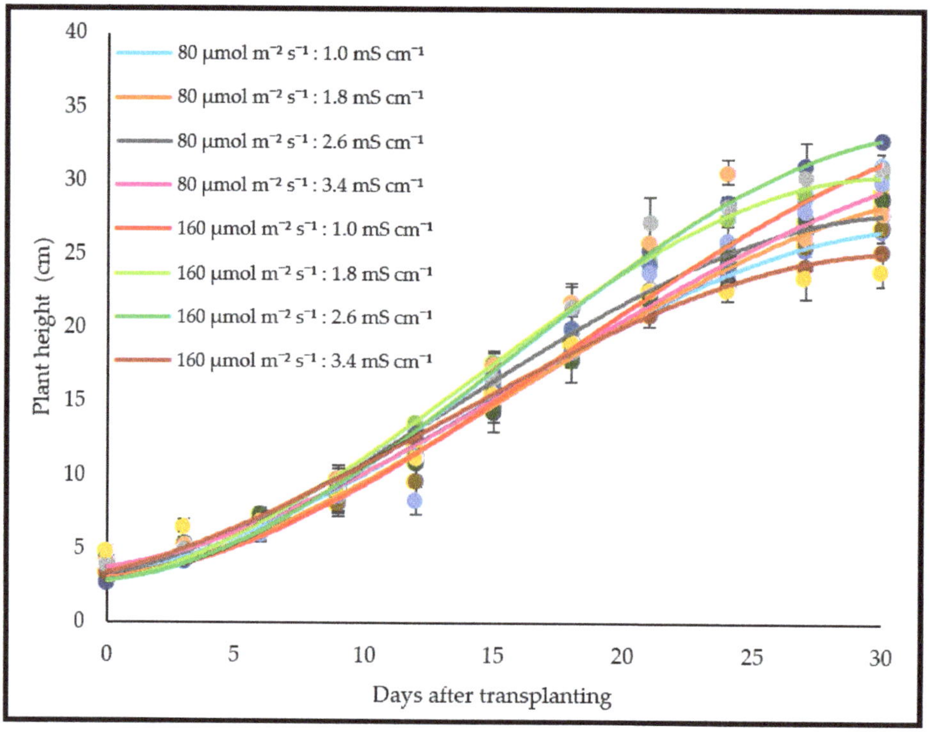

Figure 1. Plant height of lemon basil for a growing duration of 30 days under different LEDs and nutrient solution ECs.

Meanwhile, on Day 21, the plant height of the lemon basil was affected by the red and blue spectra irradiance levels of the LEDs and the EC of the nutrient solution, but no interaction effect was observed between both. The interaction effects between the irradiance levels and EC of nutrient solution were observed on Days 24 and 27 after transplant. The plants raised on irradiance levels of 160 μmol m^{-2} s^{-1} and irrigated at an EC of 1.8 mS cm^{-1} and 2.6 mS cm^{-1} were higher in height compared to other treatments.

3.1.2. Canopy Diameter

The interactions on the effects of the red and blue spectra irradiance levels of the LEDs and the nutrient solution electrical conductivity on the expansion of the canopy were recorded. The interactions were revealed by the changes in canopy diameter of the plants at all measurement dates. The canopy diameter of lemon basil grown under irradiance levels of 160 μmol m^{-2} s^{-1} and irrigated with an EC of 2.6 mS cm^{-1} and 160 μmol m^{-2} s^{-1}: 1.8 mS cm^{-1} were prominently wider than that of the plants grown under other treatments (Figure 2).

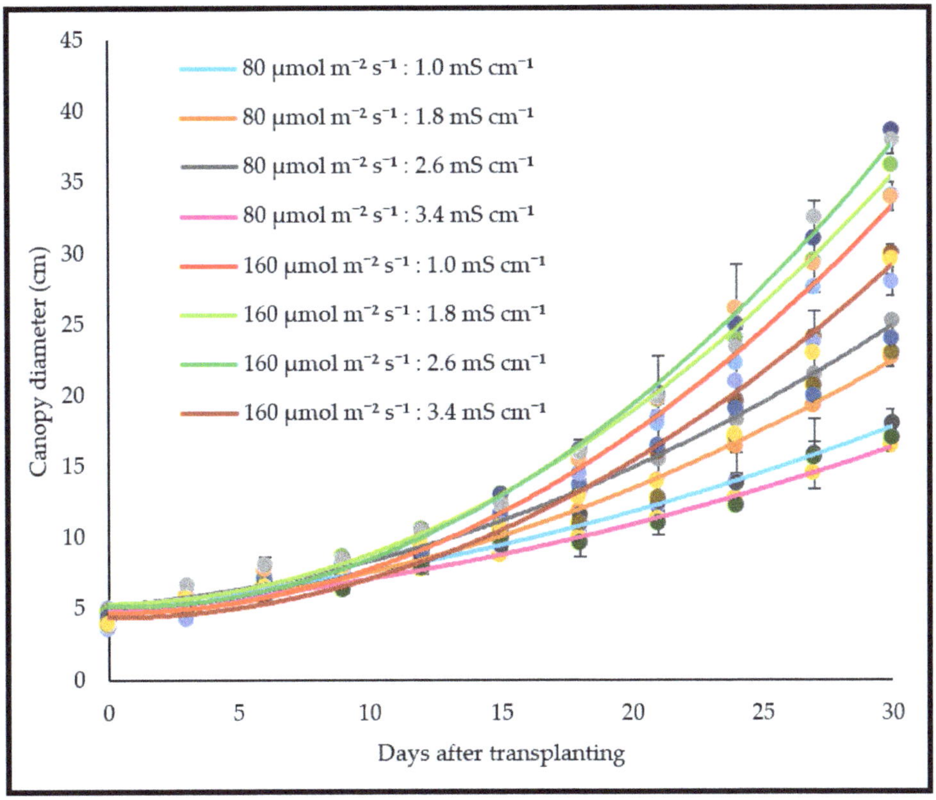

Figure 2. Canopy diameter of lemon basil for 30 days growing duration under different LEDs and nutrient solution ECs.

3.1.3. Number of Leaves

The leaves number of the lemon basil expanded exponentially resulting from the function $y = Ae^{bx}$ days after transplanting (Figure 3). The number of lemon basil leaves cultivated on red and blue spectra irradiance levels of 160 μmol m^{-2} s^{-1} and irrigated with an EC of 3.4 mS cm^{-1}, 160 μmol m^{-2} s^{-1}:2.6 mS cm^{-1} and 160 μmol m^{-2} s^{-1}:1.8 mS cm^{-1} were higher compared to other treatments due to higher rates in the rise of leaf production. The number of lemon basil leaves cultivated on irradiance levels of 160 μmol m^{-2} s^{-1} and irrigated with an EC of 2.6 and 3.4 mS cm^{-1} were both were recorded at 180, while 1.8 mS cm^{-1} was recorded at 160, with the other treatments in the range of 70 to 118.

3.2. Yield Parameters

3.2.1. Fresh Weight

Table 2 shows the fresh weight of the three plant parts: leaves, stem, and roots of the lemon basil, as affected by LED irradiance levels and nutrient solution electrical conductivity (EC). The fresh weight of the leaves, stem, and roots increased with increasing irradiance levels and nutrient solution EC. Interaction effects were observed between the irradiance levels and the EC of the nutrient solution on the fresh stem weight of the lemon basil (Figure 4). However, there were significant differences for both treatments on the leaves and roots. Plants cultivated at an irradiance level of 160 μmol m^{-2} s^{-1} produced a higher fresh weight for the leaves and roots. Plants supplemented with an EC of 2.6 mS cm^{-1} also produced a higher fresh weight for both parts when compared with other ECs. Plants grown under 160 μmol m^{-2} s^{-1} produced 60% and 55% higher leaf and root fresh weights than those from 80 μmol m^{-2} s^{-1}.

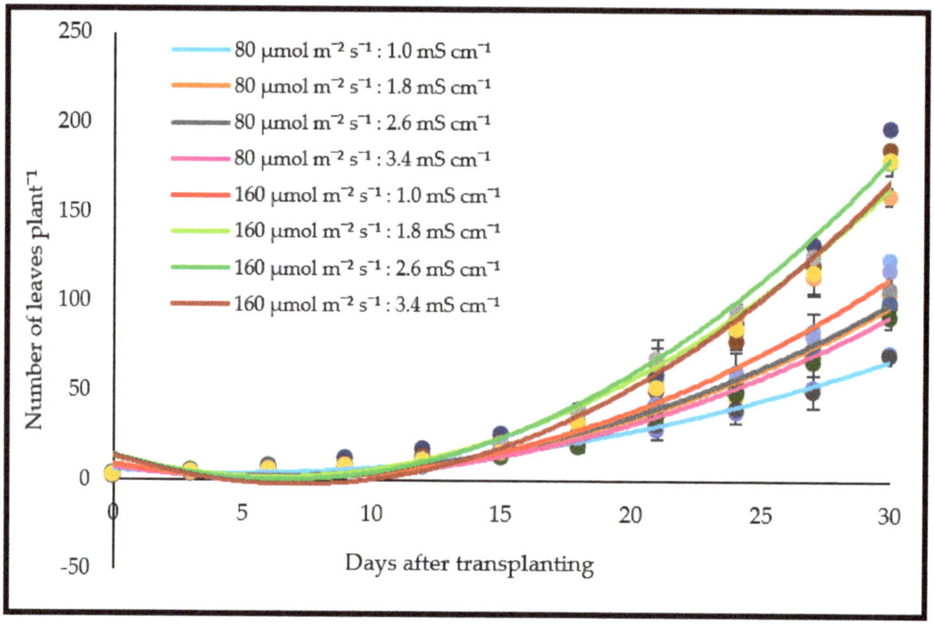

Figure 3. Number of lemon basil leaves after 30 days growing duration under different LEDs and nutrient solution ECs.

Table 2. Effects of LED irradiance levels and nutrient solution EC on the fresh weight (g) of three different parts (leaves, stem, and roots) of lemon basil.

Factor	Leaves	Stem	Roots
Irradiance levels of LEDs (μmol m^{-2} s^{-1})			
80	15.90 [b]	6.19 [b]	4.82 [b]
160	25.44 [a]	12.45 [a]	7.47 [a]
EC of nutrient solution (mS cm^{-1})			
1.0	16.75 [c]	7.04 [c]	4.60 [c]
1.8	21.14 [b]	10.21 [b]	5.862 [b]
2.6	23.77 [a]	11.84 [a]	8.72 [a]
3.4	21.02 [b]	8.21 [c]	5.40 [b]
Irradiance levels of LEDs	***	***	***
EC of nutrient solution	***	***	***
Irradiance levels of LEDs × EC of nutrient solution	ns	***	ns

*** Significant at $p < 0.001$ probability level, ns = not significant. Means in each column with different letters within each factor indicate significant differences at $p \leq 0.05$ level according to DMRT.

There is an interaction effect observed on the application of various LEDs and nutrient solution ECs on fresh stem weight (Figure 4). The application of various nutrient solution ECs at 80 μmol m^{-2} s^{-1} did not produce a significant difference in the stem weight of the lemon basil. However, the fresh weight of the stem increased when the EC of the nutrient solution increased at 160 μmol m^{-2} s^{-1}. The highest fresh stem weight was observed at irradiance levels of 160 μmol m^{-2} s^{-1} in combination with a nutrient solution EC of 2.6 mS cm^{-1}. The application of a higher nutrient solution EC of 3.4 mS cm^{-1} decreased the fresh weight of the lemon basil stem.

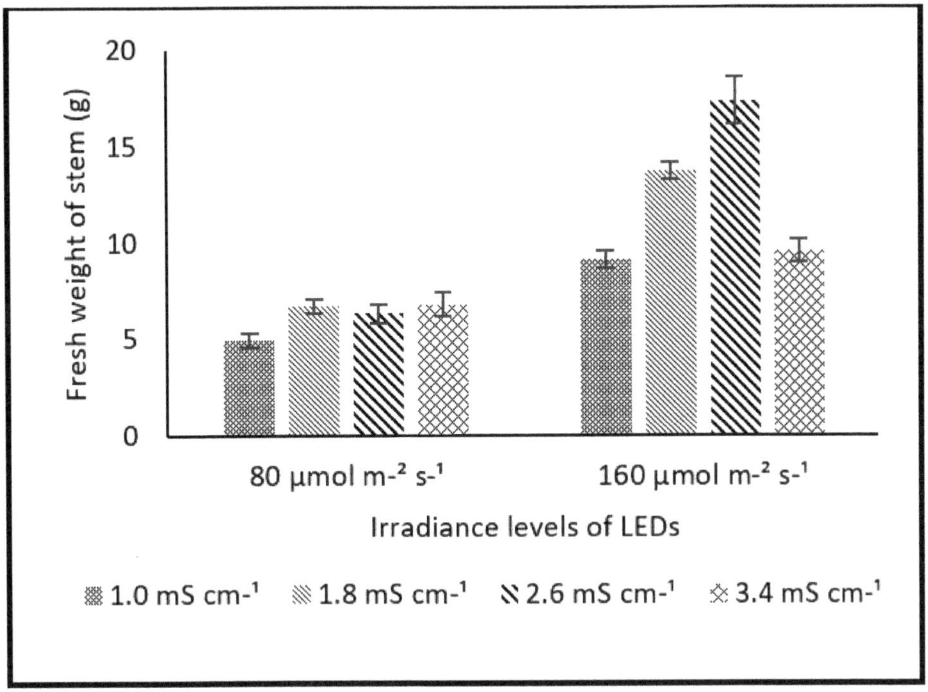

Figure 4. Interaction effect of LED irradiance levels and nutrient solution EC on fresh stem weight of lemon basil.

3.2.2. Dry Weight

The consequences of LED irradiance levels and the nutrient solution EC on the dry weight of lemon basil (leaves, stem, and roots) are shown in Table 3. The interaction effects observed between the irradiance levels and the nutrient solution EC on the dry weight of the three different parts of lemon basil are shown in Figure 5. An increase in nutrient solution EC caused an increase in the dry weight of the leaves, stem, and roots at both LED irradiance levels. However, further increases in nutrient solution EC up to 3.4 mS cm^{-1} at 160 µmol m^{-2} s^{-1} reduced the dry weight of all parts of the plants.

Table 3. Effects of LED irradiance levels and nutrient solution EC on the dry weight (g) of three different parts (leaves, stem, and roots) of lemon basil.

Factor	Leaves	Stem	Roots
Irradiance levels of LEDs (µmol m^{-2} s^{-1})			
80	0.85 ^{b}	0.64 ^{b}	0.32 ^{b}
160	1.64 ^{a}	1.29 ^{a}	0.60 ^{a}
EC of nutrient solution (mS cm^{-1})			
1.0	1.10 ^{b}	0.69 ^{d}	0.45
1.8	1.41 ^{a}	1.89 ^{c}	0.48
2.6	1.26 ^{a,b}	1.22 ^{a}	0.48
3.4	1.21 ^{b}	1.05 ^{b}	0.43
Irradiance levels of LEDs	***	***	***
EC of nutrient solution	***	***	ns
Irradiance levels of LEDs × EC of nutrient solution	***	***	***

*** Significant at $p < 0.001$ probability level, ns = not significant. Means in each column with different letters within each factor indicate significant differences at $p \leq 0.05$ level according to DMRT.

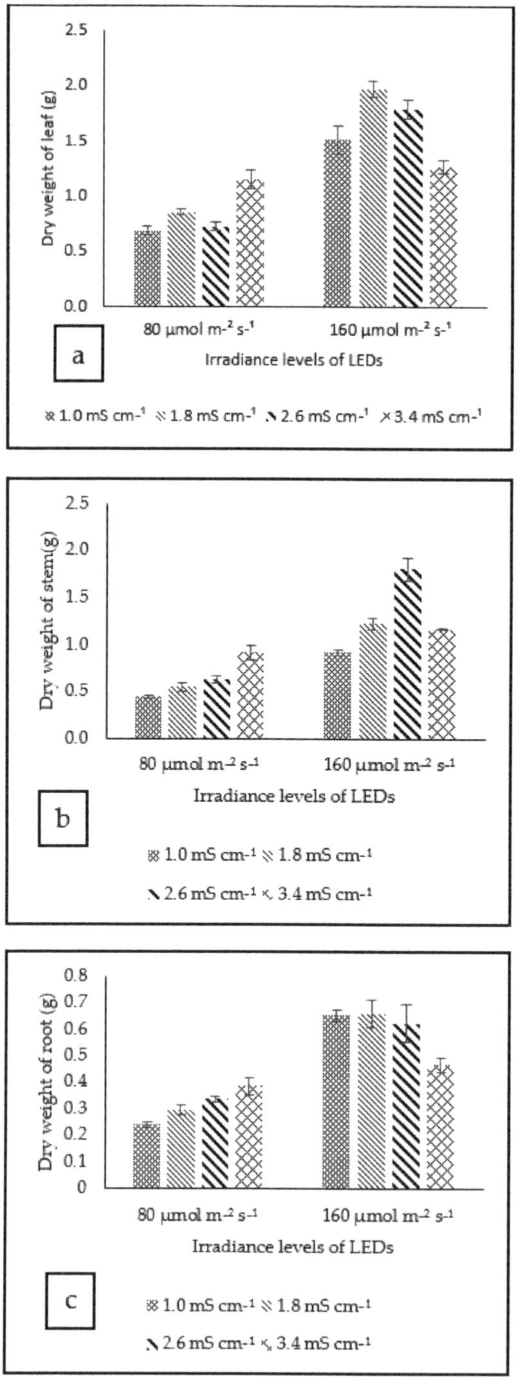

Figure 5. Interaction effect of LED irradiance levels and nutrient solution EC on the dry weight of the leaves (**a**), stems (**b**), and roots (**c**) of lemon basil.

The highest dry weight of the leaves was observed at 160 μmol m^{-2} s^{-1} in combination with 1.8 and 2.6 mS cm^{-1} with 1.97 and 1.79 g (Figure 5a), respectively; the dry weight of the stems at 160 μmol m^{-2} s^{-1} with 2.6 mS cm^{-1} was 1.82 g (Figure 5b), and the dry weight of the roots at 160 μmol m^{-2} s^{-1} in combination with 1.0, 1.8, and 2.6 mS cm^{-1} produced 0.65, 0.66, and 0.63 g (Figure 5c), respectively. In contrast, the lowest dry weight of all parts was observed at LED irradiance levels of 80 μmol m^{-2} s^{-1} and in combination with all EC levels except for 3.4 mS cm^{-1}.

3.2.3. Leaf Area

Table 4 illustrates the effects of red and blue spectra irradiance levels of LEDs and the nutrient solution ECs on leaf area in lemon basil. No interaction was observed between the irradiance levels and nutrient solution EC, but there were significant differences in both of the main effects. The leaf area of lemon basil grown under an irradiance level of 160 μmol m^{-2} s^{-1} was significantly ($p < 0.001$) the highest with 1020.62 cm^2/plant, which is 65% higher than 80 μmol m^{-2} s^{-1}. In addition, the lemon basil irrigated with a nutrient solution EC of 1.8 mS cm^{-1} also produced a greater leaf area ($p < 0.001$) compared with other EC levels with 943.23 cm^2/plant.

Table 4. Effects of LED irradiance levels and nutrient solution EC on leaf area (cm^2/plant) in lemon basil.

Factor	Leaf Area
Irradiance levels of LEDs (μmol m^{-2} s^{-1})	
80	617.14 ^{b}
160	1020.62 ^{a}
EC of nutrient solution (mS cm^{-1})	
1.0	772.84 ^{b}
1.8	943.23 ^{a}
2.6	831.26 ^{b}
3.4	848.19 ^{b}
Irradiance levels of LEDs	***
EC of nutrient solution	***
Irradiance levels of LEDs × EC of nutrient solution	ns

*** Significant at $p < 0.001$ probability level, ns = not significant. Means in each column with different letters within each factor indicate significant differences at $p \leq 0.05$ level according to DMRT.

3.3. *Phytochemical Contents*

3.3.1. Ascorbic Acid

Table 5 shows the ascorbic acid contents of the lemon basil grown under varying LED red and blue spectra irradiance levels and nutrient solution ECs. Significant interactions were not observed between the irradiance levels and the nutrient solution EC on the ascorbic acid contents in lemon basil. However, the ascorbic acid contents in the lemon basil grown under LED red and blue spectra irradiance levels at 160 μmol m^{-2} s^{-1} were significantly higher than those obtained at 80 μmol m^{-2} s^{-1}. In addition, the lemon basil fertilized with an EC of 2.6 mS cm^{-1} was also higher in ascorbic acid contents than with other ECs. The lowest contents were observed at 1.0 (67.43 mg 100^{-1} g FW).

3.3.2. Total Phenolic Content

Table 5 shows the lemon basil's TPC as affected by LED red and blue spectra irradiance levels and nutrient solution EC. The interaction between the irradiance levels and the solution EC significantly influenced TPC (Figure 6). Increasing nutrient solution EC increased the TPC at both irradiance levels of 80 and 160 μmol m^{-2} s^{-1}. Plants grown at irradiance levels of 160 μmol m^{-2} s^{-1} in combination with a nutrient solution EC of 2.6 mS cm^{-1} significantly increased in TPC compared to other levels with 1440.62 mg GAE 100 g^{-1} DW (Figure 6), and the lowest was observed at an irradiance level of

80 μmol m^{-2} s^{-1} and in combination with an EC level of 1.0 mS cm^{-1} at 491.56 mg GAE 100 g^{-1} DW. However, there was no difference with 160 μmol m^{-2} s^{-1}:3.4 mS cm^{-1}.

Table 5. Effects of LED irradiance levels and nutrient solution EC on ascorbic acid, total phenolic, and total flavonoid contents in lemon basil.

Factor	Ascorbic Acid (mg 100 g^{-1} FW)	Total Phenolic Content (mg GAE 100 g^{-1} DW)	Total Flavonoid Content (mg QE 100 g^{-1} DW)
Irradiance levels of LEDs (μmol m^{-2} s^{-1})			
80	72.82 ^{b}	620.30 ^{b}	907.87 ^{b}
160	77.59 ^{a}	1346.28 ^{a}	982.72 ^{a}
EC of nutrient solution (mS cm^{-1})			
1.0	67.43 ^{c}	866.27 ^{b}	907.34 ^{b,c}
1.8	68.54 ^{c}	995.49 ^{a}	966.63 ^{a,b}
2.6	88.37 ^{a}	1039.30 ^{a}	1022.52 ^{a}
3.4	76.48 ^{b}	1032.10 ^{a}	884.98 ^{b,c}
Irradiance levels of LEDs	***	***	**
EC of nutrient solution	***	***	***
Irradiance levels of LEDs × EC of nutrient solution	ns	*	***

*** significant at $p < 0.001$ probability level, ** significant at $p < 0.01$ probability level, * significant at $p < 0.05$ probability level, ns = not significant. Means in each column with different letters within each factor indicate significant differences at $p \leq 0.05$ level according to DMRT.

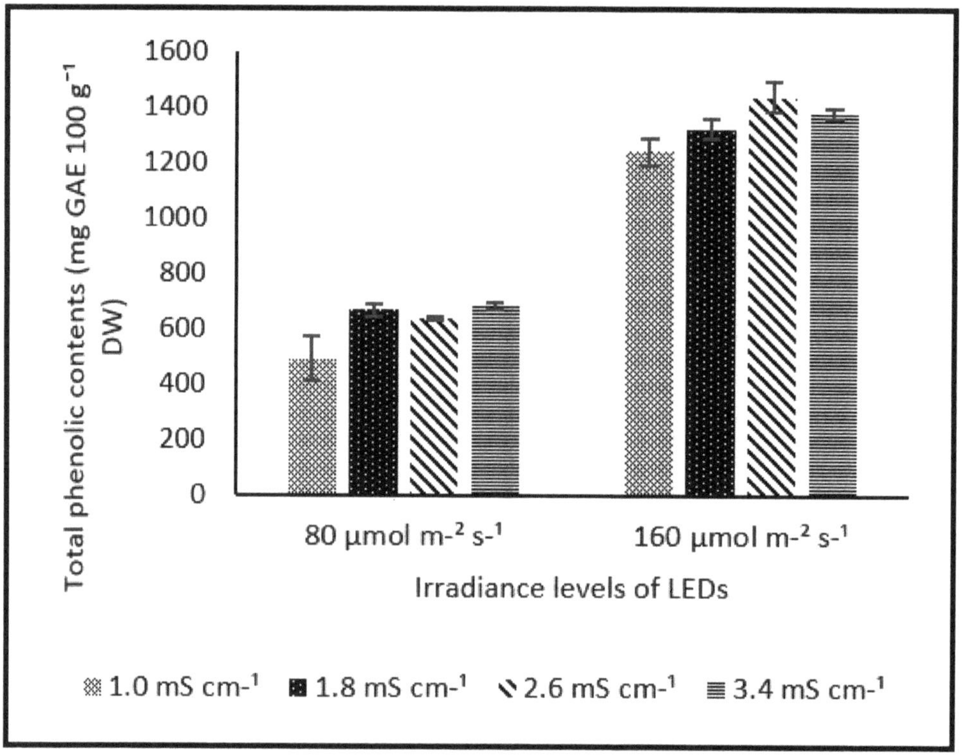

Figure 6. Interaction effect of LED irradiance levels and nutrient solution EC on total phenolic content of lemon basil.

3.3.3. Total Flavonoid Content

The total flavonoid content (TFC) of lemon basil was significantly affected by the LED red and blue spectra irradiance levels and the nutrient solution EC, and there is interaction between both (Table 5 and Figure 7). TFC decreased with increasing nutrient solution EC at 80 μmol m^{-2} s^{-1}. While at 160 μmol m^{-2} s^{-1}, increasing the nutrient solution EC up to 2.6 mS cm^{-1} increased TFC. A further increase to 3.4 mS cm^{-1} degraded the flavonoid content. The highest TFC was observed at 160 μmol m^{-2} s^{-1} irradiance supplemented with an EC of 2.6 mS cm^{-1} with 1148.79 mg quercetin equivalent 100 g^{-1} DW.

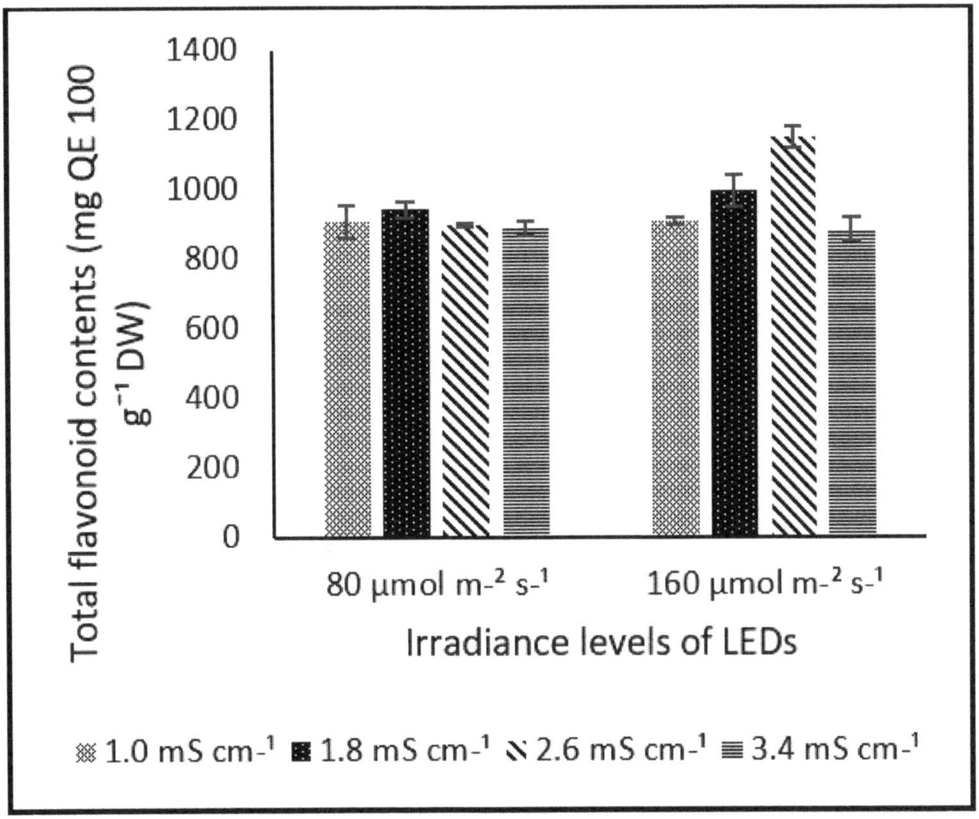

Figure 7. Interaction effect of LED irradiance levels and nutrient solution EC on total flavonoid content of lemon basil.

3.4. Individual Phenolic Compound

Table 6 shows the effects of the LEDs and EC on the concentrations of the prominent individual phenolic compounds (caftaric, rosmarinic, chicoric, and gentisic acid) of lemon basil. The interaction effects between the LED irradiance levels and nutrient solution EC on caftaric acid (Figure 8) and rosmarinic acid (Figure 9) were significantly different. In contrast, no interaction or significant differences were observed in either of the main effects on lemon basil's chicoric and gentisic acid concentrations.

Table 6. Effects of LED irradiance levels and nutrient solution EC on the concentration of the individual phenolic compounds of lemon basil.

Factor	Caftaric Acid (mg 100 g^{-1} DW)	Rosmarinic Acid (mg 100 g^{-1} DW)	Chicoric Acid (mg 100 g^{-1} DW)	Gentisic Acid (mg 100 g^{-1} DW)
Irradiance levels of LEDs (μmol m^{-2} s^{-1})				
80	2341.30	36.44 ^{b}	23.27	2.96
160	2205.70	45.41 ^{a}	23.31	3.14
EC of nutrient solution (mS cm^{-1})				
1.0	2167.90 ^{b}	32.03 ^{b}	23.40	2.35
1.8	2157.70 ^{b}	35.83 ^{b}	23.30	2.60
2.6	2677.10 ^{a}	45.28 ^{a}	23.36	4.08
3.4	2091.20 ^{b}	50.58 ^{a}	23.09	3.17
Irradiance levels of LEDs	ns	**	ns	ns
EC of nutrient solution	**	***	ns	ns
Irradiance levels of LEDs × EC of nutrient solution	***	**	ns	ns

*** significant at $p < 0.001$ probability level, ** significant at $p < 0.01$ probability level, ns = not significant. Means in each column with different letters within each factor indicate significant differences at $p \leq 0.05$ level according to DMRT.

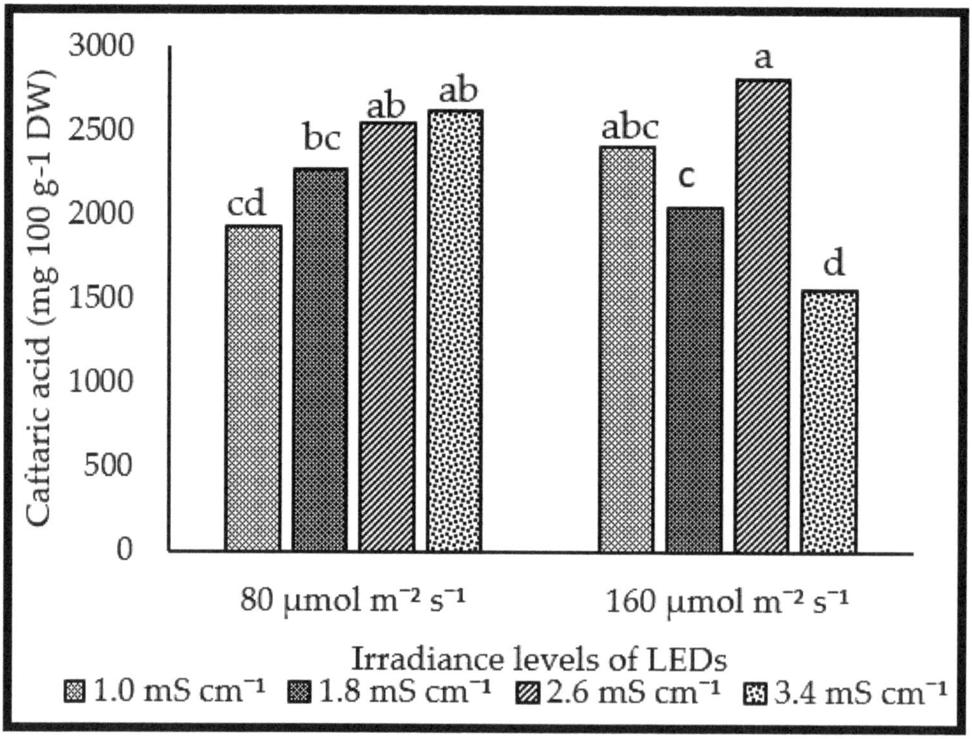

Figure 8. Interaction effect of LED irradiance levels and nutrient solution EC on caftaric acid in lemon basil.

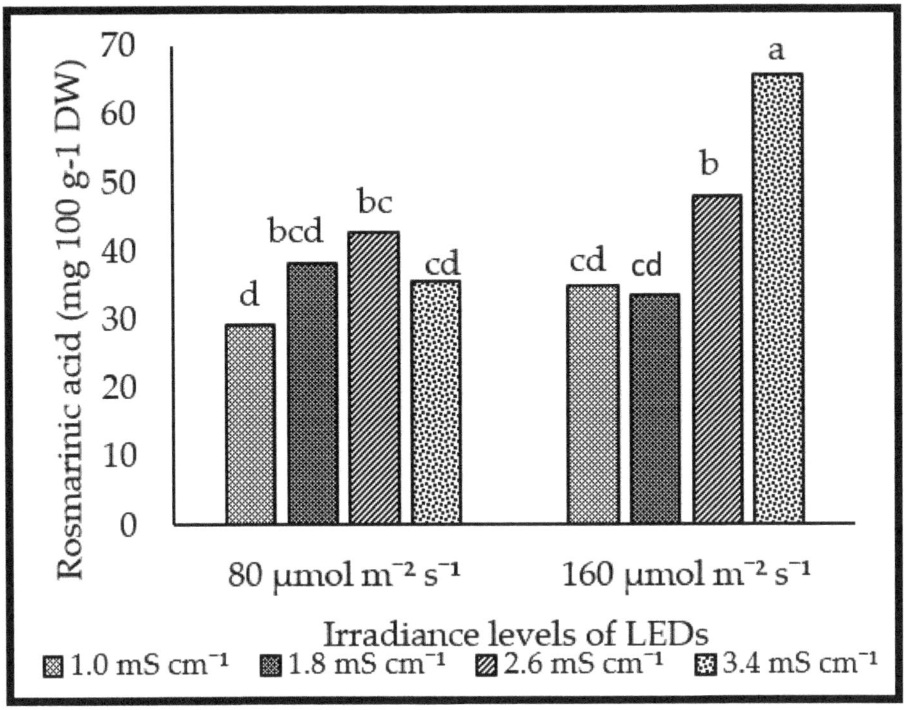

Figure 9. Interaction effect of LED irradiance levels and nutrient solution EC on rosmarinic acid in lemon basil.

The caftaric acid concentration increased with increasing nutrient solution EC at 80 μmol m^{-2} s^{-1}, whereas it fluctuated at 160 μmol m^{-2} s^{-1} with rising nutrient solution EC. The highest caftaric acid concentration was observed at an irradiance level of 160 μmol m^{-2} s^{-1} in combination with 2.6 mS cm^{-1} EC when compared to other combinations but had no differences with 80 μmol m^{-2} s^{-1}:2.6 mS cm^{-1}, 80 μmol m^{-2} s^{-1}:1.8 mS cm^{-1}, and 160 μmol m^{-2} s^{-1}:1.0 mS cm^{-1}. In contrast, the lowest caftaric acid was recorded at an LED irradiance level of s at 160 μmol m^{-2} s^{-1} supplied with a nutrient solution EC of 3.4 mS cm^{-1}, but this had no differences at 80 μmol m^{-2} s^{-1}:1.0 mS cm^{-1}.

For rosmarinic acid, significant differences were not examined at 80 μmol m^{-2} s^{-1} with increasing nutrient solution EC, whereas rosmarinic acid concentration was elevated with increasing nutrient solution EC at 160 μmol m^{-2} s^{-1}. The highest concentration was observed at 160 μmol m^{-2} s^{-1} in combination with a nutrient solution EC of 3.4 mS cm^{-1}, though the lowest was at 80 μmol m^{-2} s^{-1}:1.0 mS cm^{-1}.

4. Discussion

4.1. Plant Growth Performance

The present study reveals that the growth parameters of lemon basil (*Ocimum citriodorum* Vis.) were strongly influenced by LED red and blue spectra irradiance levels and nutrient solution EC. The growth parameters, such as plant height, canopy diameter, and the number of leaves, showed significant effects between the different treatments regarding the levels of LED irradiance and the EC of the nutrient solution. Plants cultivated at an irradiance level of 160 μmol m^{-2} s^{-1} and supplemented with an EC of 2.6 mS cm^{-1} produced optimum growth performance compared to other treatments. A previous study by Morano et al. [36] stated that a nutrient solution EC of 2.8 mS cm^{-1}

in the shortest crop cycle increased basil yield (whole plants and leaves), while an EC of 2.2 mS cm^{-1} exhibited the worst performance.

4.2. Yield Production

Leafy herbs, such as basil, are very popular crops among farmers because they are easy to grow, have a high yield index, are suitable for hydroponic and closed farming, and simultaneously have a high margin for profitability [37]. Considering our results, LED irradiance levels and nutrient solution EC significantly affected the yield production of both the fresh and dry products (Tables 2 and 3). Increases in yield for the fresh and dry weight were optimized by the appliance of LED irradiance levels of 160 μmol m^{-2} s^{-1} in combination with an EC of 2.6 mS cm^{-1}. However, the fresh and dry weights were severely inhibited at irradiance levels of 80 μmol m^{-2} s^{-1} supplemented at all nutrient solution ECs. Nemali and van Iersel [38] found similar results, reporting on the interaction effect of fertilizer concentration and photosynthetic photon flux (PPF) on the dry weight of petunia and wax begonia plants. The dry mass of the fertilized plants was higher for those plants grown at higher irradiance levels (268 μmol m^{-2} s^{-1}) when compared to those grown at low irradiance levels (113 μmol m^{-2} s^{-1}). In addition to the irradiance level treatments employed, the optimal range of nutrient concentration for dry mass varied from 0.65 to 2.0 mS cm^{-1} (wax begonia) and 1.18 to >2.77 mS cm^{-1} (petunia). In addition, no interaction was observed between the LED irradiance levels and EC on the leaf area of lemon basil. However, there were significant effects observed for both parameters on leaf area in lemon basil. Lemon basil cultivated under 160 μmol m^{-2} s^{-1} and irrigated at 1.8 mS cm^{-1} produced a higher leaf area.

4.3. Phytochemical Contents

The interaction between the LED red and blue spectra irradiance levels and nutrient solution EC did not significantly influence the ascorbic acid contents of lemon basil. However, the main effects of the LED irradiance levels and nutrient solution EC were significant. According to Fraszczak et al. [39], the amount of ascorbic acid can vary according to light conditions. The current study is in concurrence with this statement. It was observed that the ascorbic acid contents of lemon basil were significantly higher when grown at higher irradiance levels of 160 μmol m^{-2} s^{-1} than at lower irradiance levels of 80 μmol m^{-2} s^{-1}. The study by Ohashi-Kaneko et al. [40] and Fraszczak et al. [39] also revealed that plants cultivated under red and blue light saw significant increases in the content of ascorbic acid. In addition, Lee et al. [41] discovered that plant tissues would accumulate more ascorbic acid content when irradiated with high light intensity. The high light intensity would strengthen the plants so as to acquire greater photosynthesis, which would relocate more assimilation products for growth and metabolism. According to Ding et al. [42], the ascorbic acid of *pak choi* plants showed an increasing trend of up to 2.4 mS cm^{-1} but did not significantly increase over this nutrient concentration level. This divergence between different vegetable crops may be due to the differences in growth habitats and growing conditions.

The effect of LED irradiance levels on TPC and TFC varied depending on the nutrient solution EC (Figures 6 and 7). Increasing nutrient solution EC increased TPC and TFC at both irradiance levels of 80 and 160 μmol m^{-2} s^{-1}. Higher TPC and TFC were observed at irradiance levels of 160 μmol m^{-2} s^{-1} in combination with s nutrient solution EC of 2.6 mS cm^{-1}. It was revealed that LED red and blue spectra irradiance levels and nutrient solution EC influenced the accumulations of phenolic and flavonoid contents in lemon basil. The previous study by Son et al. [43] showed that blue LEDs could significantly increase the accumulation of phenolic compounds in plants. The phenolic compound content was similar between the control and white LEDs treatments in red curled lettuce, but the white LEDs accelerated the accumulation of phenolic compounds in green curled lettuce. Few studies have concluded that red light can stimulate an increase in phenolics in plants [39,44,45]. Although the mechanism is still unknown, one assumption is that red light increases the cytokinin level and thus stimulates the synthesis of phenolics [46]. A previous study by Kiferle et al. [47] stated that a lower amount of fertilizer had a significant effect on the accumulation of rosmarinic acid. In contrast, our results showed that basil fertilized at a higher nutrient solution EC accumulated significantly more rosmarinic acid compared to other ECs (Table 6).

5. Conclusions

An LED irradiance level of 160 μmol m^{-2} s^{-1} in combination with a nutrient solution EC of 2.6 mS cm^{-1} produced higher plant growth performance in terms of plant height, canopy diameter, the number of leaves, higher fresh and dry yield production, and higher phytochemical contents, such as total phenolic and total flavonoid content, along with caftaric acid concentration. The present study's findings have improved the comprehension of the effects of LED red and blue spectra irradiance levels and nutrient solution EC and its interaction on lemon basil plants that are cultivated on a vertical structure by using the hydroponic system in a controlled environment facility.

Author Contributions: Conceptualization, Z.M.D. and M.F.I.; methodology, Z.M.D. and M.F.I.; resources, Z.M.D.; data curation, Z.M.D.; writing—original draft preparation, Z.M.D.; writing—review and editing, Z.M.D., M.F.I. and M.H.; supervision, M.F.I. and M.H.; funding acquisition, M.F.I. and M.H. All authors have read and agreed to the published version of the manuscript.

Funding: This research was funded by Ministry of Higher Education Malaysia (Translational Research Program (TR@M) and Universiti Putra Malaysia (Putra Grant), grant number 5526700 and 9433800, respectively and the APC was funded by Universiti Putra Malaysia.

Institutional Review Board Statement: Not applicable.

Informed Consent Statement: Not applicable.

Data Availability Statement: The data presented in this study are available in the article.

Acknowledgments: Authors would like to thank the Ministry of Higher Education, Malaysia, for the scholarship.

Conflicts of Interest: The authors declare no conflict of interest.

References

1. Makri, O.; Kintzios, S. *Ocimum* sp. (basil) botany, cultivation, pharmaceutical properties and biotechnology. *J. Herbs Spices Med. Plants* **2008**, *13*, 123–150. [CrossRef]
2. Kwee, E.M.; Niemeyer, E.D. Variations in phenolic compositions and antioxidant properties among 15 basil (*Ocimum basilicum* L.) cultivars. *J. Food Chem.* **2011**, *128*, 1044–1050. [CrossRef]
3. Lee, J.; Scagel, C.F. Chicoric acid found in basil (*Ocimum basilicum* L.) leaves. *J. Food Chem.* **2009**, *115*, 650–656. [CrossRef]
4. Javanmardi, J.; Khalighi, A.; Kashi, A.; Bais, H.P.; Vivanco, J.M. Chemical characterization of basil (*Ocimum basilicum* L.) found in local accessions and used in traditional medicines in Iran. *Agric. Food Chem.* **2002**, *50*, 5878–5883. [CrossRef] [PubMed]
5. Department of Agriculture, Forestry and Fisheries of Republic of South Africa. *Basil Production*; Department of Agriculture, Forestry and Fisheries: Pretoria, South Africa, 2012.
6. Fischer, R.; Nitzan, N.; Chaimovitsh, D.; Rubin, B.; Dudai, N. Variation in essential oil composition within individual leaves of sweet basil (*Ocimum basilicum* L.) is more affected by leaf position than by leaf age. *J. Agric. Food Chem.* **2011**, *59*, 4913–4922. [CrossRef] [PubMed]
7. Hassanpouraghdam, M.B.; Gohari, G.R.; Tabatabaei, S.J.; Dadpour, M.R. Inflorescence and leaves essential oil composition of hydraponically grown *Ocimum basilicum* L. *J. Serbian Chem. Soc.* **2010**, *75*, 1361–1368. [CrossRef]
8. Pushpangadam, P.; George, V. Basil. In *Handbook of Herbs and Spices*; Peter, K.V., Ed.; Elsevier: Atlanta, GA, USA, 2012; pp. 55–72.
9. Despommier, D. Farming up the city: The rise of urban vertical farms. *Trends Biothechnol.* **2013**, *31*, 388–389. [CrossRef]
10. Kozai, T.; Niu, G.; Takagaki, M. *Plant Factory: An Indoor Vertical Farming System for Efficient Quality Food Production*; Academic Press: San Diego, CA, USA, 2015.
11. Chang, X.; Alderson, P.G.; Wright, C.J. Solar irradiance level alters the growth of basil (*Ocimum basilicum* L.) and its content of volatile oils. *Environ. Expt. Bot.* **2008**, *63*, 216–223. [CrossRef]
12. Dou, H.; Niu, G.; Gu, M.; Masabni, J.G. Effects of light quality on growth and phytonutrient accumulation of herbs under controlled environments. *Horticulturae* **2017**, *3*, 36. [CrossRef]
13. Figueiredo, A.C.; Barroso, J.G.; Pedro, L.G.; Scheffer, J.J. Factors affecting secondary production in plants: Volatile components and essential oils. *J. Flav. Frag.* **2008**, *23*, 213–226. [CrossRef]

14. Shaffie-Hajiabad, M.; Novak, J.; Honermeier, B. Content and composition of essential oil of four *Origanum vulgare* L. accessions under reduced and normal light intensity conditions. *J. Appl. Bot. Food Qual.* **2016**, *89*, 126–134.
15. Olle, M.; Virsille, A. The effect of light emitting diode lighting on greenhouse plant growth and quality. *J. Agric. Food Sci.* **2013**, *22*, 223–234. [CrossRef]
16. Darko, E.; Heydarizadeh, P.; Shoefs, B.; Sabzalian, M.R. Photosynthesis under artificial light: The shift in primary and secondary metabolism. *J. Philos. Trans. R. Soc. Biol. Sci.* **2014**, *369*, 20130243. [CrossRef] [PubMed]
17. Shiga, T.; Shoji, K.; Shimada, H.; Hashida, S.N.; Goto, F.; Yoshihara, T. Effect of light quality on rosmarinic acid content and antioxidant activity of sweet basil, *Ocimum basilicum* L. *J. Plant Biotechnol.* **2009**, *26*, 255–259. [CrossRef]
18. Liao, H.L.; Alferez, F.; Burns, J.K. Assessment of blue light treatments on citrus postharvest diseases. *J. Postharvest Biol. Technol.* **2013**, *81*, 81–88. [CrossRef]
19. Johkan, M.; Shoji, K.; Goto, F.; Hashida, S.; Yoshihara, T. Blue light-emitting diode light irradiation of seedlings improves seedling quality and growth after transplanting in red leaf lettuce. *J. Hortic. Sci.* **2010**, *45*, 1809–1814. [CrossRef]
20. Kook, H.S.; Park, S.H.; Jang, Y.J.; Lee, G.W.; Kim, J.S.; Kim, H.M.; Oh, B.T.; Chae, J.C.; Lee, K.J. Blue LED (light-emitting diodes)-mediated growth promotion and control of Botrytis disease in lettuce. *Acta Agric. Scand. Sect. B-Soil Plant Sci.* **2013**, *63*, 271–277.
21. Samuoliene, G.; Brazaityte, A.; Jankauskiene, J.; Virsile, A.; Sirtautas, R.; Novickovas, A.; Sakalauskiene, S.; Sakalauskaite, J.; Duchovskis, P. LED irradiance level affects growth and nutritional quality of Brassica microgreens. *J. Century Eur. Biol.* **2013**, *8*, 1241–1249. [CrossRef]
22. Bantis, F.; Ouzounis, T.; Radoglou, K. Artificial LED lighting enhances growth characteristics and total phenolic content of *Ocimum basilicum*, but variably affects transplant success. *J. Hortic. Sci.* **2016**, *198*, 277–283. [CrossRef]
23. Bekhradi, F.; Delshad, M.; Marin, A.; Luna, M.C.; Garrido, Y.; Kashi, A.; Babalar, M.; Gil, M.I. Effects of salt stress on physiological and postharvest quality characteristics of different Iranian genotypes of basil. *Hortic. Environ. Biotechnol.* **2015**, *56*, 777–785. [CrossRef]
24. De Pascale, S.; Maggio, A.; Orsini, F.; Barbieri, G. Nutrient influence on ready to eat sweet basil quality. *Acta Hortic.* **2006**, *718*, 523–530. [CrossRef]
25. Fallovo, C.; Rouphael, Y.; Rea, E.; Battistelli, A.; Colla, G. Nutrient solution concentration and growing season affect yield and quality of *Lactuca sativa* L. var. *acephala* raft culture. *J. Food Sci. Agric.* **2009**, *89*, 1682–1689. [CrossRef]
26. Vendrame, W.; Moore, K.K.; Broschat, T.K. Interaction of light intensity and controlled release fertilization rate on growth and flowering of two New Guinea impatiens cultivars. *HortTechnology* **2004**, *14*, 491–495.
27. Poorter, H.; Nagel, O. The role of biomass allocation in the growth response of plants to different levels of light, CO_2, nutrients andwater: A quatitative review. *J. Funct. Plant Biol.* **2000**, *27*, 1191. [CrossRef]
28. Lu, N.; Bernardo, E.L.; Tippayadarapanich, C.; Takagaki, M.; Kagawa, N.; Yamori, W. Growth and accumulation of secondary metabolites in Perilla as affected by photosynthetic photon flux density and electrical conductivity of the nutrient solution. *Front. Plant Sci.* **2017**, *8*, 708. [CrossRef] [PubMed]
29. Samarakoon, U.C.; Weerasinghe, P.A.; Weerakkody, A.P. Effect of electrical conductivity (EC) of the nutrient solution on nutrient uptake, growth, and yield of leaf lettuce (*Lactuca sativa* L.) in stationary culture. *J. Trop. Agric. Res.* **2006**, *18*, 13–21.
30. Jagota, S.K.; Dani, H.M. A new colorimetric technique for the estimation of vitamin C using Folin phenol reagent. *Anal. Biochem.* **1982**, *127*, 178–182. [CrossRef]
31. Nguyen, P.M.; Niemeyer, E.D. Effects of nitrogen fertilization on the phenolic composition and antioxidant properties of basil (*Ocimum basilicum* L.). Agric. *Food Chem.* **2008**, *56*, 8685–8691. [CrossRef] [PubMed]
32. Ghasemzadeh, A.; Jaffar, H.Z.; Rahmat, A. Synthesis of phenolics and flavonoids in ginger (*Zingiber officinale Roscoe*) and their effects on photosynthesis rate. *Int. J. Mol. Sci.* **2010**, *11*, 4539–4555. [CrossRef]
33. Ismail, H.I.; Chan, K.W.; Mariad, A.A.; Ismail, M. Phenolic content, and antioxidant activity of cantaloupe (*Cucumis melo*) methanolic extracts. *J. Food Chem.* **2010**, *119*, 643–647. [CrossRef]
34. Flanigan, P.M.; Niemeyer, E.D. Effect of cultivar on phenolic levels, anthocyanin composition, and antioxidant properties in purple basil (*Ocimum basilicum* L.). *Food Chem.* **2014**, *164*, 518–526. [CrossRef] [PubMed]
35. SAS Institute. *SAS User's Guide: Statistics*; SAS Institute: Cary, NC, USA, 1999.
36. Morano, G.; Amalfitano, C.; Sellitto, M.; Cuciniello, A.; Maiello, R.; Caruso, G. Effects of nutritive solution electrical conductivity and plant density on growth, yield, and quality of sweet basil grown in gullies by subirrigation. *J. Adv. Hortic. Sci.* **2017**, *31*, 25–30.
37. Ruta, S.; Lauzike, K.; Pukas, T.; Samuoliene, G. Effect of light intensity on the growth and antioxidant activity of sweet basil and lettuce. *J. Plants* **2022**, *11*, 1709.
38. Nemali, K.S.; van Iersel, M.W. Light intensity and fertilizer concentration: I. Estimating optimal fertilizer concentration from water-use efficiency of wax begonia. *HortScience* **2004**, *39*, 1287–1292. [CrossRef]
39. Fraszczak, B.; Golcz, A.; Zawirska-Wojtasiak, R.; Janowska, B. Growth rate of sweet basil and lemon balm plants grown under fluorescent lamps and LED modules. *Acta Sci. Pol. Hortorum Cultus* **2014**, *13*, 3–13.
40. Ohashi-Kaneko, K.; Takse, M.; Kon, N.; Fujiwara, K.; Kurata, K. Effect of light quality on growth and vegetable quality in leaf lettuce, spinach and komatsuna. *J. Environ. Control Biol.* **2007**, *45*, 189–198. [CrossRef]
41. Lee, S.K.; Kader, A.A. Pre-harvest and postharvest factors influencing vitamin C content of horticultural crops. *J. Postharvest Biol. Technol.* **2000**, *20*, 207–220. [CrossRef]

42. Ding, X.; Jiang, Y.; Zhao, H.; Guo, D.; He, L.; Liu, F.; Zhou, Q.; Nandwani, D.; Hui, D.; Yu, J. Electrical conductivity of nutrient solution influenced photosynthesis, quality and antioxidant enzyme activity of pakchoi (*Brassica campestris* L. ssp. *chinensis*) in a hydroponic system. *PLoS ONE* **2018**, *13*, e0202090. [CrossRef]
43. Son, K.H.; Park, J.H.; Kim, D.; Oh, M.M. Leaf shape, growth and phytochemicals in two leaf lettuce cultivars grown under monochromatic light-emitting diodes. *Korean J. Hortic. Sci. Technol.* **2012**, *30*, 664–672. [CrossRef]
44. Lee, Y.J.; Ha, J.Y.; Oh, J.E.; Cho, M.S. The effect of LED irradiation on the quality of cabbage stored at a low temperature. *J. Food Sci. Biotechnol.* **2014**, *23*, 1087–1093. [CrossRef]
45. Li, Q.; Kubota, C. Effects of supplemental light quality on growth and phytochemicals of baby leaf lettuce. *J. Environ. Exp. Bot.* **2009**, *67*, 59–64. [CrossRef]
46. Galuszka, P.; Frebortova, J.; Luhova, L.; Bilyeu, K.D.; English, J.T.; Frebort, I. Tissue localization of cytokinin dehydrogenase in maize: Possible involvement of quinone species generated from plant phenolic by other enzymatic systems in the catalytic reaction. *J. Plant Cell Physiol.* **2005**, *46*, 716–728. [CrossRef] [PubMed]
47. Kiferle, C.; Maggini, R.; Pardossi, A. Influence of nitrogen nutrition on growth and accumulation of rosmarinic acid in sweet basil (*Ocimum basilicum* L.) grown in hydroponic culture. *J. Aus. Crop Sci.* **2013**, *9*, 3–12.

horticulturae

Review

Modification of Light Characteristics Affect the Phytochemical Profile of Peppers

Yamir Jiménez-Viveros [1], Héctor Gordon Núñez-Palenius [2], Grisel Fierros-Romero [3] and Juan Ignacio Valiente-Banuet [1,*]

[1] Tecnologico de Monterrey, Escuela de Ingeniería y Ciencias, Epigmenio González 500, Querétaro 76130, Mexico
[2] División de Ciencias de la Vida, Campus Irapuato-Salamanca, Universidad de Guanajuato, Agronomía, Ex Hacienda el Copal km 9, Carretera Irapuato-Silao, Irapuato 36500, Mexico
[3] School of the Environment, Florida Agricultural and Mechanical University, 1601 S. Martin Luther King Jr. Blvd, Tallahassee, FL 32307, USA
* Correspondence: valiente@tec.mx; Tel.: +52-442-592-6268

Abstract: *Capsicum* is one of the most economically important genera in the Solanaceae family. *Capsicum* fruits (peppers) are rich in phytochemicals with high nutritional value and significant health-promoting characteristics. The phytochemical profile of peppers consists of capsaicinoids, carotenoids, and phenolics, primarily. Currently, most of the pepper production is carried out under protected horticulture conditions. The objective of this article was to provide a comprehensive review on how light characteristics and manipulation by different horticultural technologies can affect the biosynthesis and accumulation of phytochemicals in *Capsicum* fruits. The use of shade nets or plastic covers to reduce light intensity does not seem to yield consistent responses on the phytochemical profile, as the final profile results from the interaction of several factors. Other factors involved in the accumulation of phytochemicals include temperature, water availability and plant nutrition. Exposure of plants to supplemental light with specific wavelengths (using LEDs) seems to result in a more precise stimulation of specific metabolites. In this article, we examine the effects of light irradiance and spectrum on the specific phytochemicals of *Capsicum* fruits.

Keywords: capsaicinoids; carotenoids; irradiance; phenolic compounds; plant secondary metabolites; spectrum light; solar radiation

1. Introduction

Capsicum is one of the most economically important genera in the *Solanaceae* family. This genus encompasses five domesticated species with more than 50,000 cultivars [1]. The fruits of *Capsicum* (peppers) are associated with significant health-promoting properties attributable to their nutritional composition and metabolite contents. These properties include analgesic, anti-obesity, cardioprotective, pharmacological, neurological, and dietetic, among others [2]. The specific phytochemicals associated with these properties include carotenoids (provitamin A), phenolic compounds, and capsaicinoids, primarily [3].

The phytochemical and secondary metabolite profiles of peppers are also a good source of nutrients and bioactive compounds [4,5]. Secondary metabolites are a large group of organic compounds with low molecular weight and specific physiological functions. These metabolites serve as chemical adaptations to stress conditions, or as defensive, protective, or offensive chemical agents against micro-organisms, insects, and herbivores [6].

The chemical composition of peppers is closely related to genotype, the process of fruit ripening [3,7], and environmental conditions [8,9]. The environmental factors that affect the biosynthesis, metabolism, and accumulation of phytochemicals in peppers include light, temperature, soil-water availability, and plant nutrition [10]. Thus, changes in environmental conditions can affect the biosynthesis of bioactive compounds in peppers [8].

Peppers vary in color, shape, and chemical composition [7]. Color properties vary by genotype and cultivar. Color changes occur during fruit maturation when the plastids transition from chloroplast to chromoplast in the fruits' pericarp [3].

Currently, the production of peppers is carried out predominantly under protected horticulture conditions [11]. In particular, the manipulation of natural light by photo-selective netting or plastics, and supplemental lighting (artificial light) can be used to reduce heat and light stress and improve the yield and quality of horticultural crops [12]. These horticultural practices modify the light intensity and spectrum intercepted by the plants and may also affect the production levels of total phenols, ascorbic acid, and antioxidants due to the influence of modified light conditions on the metabolic pathways that lead to the formation of the phytochemicals [13]. Controlled growing conditions in glasshouses impacted the carotenoid contents in sweet peppers [14]. Thus, light intensity (irradiance) and spectrum are environmental factors that affect the phytochemical contents of peppers [15].

Even though the pathways for the biosynthesis of the secondary metabolites of peppers have been described, limited information is currently available on the interaction between the effects of light on the synthesis and accumulation of bioactive compounds in *Capsicum* species. The objective of this review article is to examine how changes in light characteristics affect the biosynthesis and accumulation of metabolites of *Capsicum* fruits, and, in turn, alter the phytochemical profile of peppers.

2. Light Interactions with *Capsicum* Plants

The growth and productivity of pepper crops are affected by environmental factors [16]. Among these factors, light is the principal source of energy that drives physiological processes, which include: photosynthesis, photomorphogenesis, fruit development, and maturation [17,18]. Plants interact with light through specific pigments that acquire light energy, and photoreceptors which are proteins that elicit different responses based on light conditions [19]. The most important plant photoreceptors reported for pepper plants include phytochromes, cryptochromes, phototropins, and UV-B-Resistance 8 (UVR8) photoreceptors (Figure 1) [20]. These photoreceptors have peak absorbance wavelengths for the induction of the responses.

Currently, most of the horticultural production of peppers is carried out under protected agriculture conditions [21] primarily by the implementation of photo-selective shading nets [22], plastics [23], and, in some cases, artificial lighting [9,24] which includes ultraviolet radiation (UV), fluorescent lamps, and light-emitting diodes (LEDs) [25]. The active manipulation of light can improve plant productivity and the quality of peppers [26,27].

The biosynthesis of phytochemicals changes depending on light intensity and spectral quality. Plants accumulate phenolic compounds and other antioxidants such as carotenoids, flavonoids, and anthocyanins to protect against damaging high irradiance and UV radiation. Thus, spectral and irradiance manipulation could promote morphological and physiological responses and influence the biosynthesis, accumulation, and retention of phytochemicals [28,29]. UV radiation and excessive irradiance produced by different light sources may cause stress conditions and activate the defense response, changing a variety of bioactive compounds [25].

Shade nets and plastic covers reduce the light intensity (irradiance) and alter the light spectra that reach the crops. Reduced light intensity affects the physiological responses by decreasing photosynthetic rate and promoting an increase in leaf area [12], while scattering improves the penetration of spectrally modified light into the inner canopy of the crop [28,30]. Currently, the use of black shade nets is the predominant practice in the horticultural production of peppers. Black nets reduce light intensity and have a limited effect on light quality [31,32]. By contrast, colored shading nets selectively filter the solar radiation and promote specific wavelengths [33]. Colored shading nets could promote plants' physiological and morphological responses [34]. Colored shading nets can selectively change the red to far-red ratios that are detected by the phytochromes, enhance the radiation available to activate the blue/ultraviolet-A photoreceptors, alter the blue light involved in phototropic responses mediated by phototropins, or enhance radiation at other wavelengths that influence plant response [35].

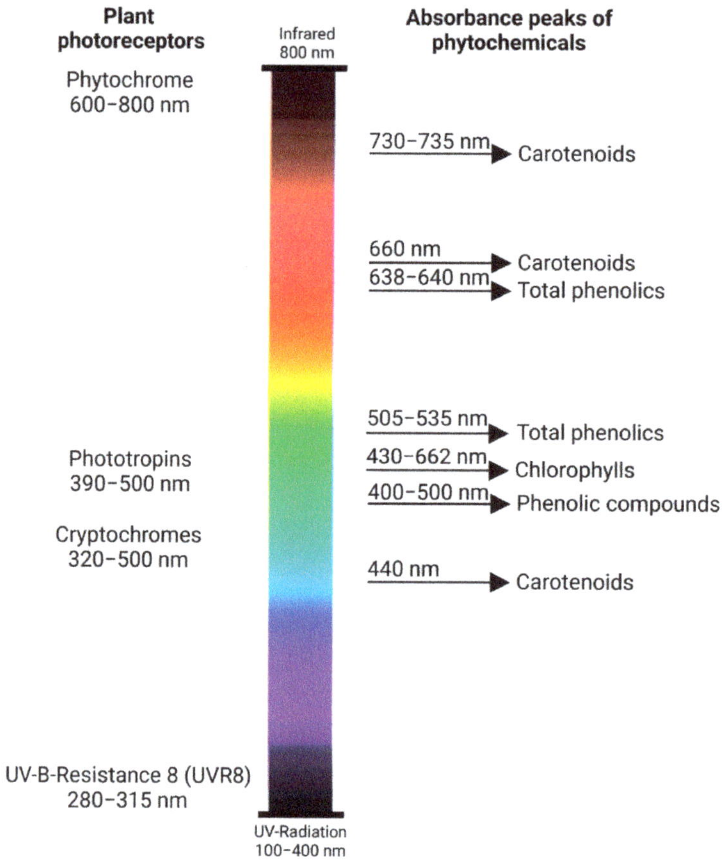

Figure 1. Plant photoreceptors (phytochrome, phototropins, cryptochromes, and UV-B-Resistance 8 (UVR8)) with the corresponding absorbance peaks (wavelengths of the electromagnetic spectrum) for each light-sensing photoreceptor protein. The light-responding groups of phytochemicals in plants in the specific wavelength ranges are provided on the right.

The traditional supplemental light sources used for greenhouse and in vitro applications include fluorescent, metal halide, high-pressure sodium, and incandescent lamps. These light sources have certain limitations as they produce an impractical mixture of wavelengths for plant growth [36], and their electricity consumption is high [37]. LEDs are considered improved light sources for greenhouse production as they can emit specific wavelengths aimed at increasing crop yield, higher quality yield, manipulation of harvest dates, and enhanced nutritional value in cultured plants [38]. Currently, these technologies are preferred for in vitro propagation and indoor plant growth, which are effective for the stimulation of plant phytochemicals during fruit development and postharvest [39].

3. Effects of Light Characteristics on the Phytochemicals of *Capsicum* Fruits

The most abundant secondary metabolites in *Capsicum* fruits include capsaicinoids, carotenoids, phenolic compounds, flavonoids, and a wide range of volatile compounds.

The accumulation of phytochemicals in peppers is light-dependent, and the high variability of these compounds determines the diversity of aroma and flavor of peppers [40].

3.1. Capsaicinoids

Capsaicinoids are secondary metabolites biosynthesized exclusively by the fruits of *Capsicum* plants [41]. These metabolites are the bioactive compounds responsible for the pungent taste of peppers [42]. Capsaicinoids may occur in peppers in a wide range of contents from 'Bell peppers', where they are practically non-existent, to other high-pungency cultivars such as 'Naga peppers' [43]. Capsaicinoids are considered natural defense mechanisms against herbivores ranging from insects to rodents [1]. Capsaicinoids also mediate interactions with birds, who act as seed dispersers for wild peppers [44].

In recent years, capsaicinoid research has been influential in the development of innovative applications in the food and pharmaceutical industries [41] due to their value as antioxidants (free radical scavengers) [45], anti-arthritic [46], gastroprotective [47,48], anti-cancer [49], and analgesic agents [50], among others.

The most abundant capsaicinoids in peppers are capsaicin and dihydrocapsaicin [51,52]. Together, these compounds encompass more than 90% of the total capsaicinoid content of peppers [53]. Nonetheless, at least nine other capsaicinoids including nordihydrocapsaicin, homodihydrocapsaicin, and homocapsaicin have also been identified [43]. Capsaicinoid levels are influenced by the ontogenetic development of the peppers. The accumulation of capsaicinoids starts at the early stages of fruit development, followed by a high peak and a rapid decline [54].

3.1.1. Biosynthesis of Capsaicinoids

Capsaicinoid biosynthesis is derived from the phenylpropanoid pathway (Figure 2) [54–56] and occurs after the enzymatic condensation of a molecule of vanillylamine derived from phenylalanine, valine, or leucine to a branched-chain amino acid. The enzymes whose alleles determine pungency levels in peppers are CaMYB31, *pAMT*, *CS/AT3/Pun1*, and *CaKR1* [57]. Capsaicin synthase (*CS*) is the last enzyme (encoded by the *Pun1* gen) responsible for the condensation between vanillylamine and a fatty acid-CoA while the aromatic vanillylamine moiety is paired with many acyl groups, mostly medium-length (from 9 to 11 carbon atoms), giving the immediate reaction of capsaicin biosynthesis [58,59]. Capsaicinoids differ in their chemical structures, specifically in the side chain with a variable number of double bonds placed in different positions; the type of capsaicinoid depends on the products obtained from the different fatty acids in the dehydration synthesis reaction [55].

Differences in capsaicinoid contents can be attributed to changes in the gene expression of the phenylpropanoid pathway. This biosynthetic pathway depends on the genotype and is affected by environmental conditions that include light, temperature, soil-water availability, and mineral nutrition [36,41]. Light intensity directly affects the biosynthesis and accumulation of capsaicinoids in peppers. Light exposure has a positive influence on the expression of the capsaicin synthase gene (*CS*) that has light-responsive motifs in its promoter region *KAS* (keto-acyl ACP synthase) and *AMT* (aminotransferase), with a negative effect through the induction of peroxidases that can degrade capsaicin. Currently, it is not well understood how this balance is controlled and adjusted [54]. The expression of the CaMYB31, *KAS*, and *pAMT* is affected in peppers of the *C. annuum* genus mainly by light but also by temperature, mechanical stress, and plant hormones [60]. The promoter of the *Pun1* gene has light-responsive motifs and consensus elements that promote capsaicinoid biosynthesis [61].

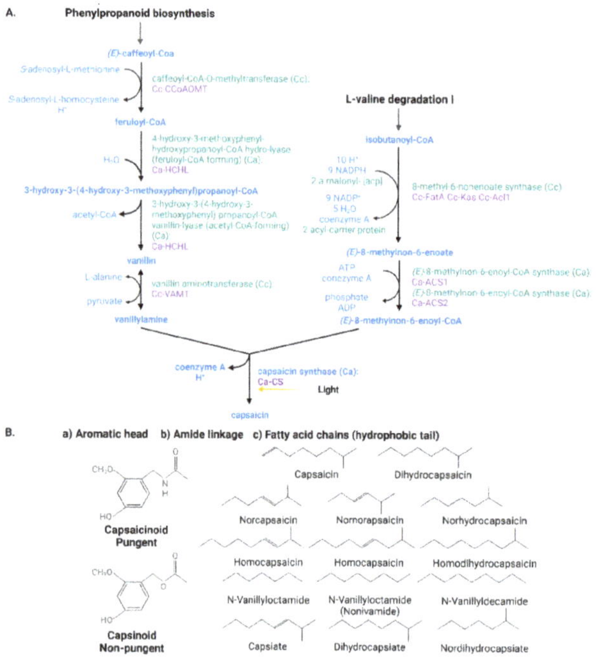

Figure 2. (**A**) Capsaicinoid biosynthetic pathway in peppers (*Capsicum* spp.) via phenylpropanoid and L-valine Degradation I. The yellow arrow indicates the light signal that regulates transcription factors at the molecular level. (**B**) Chemical structure of the most abundant capsaicinoids (pungent) and capsinoids (non-pungent) molecules of *Capsicum* fruits. Capsaicinoids and capsinoids differ in the R group (fatty acids) present.

3.1.2. Effects of Light on Capsaicinoids

In a study on bell pepper production, the optimum light intensity reported to obtain maximum fruit yield was estimated in the range of 1365 to 1470 $\mu mol \cdot m^{-2} \cdot s^{-1}$ [62]. Horticultural practices that modify irradiance may result in the enhancement or reduction of capsaicinoid contents (Table 1), depending on the species and the light modification mechanisms (e.g., color and degree of shading, or quality of light emitted by artificial illumination) [63].

Capsaicinoid accumulation is affected by the interaction of light intensity with temperature and relative humidity. In high-pungency peppers (*C. chinense* Jacq.), reduced light intensity and temperature caused lower capsaicinoid production of 4.82 and 3.49 mg plant^{-1} when plants were grown under 50% and 70% shade, respectively [63]. Reduced capsaicinoid accumulation also occurred at high irradiance levels and high temperatures. In addition, environments with reduced light intensity (713–783 $\mu mol \cdot m^{-2} \cdot s^{-1}$) and higher relative humidity increased capsaicinoid production [64]. Thus, the authors suggest an optimum light intensity of 700 to 950 $\mu mol \cdot m^{-2} \cdot s^{-1}$ for capsaicinoid production in these cultivars [63].

Total capsaicinoid contents were significantly affected by the interaction of reduced light intensity using different color shades and harvest time in *C. annuum* 'Star flame' and 'Fire flame' [65]. The capsaicinoid contents of peppers grown under colored shading net treatments (white, red, and green) were higher than the unshaded treatment. Of those, the green shade treatment had a considerably higher capsaicinoid content at the first harvest time. This effect could be related to a higher average temperature (22–28 °C) during the cycle. However, other studies showed that higher average temperature and increased solar radiation were associated with lower capsaicinoid contents [41].

Exposure of pepper plants (*C. chinense* Jacq.) to reduced light intensities using shade nets increased the contents of secondary metabolites, including capsaicinoids and other phenolic compounds [63]. Reduced light intensities increased the contents of the phenylalanine ammonia-lyase (PAL) enzyme, which plays a vital role in capsaicinoid biosynthesis. Thus, an increase in the contents of PAL may also cause an increase in capsaicinoids in peppers [66]. Currently, there is not a full understanding of how capsaicinoid accumulation relates to the relevant biochemical reactions with precursors and environmental factors [58].

As for supplemental light, pepper fruits accumulated more capsaicinoids in plants grown in a closed environment under continuous fluorescent illumination (150–350 $\mu mol \cdot m^{-2} \cdot s^{-1}$) and constant temperature (28 °C) than pepper fruits grown under greenhouse conditions during the summer season [67].

Table 1. Effect of light condition treatments on the capsaicinoid content in *Capsicum* species.

***Capsicum* spp.**	**Light Treatment**	**Effects on Capsaicinoids Compared to Control**	**Biosynthetic Effect**
C. chinense Jacq. Seven hot hybrid peppers	Light intensities (1200, 1313, 713, 1112, 774, and 783 $\mu mol \cdot m^{-2} \cdot s^{-1}$) in different locations with shading net with 50% shade	Reduced light intensity (713–783 $\mu mol \cdot m^{-2} \cdot s^{-1}$) and higher relative humidity increased capsaicinoid production in cultivars	Not reported [64]
C. chinense Jacq. 'Bhut Jolokia' 'Akanee Pirote' 'Habanero'	Shading nets with 50%, and 70% shade, and unshaded as control	'Bhut Jolokia' showed the highest capsaicinoid yield under 70% shading, 'Akanee Pirote' under 50% shading, and habanero peppers showed the lowest capsaicinoid content under shading treatments	Levels of phenylalanine ammonia-lyase (PAL) increased under low light intensities [63]
C. annuum 'Star flame' 'Fire flame'	Colored shading nets: white, red, and green with 40% shade, and unshaded as control	Capsaicinoid content increased in color-shading treatments, specifically in green treatment in both cultivars	A high average temperature of 22–28 °C may have promoted capsaicinoid biosynthesis [65]
C. annuum 'Super hot'	Greenhouse conditions with LED lighting treatments: blue, red, and a mixture of blue and red light, and 12 h of sunlight as control	Blue LEDs significantly increased nordihydrocapsaicin, capsaicin, dihydrocapsaicin, homocapsaicin, and homodihydrocapsaicin contents by 57, 43, 56, 28, and 54%, respectively	Capsaicin and dihydrocapsaicin accumulation helped in oxidative stress defense. Valine and phenylalanine increased in blue LED lights contributing to a higher content of capsaicinoids [68]
C. annuum 'Cheonyang'	LED lighting treatments: red, blue, and red plus blue, and fluorescent lamps as control	Blue LEDs increased capsaicinoid contents, red LEDs reduce two times the capsaicinoid content compared to fluorescent light	Not reported [36]
C. annuum 'Shishito pepper'	Continuous fluorescent illumination (150–350 $\mu mol \cdot m^{-2} \cdot s^{-1}$) at constant temperature (28 °C), and greenhouse conditions as control	Fewer seeds and higher concentration of capsaicin in fruits under continuous fluorescent illumination	There is a negative correlation between seed formation and capsaicin biosynthesis [67]

Table 1. *Cont.*

Capsicum spp.	Light Treatment	Effects on Capsaicinoids Compared to Control	Biosynthetic Effect
C. annuum Serrano 'Tampiqueño 74′ Sweet pepper 'California wonder'	Artificial light in postharvest (50 μmol·m^{-2}·s^{-2}) and dark conditions as control	Light factors increased capsaicin content in 'Tampiqueño 74′	CaMYB31-expression analysis from placental tissue of pungent and non-pungent fruits showed a positive correlation with the structural genes *Ca4H*, *Comt*, *KAS*, *pAMT*, and *AT3* expression, and with the content of capsaicin and dihydrocapsaicin during fruit development [60]

Differences in light spectral quality can also affect the accumulation of capsaicinoids in peppers. Peppers produced under blue spectrum light-emitting diodes (LEDs) increased capsaicinoid contents in comparison to plants exposed to fluorescent lights [36]. In a similar study under greenhouse conditions, supplemental blue light LEDs placed at the top and between plant rows also increased capsaicinoid levels in peppers. This was attributed to the blue wavelength, which is near the UV spectra, and causes the same oxidative stress response during the biosynthesis of capsaicin. Blue light also plays a role in chloroplast development, chlorophyll formation, and stomatal opening [68]. In postharvest, Serrano pepper fruits ('Tampiqueño 74′) treated with light or dark conditions with varying exposure times, the expression of the structural genes *KAS*, *pAMT*, and the transcription factor gene CaMYB31 was higher under the light stimulus than fruits stored in the dark [60].

3.2. Carotenoids

Carotenoids are a numerous family of more than 850 naturally occurring lipophilic isoprenoid compounds widely distributed in nature [69]. All photosynthetic organisms, including plants, algae, and cyanobacteria, and some non-photosynthetic micro-organisms, including fungi and bacteria, synthesize carotenoids [70]. In plants, the principal function of carotenoids is the protection of cells and organelles against oxidative damage. Carotenoids prevent the accumulation of harmful oxygen species by interacting with singlet oxygen molecules and scavenging peroxy radicals [71]. Carotenoids are also involved in the photosynthetic process and play a role in photo-protection, photo-morphogenesis, and plant development. Carotenoids also promote the biosynthesis of other essential compounds and play a role in the attraction of insects for pollination and seed dispersal [4,71,72].

Carotenoids have several important essential functions in human nutrition and health. This group of compounds can prevent and protect from cardiovascular diseases, inhibit carcinogenic cells, macular degeneration, and cataracts [73]. Carotenoids are considered the most effective antioxidant compounds found in peppers, besides phenolic and flavonoid compounds, which act synergistically as efficient free radical scavengers [74,75]. Carotenoids deactivate free radicals and quench reactive oxygen species due to the presence of conjugated double bonds [42,76]. In addition, plant carotenoids are endogenous isoprenoid precursors of vitamin A, β-carotene, α-carotene, γ-carotene, and β-cryptoxanthin which can be converted into retinol, the assimilable form of vitamin A in the human body [77].

Capsicum fruits are rich sources of carotenoids. The wide range of colors in peppers is related to the stage of maturation and the differential accumulation of carotenoids [78,79]. Specifically, oxygenated carotenoids are responsible for the yellow, orange, and red colors of pepper fruits [80].

3.2.1. Biosynthesis of Carotenoids

Carotenoids are derived from the universal five-carbon precursor isopentenyl pyrophosphate (IPP, C5) [7]. In *Capsicum*, the plastidial isoprenoid biosynthesis pathway starts with the mevalonic acid which is entered into several reactions to produce the C5 building block precursors—isopentenyl diphosphate and dimethylallyl pyrophosphate. In plants, carotenoids are synthesized in the plastid using IPP generated from the methylerythritol-4-phosphate (MEP) pathway (Figure 3) [4,81]. The MEP pathway receives substrates, G3P and pyruvate, from primary metabolism and delivers IPP to the prenyl lipid pathway. Phytoene, the first carotenoid in the pathway, is synthesized from eight IPP units in the prenyl lipid pathway [72]. The carotenoid biosynthesis pathway is split into the α and β branches. The addition of a hydroxyl group to the end rings characterizes the transition from carotene to xanthophyll. The end-products found in red *Capsicum* fruits are the red pigments capsorubin and capsanthin with κ end groups, the latter being the most abundant [7].

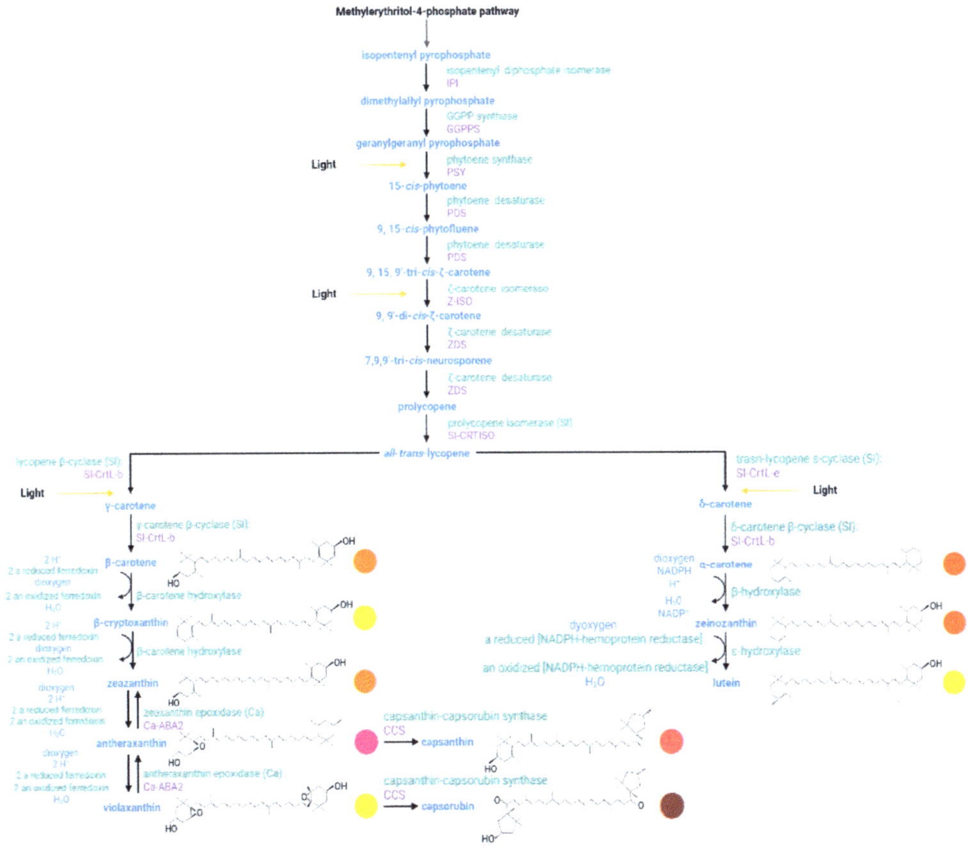

Figure 3. Carotenoid biosynthetic pathway in peppers (*Capsicum* spp.). Yellow arrows indicate the specific reaction steps at which light signal regulates the transcription factors at the molecular level. Chemical structures of the most abundant carotenoids present in *Capsicum* fruits. The circles indicate the color to which each carotenoid is associated in plant tissue.

In *Capsicum* fruits, carotenoid accumulation has been associated with the esterification of xanthophylls to allow for more efficient storage and increased stability, with the expression of a putative carotenoid acyl transferase, and an increased fibril content within the plastid [7].

3.2.2. Effects of Light on Carotenoids

Light signaling regulates the biosynthesis and accumulation of carotenoids through molecular mechanisms by which photoreceptors detect light signals in different plant organs [82]. Light regulates *Psy* (Figure 2) to modulate carotenoid biosynthesis during photomorphogenesis or de-etiolation, which is the process that occurs by the transition from the etioplast to the chloroplast [69]. Phytoene (15-cis-phytoene) has two sequential desaturations by *PDS* to produce 9,15-cis-phytofluene and 9,15,9′-cis-ζ-carotene, which can isomerize to ζ-carotene by light [71]. In peppers under protected cultivation, carotenoid content and *Psy* expression decreased compared to fruits grown under direct white light [83]. The expression of the *Psy* gene has also been reported in other plants including tomato exposed to blue LEDs [84]. The similarities between these two crops include the transition of tissues from chloroplast to chromoplasts during ripening and the high content of carotenoids in these chromoplast-containing fruits, resulting in the characteristic red color [84].

The biosynthesis and final contents of carotenoids are related to the fruit maturation process. Carotenoid accumulation is associated with a reduction in chlorophyll content. In immature fruits, chlorophylls are abundant and contribute to the characteristic green color. As the pepper fruits mature and the chloroplasts differentiate into chromoplasts, the chlorophyll contents of the epicarp lower significantly, and the biosynthesis of carotenoids occurs. During this process, carotenoids start to accumulate and contribute to fruit color [42,76]. The final carotenoid concentration is diverse, and the carotenoid profile is related to fruit color at harvest [74]. Color changes in response to more than thirty types of carotenoids [42]. In mature peppers, the most diverse carotenoid profile consisted of β-carotene, violaxanthin, antheraxanthin, zeaxanthin, and the intense red ketocarotenoids (capsanthin, capsorubin, and capsanthin-5,6-epoxide) [74].

In addition to the maturation process, other factors that affect carotenoid contents in peppers include genotype differences [85,86], environmental conditions during agricultural production [87,88], postharvest handling [9], processing [89], and storage (Table 2) [76].

Light is an important environmental factor involved in carotenoid biosynthesis. The quality and intensity of the light intercepted by the crop have a direct effect on the production and accumulation of carotenoids in peppers [15].

In sweet pepper cultivars, enhanced accumulation of carotenoids was obtained by a reduction in light intensity on the crop using shade nets. The five identified carotenoids were capsanthin, lutein, β-cryptoxanthin, β-carotene, and phytoene. Of these, capsanthin was the major carotenoid compound [8]. Similarly, reduced light stress in a shaded greenhouse also promoted carotenoid accumulation in three orange-fruited pepper cultivars. For these cultivars, the primary carotenoids present at the highest concentrations were lutein, zeaxanthin, and violaxanthin [90]. The increase in carotenoid contents caused by shaded conditions was also observable in postharvest studies. The use of black nets increased the carotenoid contents of β-carotene and lycopene in two different red and yellow sweet pepper cultivars [27].

The use of shading nets (black or colored) affects the accumulation of carotenoids in peppers. Plants cultivated in unshaded conditions (open field) produced peppers with the lowest levels of carotenoids in comparison to plants covered by black or colored shading nets [91]. Unshaded plants yielded fruits with less than 50% of the carotenoid contents in comparison to those grown under white nets. As for colored nets, peppers grown under yellow and red nets contained the lowest amounts of carotenoids (except for the unshaded control plants). However, 'Kapia'-type red sweet peppers grown under white shading nets resulted in significantly higher carotenoid contents in comparison to the green and yellow shades [92].

In postharvest studies of peppers, the exposure of green 'Takanotsume' peppers to different light wavelengths affected the carotenoid profile (including β-carotene, free-capsanthin, and total carotenoids). Peppers treated with red LEDs (660 nm) presented the highest increase in carotenoid contents, followed by those exposed to blue LEDs (470 nm). This response was associated with a reduction of chlorophyll in the fruits [9].

Accumulation of carotenoids can be induced by UV radiation (wavelengths from 100 to 400 nm). Of these, UV-A ranges from 315 to 400 nm, UV-B from 280 to 315 nm, and UV-C from 100 to 280 nm [93]. UV-C wavelengths do not reach the Earth's surface but can be applied in horticulture by artificial illumination to enhance the biosynthesis of metabolites. UV-C radiation has shown increased carotenoid levels when applied at low intensities. Nonetheless, high intensities can negatively affect photosynthesis and damage plant tissues [92].

The application of UV radiation to red sweet peppers during postharvest increased the levels of carotenoids after 14 days at 7 °C. Carotenoids increased exponentially by exposure to UV-C and UV-B in comparison to the non-UV treatment [94]. The UVR8 protein may be the principal UV-B receptor, and its action spectrum also includes the UV-C region. Thus, the application of low levels of single UV-C can also stimulate carotenoids and other phytochemicals. Exposure to red and blue (RB) LEDs light and RB with far-red wavelengths in red and yellow sweet pepper fruits increased the carotenoid content when compared to natural light exposure. The major carotenoids found in red fruits were capsanthin and capsorubin, whereas in yellow fruits, they were violaxanthin and lutenin [95]. In peppers that accumulate plastids after the breaker, the far-red wavelengths can act as a signal for the initiation of plastid accumulation [84]. Storage of habanero fruits in closed packages at low temperatures under blue and UV-C treatments affected carotenoid biosynthesis. During the first five days, the contents of chlorophylls and total carotenoids were reduced in comparison to the untreated peppers. This response could be attributed to the synthesis of photosynthetic pigments in chloroplasts to protect the photosystems [96].

Table 2. Effect of light-condition treatments on the carotenoid content in *Capsicum* species.

Capsicum spp.	Light Treatment	Effects on Carotenoids Compared to Control	Biosynthetic Effect
C. annuum Sweet pepper	Colored shading net: white with 40% shade and controlled-temperature plastic tunnel environment	Controlled temperature plastic tunnel enhanced the accumulation of carotenoid components	Capsanthin biosynthesis was not affected by treatments in most of the cultivars; peppers showed a homogeneous behavior in β-cryptoxanthin biosynthesis, which was not significantly affected in most cultivars in any of the treatments. Shading effect influences a change in the active form of phytochrome, facilitating the degradation of phytochrome interacting factor (PIF1a) and activating *PSY1* expression and carotenoid biosynthesis [8]
C. annuum Sweet pepper 'Cameleon'	Plastic tunnel plus colored shading nets: red, black, pearl, and blue shading nets with 40% shade, and open field as control	Black nets increased the carotenoid contents of β-carotene and lycopene	Not reported [11]

Table 2. *Cont.*

Capsicum spp.	Light Treatment	Effects on Carotenoids Compared to Control	Biosynthetic Effect
C. annuum Sweet pepper 'Karpex'	Colored shading nets: red, yellow, red, green, and white with 40% shade and unshaded as control	The unshaded control produced more than 50% less carotenoid than that under the white net. Peppers under the yellow and red nets produced the lowest content of carotenoids	Exposure to high temperature and radiation can lead to inhibition of carotenoid biosynthesis [91]
C. annuum Sweet pepper 'Kapia'	Colored shading nets: white, green, yellow, red, and unshaded as control	White shade net resulted in significantly higher carotenoid content compared to the green and the yellow nets	Not reported [92]
C. annuum 'Fogo' 'NuMex' 'Sunset' 'Orange Grande'	Shaded greenhouse with 40–50% shade, greenhouse conditions, and open field as control	Carotenoid concentrations decreased in fruits grown under increased light levels and increased in treatments with lower light intensity level	Not reported [90]
C. annuum Red and yellow sweet pepper	LED lighting treatments: natural light with red and blue LED, red and blue LED with far-red light, and natural light as control	In both colored fruits, carotenoid content was higher in LED treatments	Far-red light can act as a signal for starting plastid accumulation. Carotenoids changed by adding far-red light to the red and blue lighting [95]
C. annuum Red sweet pepper 'Angus'	UV lighting: UV-C, UV-B, UV-B+C, and no UV treatment as control	UV treatments induced carotenoid accumulation; after 14 days at 7 °C, UV-B and UV-C increased by 59% the total carotenoid content, and UVB + C by 94%	The active form of UVR8, a UV photoreceptor specific for UV-C and UV-B wavelengths, directly interacts with COP1 and regulates the expression of the *HY5* gene, which promotes the production of carotenoids [94]
C. chinense Habanero pepper	Irradiation treatments: blue lamps (0, 1.5, and 3 min), and UV-C light (0, 0.5, and 1 min) at 4–5 °C	Both lights stimulated bioactive compounds. Carotenoid content increased only in the first days of storage	Blue and UV-C light may stimulate the synthesis of chlorophylls and total carotenoids [96]
C. annuum Sweet peppers	LED lighting treatments in postharvest: yellow light at a wavelength of 590 nm and dark conditions as control	LED light slightly accelerated the ripening of fruits and increased the content of β-carotene, α-tocopherol, γ-tocopherol, chlorophyll, and lutein. Fruits showed higher antioxidant potential	Not reported [97]

3.3. Phenolic Compounds

Phenolic compounds constitute another essential group of secondary metabolites in *Capsicum* fruits. This group of compounds is usually reported as total phenolic compounds (TPC) and include phenols, phenolic acids, flavonoids, anthocyanins, lignans and lignins, stilbenes, and tannins. In peppers, the highest levels of TPC are found in the pericarp of fruits [95,96]. Peppers are rich in polyphenols, such as p-coumaric, ferulic, p-hydroxybenzoic, caffeic acid, sinapic acid, and quercetin-3-glucoside (Figure 4) [8].

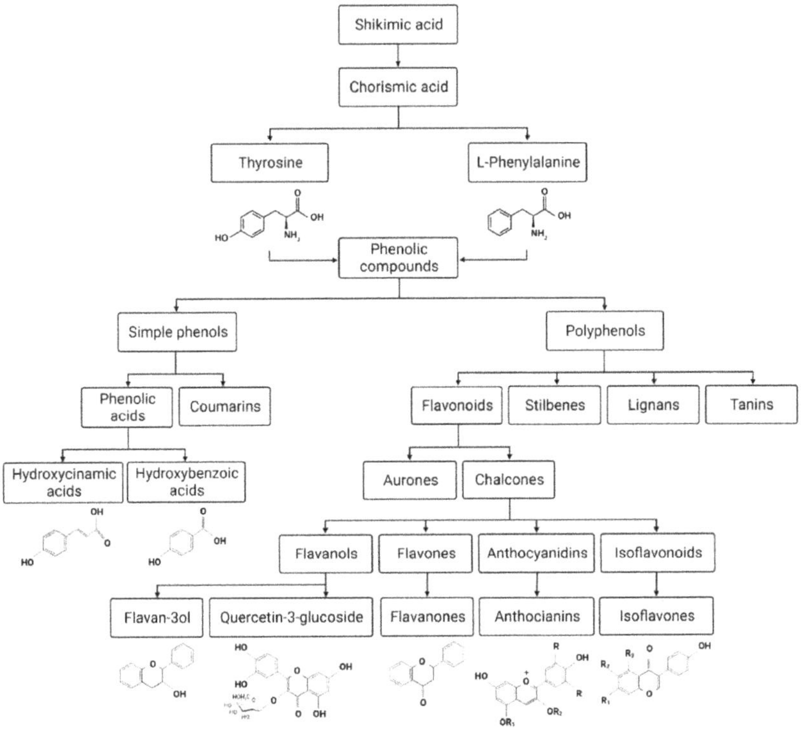

Figure 4. Phenolic compounds pathway in peppers (*Capsicum* spp.). Chemical structure of the most abundant phenolic compounds in *Capsicum* fruits (from their precursors).

Phenolic compounds result from the adaptation of plants to biotic and abiotic conditions that include infection, wounds, water, cold, and light intensity stress, among others [80,98,99]. Phenolics assist and interact as defense mechanisms with biotic and abiotic factors [52,100]. Phenolics quench the reactive oxygen species (ROS) produced during stress and protect the photosynthetic cells, and are related to the capacity of plants to absorb UV-B radiation [80,101].

Phenolic compounds are considered health-promoting metabolites [102]. Flavonoids are associated with the prevention of cancer, cardiovascular and autoimmune diseases, and are involved in the delay of the aging process [2]. These effects can be attributed to their direct role as free radical scavengers; modulators of detoxification enzymes, oxidation, and reduction processes; and strengtheners of the immune system, regulating gene expression, cell signaling, and hormone metabolism [103,104].

Phenolic compounds are phytochemicals with one aromatic ring attached to a hydroxyl group at a minimum. Phenolic compounds are divided into different classes by their chemical structure and the number of carbon atoms in their molecule [105]. The classification of phenolic compounds depends on the number of phenol units as simple phenols or polyphenols. Phenols contain one phenol unit, and polyphenols consist of two or more phenolic groups, up to polymeric structures [98]. Polyphenols rarely appear as free compounds and can be found in plants in the form of esters or glycosides with other natural compounds such as flavonoids, alcohols, and sterols [2,106].

3.3.1. Biosynthesis of Phenolic Compounds

Phenolic compounds are products of the secondary metabolism, in particular the shikimate pathway. Even though the precursors, phenylalanine or tyrosine, are the same, this pathway has different branches that lead to different compounds, which makes the biosynthetic pathway very complex [107,108]. Multiple genes are involved in the regulation of the different transcription factors involved in this pathway. Nonetheless, in *Capsicum*, only a few of the genes are known. The synthesis of flavonoids and other phenolic compounds can be regulated through a series of internal and external factors, including light [103]. The biosynthesis of phenolics is closely related to PAR irradiation and spectral quality; therefore, the manipulation of light conditions can cause changes in the content of metabolites and, consequently, alter photoprotection mechanisms [109].

The biosynthesis of flavonoids follows the phenylpropanoid pathway, which is impacted by environmental conditions. Nutrient deficiency, UV radiation, or an increase in stress levels caused by pathogens can influence the biosynthesis of flavonoids in many types of peppers [101].

In sweet pepper cultivars, the interaction between cultivar and growing conditions under protected cultivation affected the accumulation of phenolic compounds and antioxidant activity. Light intensity modified by white shade nets increased the accumulation of phenolic compounds and antioxidant activity in most of the studied cultivars. Similarly, the cultivation of peppers in plastic tunnels also favored the production of phenolics in other cultivars [8]. Similar results under white and red nets were reported, where higher R/FR ratios in spectral quality and reduced PAR increased the accumulation of phenols, quercetin, and other flavonoids in peppers [22].

3.3.2. Effects of Light on Phenolic Compounds

Light intensity and spectral quality during cultivation enhance the content of TPC in peppers [22,88] during cultivation, postharvest, and storage (Table 3) [96].

In postharvest studies, the spectral characteristics of light affect the accumulation and retention of bioactive compounds and physicochemical parameters in green peppers at harvest and during postharvest storage [22]. The antioxidant activity in peppers also increases during postharvest storage; this activity is associated with the metabolic pathways involved during the ripening and the production of lipophilic antioxidants [110]. Peppers produced under black or yellow nets showed a reduction of TPC. A further reduction was observed in fruits under black nets after postharvest storage. By contrast, peppers produced under pearl and red nets had a higher concentration of total phenols at harvest and remained high after postharvest storage. Total phenols, flavonoids, and even the antioxidant capacity in bell peppers were among the highest in unshaded conditions [89].

Exposure of pepper fruits during postharvest to red and blue LED also changed the TPC. Blue LED resulted in a significant increase in phenolic compounds in fruits when compared to the red LED and the control (fruits incubated in darkness). This effect was spectrum-specific as the red LED did not cause a significantly different response of the TPC [78]. Similar studies revealed an increase in total phenolic compounds in yellow and green sweet peppers exposed to red LED light and red peppers exposed to blue LED light during postharvest by increasing phenylalanine ammonia-lyase activity [24]. As described before, a wide variety of enzyme-catalyzed reactions are involved in the biosynthesis of phenols and flavonoids. However, only some of the genes involved in the *Capsicum* genus are known [98]. Therefore, detailed studies at the genomic and transcriptional levels are needed to elucidate the mechanism of light effects on phenolic compound production in peppers.

Table 3. Effect of light-condition treatments on the phenolic compounds content in *Capsicum* species.

Capsicum spp.	Light Treatment	Effects on Phenolic Compounds Compared to Control	Biosynthetic Effect
C. annuum Sweet peppers c.v. 'California Wonder'	Polytrench greenhouse, shaded greenhouse (Polytrench + red shade net), and open field as control	The total contents of phenols and flavonoids were reduced by 35.2 and 14.6%, respectively, in the greenhouse treatment.	Not reported [106]
C. annuum Green sweet peppers	Colored shading nets: pearl, red, and yellow with 40% shade, and black net with 25% shade as control	Fruits produced under the pearl nets showed higher ascorbic acid content, and antioxidant scavenging activity after postharvest storage	Red–far-red photon ratio under the pearl net could have improved the ascorbic acid content and the antioxidant scavenging activity in green peppers [22]
C. annuum Sweet peppers	Colored shading nets: black, red, silver, white with 30% to 46% shade, and unshaded as control	Total phenols and flavonoids were among the highest in the unshaded treatment and under the white net, and the lowest content under the black net	Not reported [87]
C. annuum Sweet peppers, eleven cultivars	Colored shading net: white with 40% shade and controlled temperature plastic tunnel	White shade nets increased the accumulation of phenolic compounds and antioxidant activity in most of the studied cultivars	Not reported [8]
C. annuum c.v. 'Takanotsume'	LED lighting treatments: red (660 nm) and blue (470 nm) light at an intensity of 50 μmol$\cdot$m$^{-2}\cdot$s^{-1}	The total phenolic, vitamin C content, and antioxidant capacity were higher in the blue LED-treated fruits	The blue LED was more effective in increasing the expression of the phytoene synthase (*Psy*) gene [78]
C. annuum Red sweet peppers	HPS and LED lighting in a glass greenhouse	LEDs at 622 nm enhanced phenolic compounds. HPS lighting supplemented with different LEDs was not efficient.	Not reported [111]
C. annuum Purple bell pepper	LED lighting treatments: white-red, and blue light	High blue-light fractions increased anthocyanin levels; white-red light is not efficient in the accumulation of anthocyanins	Increasing anthocyanin levels, via enhancing anthocyanin biosynthesis, was supported by kinetic modeling and higher expression levels of the anthocyanin biosynthetic genes *CaMYB*, *CaCHS*, *CaDFR*, *CaANS* and *CaUFGT* [85]
C. annuum Yellow, green, and red sweet peppers	LED lighting treatments: red, blue, and white light, and darkness as control	Red LED light for 8 h per day during storage at 7 °C was beneficial to retain bioactive compounds such as phenols and flavonoids	PAL activity in the yellow and green peppers exposed to red LED light increased and was correlated with the number of bioactive compounds [24]

Exposure of bell peppers to UV-C radiation in postharvest studies reduced the incidence and severity of the chilling injury and reduced the accumulation of phenolic compounds [112]. The response to UV-C radiation is highly dose-dependent as exposure to UV-C may significantly affect the enzymes involved in the biosynthesis of phytochemicals [113]. Moderate doses induce physiological responses, whereas high doses may reduce the enzymatic role, which causes a reduction in the production of bioactive phenolic compounds and other antioxidants [114].

4. Summary

Light is an elicitor of bioactive compounds in peppers and affects the biosynthesis and accumulation of phytochemicals. Current horticultural technologies that modify light intensity and spectrum aimed at improving pepper yields can also cause changes in the accumulation of bioactive compounds. The use of shade nets or plastic covers to reduce light intensity does not seem to yield consistent responses on the phytochemical profile, as the final profile results from the interaction of several factors. Exposure of plants to supplemental light with specific wavelengths seems to result in a more precise stimulation of specific metabolites. The molecular mechanisms underlying the specific effects of light on the phytochemical profile of peppers are still unclear. Further research is needed for a better understanding of the biochemical and molecular mechanisms of phytochemicals to reveal the complete effects of light on the phytochemical profile of peppers.

Author Contributions: Conceptualization, J.I.V.-B. and Y.J.-V.; methodology, J.I.V.-B. and Y.J.-V.; investigation, Y.J.-V.; resources, Y.J.-V.; writing—original draft preparation, Y.J.-V.; writing—review and editing, J.I.V.-B., G.F.-R. and H.G.N.-P.; supervision, J.I.V.-B.; project administration, J.I.V.-B.; funding acquisition, J.I.V.-B. and Y.J.-V. All authors have read and agreed to the published version of the manuscript.

Funding: This research was funded by the *Consejo Nacional de Ciencia y Tecnología* (CONACyT Spanish acronym) grant number 624964 And The APC was funded by *Tecnológico de Monterrey*.

Institutional Review Board Statement: Not applicable.

Acknowledgments: To CONACyT for providing the Ph. D. scholarship for Yamir Jiménez-Viveros (624964).

Conflicts of Interest: The authors declare no conflict of interest.

References

1. Wahyuni, Y.; Ballester, A.R.; Tikunov, Y.; de Vos, R.C.H.; Pelgrom, K.T.B.; Maharijaya, A.; Sudarmonowati, E.; Bino, R.J.; Bovy, A.G. Metabolomics and molecular marker analysis to explore pepper (*Capsicum* sp.) biodiversity. *Metabolomics* **2013**, *9*, 130–144. [CrossRef]
2. De Sá Mendes, N.; de Andrade Gonçalves, É.C.B. The role of bioactive components found in peppers. *Trends Food Sci. Technol.* **2020**, *99*, 229–243. [CrossRef]
3. Cisternas-Jamet, J.; Salvatierra-Martínez, R.; Vega-Gálvez, A.; Stoll, A.; Uribe, E.; Goñi, M.G. Biochemical composition as a function of fruit maturity stage of bell pepper (*Capsicum annum*) inoculated with *Bacillus amyloliquefaciens*. *Sci. Hort.* **2020**, *263*, 109107. [CrossRef]
4. Antonio, A.S.; Wiedemann, L.S.M.; Veiga Junior, V.F. The genus: *Capsicum*: A phytochemical review of bioactive secondary metabolites. *RSC Adv.* **2018**, *8*, 25767–25784. [CrossRef]
5. Wahyuni, Y.; Ballester, A.R.; Sudarmonowati, E.; Bino, R.J.; Bovy, A.G. Metabolite biodiversity in pepper (*Capsicum*) fruits of thirty-two diverse accessions: Variation in health-related compounds and implications for breeding. *Phytochemistry* **2011**, *72*, 1358–1370. [CrossRef]
6. Namdeo, A.G. Plant cell elicitation for production of secondary metabolites: A Review. *Pharmacogn. Rev.* **2007**, *1*, 60–79.
7. Berry, H.M.; Rickett, D.V.; Baxter, C.J.; Enfissi, E.M.A.; Fraser, P.D. Carotenoid biosynthesis and sequestration in red chilli pepper fruit and its impact on colour intensity traits. *J. Exp. Bot.* **2019**, *70*, 2637–2650. [CrossRef]
8. Lekala, C.S.; Madani, K.S.H.; Phan, A.D.T.; Maboko, M.M.; Fotouo, H.; Soundy, P.; Sultanbawa, Y.; Sivakumar, D. Cultivar-specific responses in red sweet peppers grown under shade nets and controlled-temperature plastic tunnel environment on antioxidant constituents at harvest. *Food Chem.* **2019**, *275*, 85–94. [CrossRef]
9. Pola, W.; Sugaya, S.; Photchanachai, S. Influence of postharvest temperatures on carotenoid biosynthesis and phytochemicals in mature green chili (*Capsicum annuum* L.). *Antioxidants* **2020**, *9*, 203. [CrossRef]
10. Ncise, W.; Daniels, C.W.; Nchu, F. Effects of light intensities and varying watering intervals on growth, tissue nutrient content and antifungal activity of hydroponic cultivated *Tulbaghia violacea* L. under greenhouse conditions. *Heliyon* **2020**, *6*, e03906. [CrossRef]
11. Ilić, Z.S.; Milenković, L.; Šunić, L.; Barać, S.; Mastilović, J.; Kevrešan, Ž.; Fallik, E. Effect of shading by coloured nets on yield and fruit quality of sweet pepper. *Zemdirbyste* **2017**, *104*, 53–62. [CrossRef]
12. Valiente-Banuet, J.I.; Gutiérrez-Ochoa, A. Effect of irrigation frequency and shade levels on vegetative growth, yield, and fruit quality of piquin pepper (*Capsicum annuum* L. var. *glabriusculum*). *HortScience* **2016**, *51*, 573–579. [CrossRef]
13. Selahle, K.M.; Sivakumar, D.; Jifon, J.; Soundy, P. Postharvest responses of red and yellow sweet peppers grown under photo-selective nets. *Food Chem.* **2015**, *173*, 951–956. [CrossRef] [PubMed]

14. Russo, V.M.; Howard, L.R. Carotenoids in pungent and non-pungent peppers at various developmental stages grown in the field and glasshouse. *J. Sci. Food. Agr.* **2002**, *82*, 615–624. [CrossRef]
15. Bian, Z.H.; Yang, Q.C.; Liu, W.K. Effects of light quality on the accumulation of phytochemicals in vegetables produced in controlled environments: A review. *J. Sci. Food Agric.* **2015**, *95*, 869–877. [CrossRef]
16. Shafiq, I.; Hussain, S.; Raza, M.A.; Iqbal, N.; Asghar, M.A.; Raza, A.; Fan, Y.-F.; Mumtaz, M.; Shoaib, M.; Ansar, M.; et al. Crop photosynthetic response to light quality and light intensity. *J. Integr. Agric.* **2021**, *19*, 2–21. [CrossRef]
17. Binotti, E.D.; Costa, E.; Binotti, F.F.D.S.; Batista, T.B. Chemical agents and shading levels for the production of pepper seedlings. *Eng. Agric.* **2018**, *38*, 450–456. [CrossRef]
18. Casierra-Posada, F.; Matallana-Díaz, Y.A.; Zapata-Casierra, E. Growth of bell pepper plants (*Capsicum annuum*) affected by coloured covers. *Gesunde Pflanzen.* **2014**, *66*, 149–155. [CrossRef]
19. Holopainen, J.K.; Kivimäenpää, M.; Julkunen-Tiitto, R. New light for phytochemicals. *Trends Biotechnol.* **2018**, *36*, 7–10. [CrossRef]
20. Jiao, Y.; Lau, O.S.; Deng, X.W. Light-regulated transcriptional networks in higher plants. *Nat. Rev. Genet.* **2007**, *8*, 217–230. [CrossRef]
21. Jovicich, E.; VanSickle, J.J.; Cantliffe, D.J.; Stoffella, P.J. Greenhouse-grown colored peppers: A profitable alternative for vegetable production in Florida? *Hort. Technol.* **2005**, *15*, 355–369. [CrossRef]
22. Mashabela, M.N.; Selahle, K.M.; Soundy, P.; Crosby, K.M.; Sivakumar, D. Bioactive compounds and fruit quality of green sweet pepper grown under different colored shade netting during postharvest storage. *J. Food Sci.* **2015**, *80*, H2612–H2618. [CrossRef]
23. Zermeño-González, A.; Claveria-Cigarrero, G.L.; Melendres-Alvarez, A.I.; Ramírez-Rodriguez, H.; Munguía-López, J.P.; Campos-Magaña, S.G.; Cadena-Zapata, M. Colored plastic covers and its relationship with radiation, growth and yield of a sweet yellow pepper (*Capsicum annuum* L.) crop. *Agrociencia* **2019**, *53*, 709–723.
24. Maroga, G.M.; Soundy, P.; Sivakumar, D. Different postharvest responses of fresh-cut sweet peppers related to quality and antioxidant and phenylalanine ammonia lyase activities during exposure to light-emitting diode treatments. *Foods* **2019**, *8*, 359. [CrossRef]
25. Hashim, M.; Ahmad, B.; Drouet, S.; Hano, C.; Abbasi, B.H.; Anjum, S. Comparative effects of different light sources on the production of key secondary metabolites in plants in vitro cultures. *Plants* **2021**, *10*, 1521. [CrossRef]
26. Dueck, T.; van Ieperen, W.; Taulavuori, K. Light perception, signalling and plant responses to spectral quality and photoperiod in natural and horticultural environments. *Environ. Exp. Bot.* **2016**, *121*, 1–150. [CrossRef]
27. Ilić, Z.S.; Fallik, E. Light quality manipulation improves vegetable quality at harvest and postharvest: A review. *Environ. Exp. Bot.* **2017**, *139*, 79–90. [CrossRef]
28. Ilić, Z.S.; Milenković, L.; Šunić, L.; Fallik, E. Effect of coloured shade-nets on plant leaf parameters and tomato fruit quality. *J. Sci. Food Agr.* **2015**, *95*, 2660–2667. [CrossRef]
29. Lester, G.E. Environmental regulation of human health nutrients (ascorbic acid, β-carotene, and folic acid) in fruits and vegetables. *HortScience* **2006**, *41*, 59–64. [CrossRef]
30. Shahak, Y. Photo-selective netting for improved performance of horticultural crops. A review of ornamental and vegetable studies carried out in Israel. *Acta Hortic.* **2008**, *770*, 161–168. [CrossRef]
31. Castellano, S.; Candura, A.; Scarascia Mugnozza, G. Relationship between solidity ratio, colour and shading effect of agricultural nets. *Acta Hortic.* **2008**, *801*, 253–258. [CrossRef]
32. Sivakumar, D.; Jifon, J.; Soundy, P. Spectral quality of photo-selective shade nettings improves antioxidants and overall quality in selected fresh produce after postharvest storage. *Food Rev. Int.* **2018**, *34*, 290–307. [CrossRef]
33. Castellano, S.; Mugnozza, G.S.; Russo, G.; Briassoulis, D.; Mistriotis, A.; Hemming, S.; Waajienberg, D. Plastic nets in agriculture: A general review of types and applications. *Appl. Eng. Agric.* **2008**, *24*, 799–808. [CrossRef]
34. Santana, J.Q.; Balbino, M.A.; Tavares, T.R.; Bezerra, R.S.; Farias, J.G.; Ferreira, R.C. Effect of photoselective screens in the development and productivity of red and yellow sweet pepper. *Acta Hortic.* **2012**, *956*, 493–500. [CrossRef]
35. Stamps, R.H. Use of colored shade netting in horticulture. *HortScience* **2009**, *44*, 239–241. [CrossRef]
36. Gangadhar, B.H.; Mishra, R.K.; Pandian, G.; Park, S.W. Comparative study of color, pungency, and biochemical composition in chili pepper (*Capsicum annuum*) under different light-emitting diode treatments. *HortScience* **2012**, *47*, 1729–1735. [CrossRef]
37. Al Murad, M.; Razi, K.; Jeong, B.R.; Samy, P.M.A.; Muneer, S. Light emitting diodes (LEDs) as agricultural lighting: Impact and its potential on improving physiology, flowering, and secondary metabolites of crops. *Sustainability* **2021**, *13*, 1985. [CrossRef]
38. Neugart, S.; Baldermann, S.; Hanschen, F.S.; Klopsch, R.; Wiesner-Reinhold, M.; Schreiner, M. The intrinsic quality of brassicaceous vegetables: How secondary plant metabolites are affected by genetic, environmental, and agronomic factors. *Sci. Hortic.* **2018**, *233*, 460–478. [CrossRef]
39. Jung, W.S.; Chung, I.M.; Hwang, M.H.; Kim, S.H.; Yu, C.Y.; Ghimire, B.K. Application of light-emitting diodes for improving the nutritional quality and bioactive compound levels of some crops and medicinal plants. *Molecules* **2021**, *26*, 1477. [CrossRef]
40. Vera-Guzmán, A.M.; Chávez-Servia, J.L.; Carrillo-Rodríguez, J.C.; López, M.G. Mexico evaluación fitoquímica en chile (*Capsicum annuum* L. and *C. pubescens* Ruiz & Pav.) silvestre y cultivado en Oaxaca, México. *Chil. J. Agr. Res.* **2011**, *71*, 578–585. [CrossRef]
41. Gurung, T.; Techawongstien, S.; Suriharn, B.; Techawongstien, S. Impact of environments on the accumulation of capsaicinoids in *Capsicum* spp. *HortScience* **2011**, *46*, 1576–1581. [CrossRef]

42. Batiha, G.E.S.; Alqahtani, A.; Ojo, O.A.; Shaheen, H.M.; Wasef, L.; Elzeiny, M.; Ismail, M.; Shalaby, M.; Murata, T.; Zaragoza-Bastida, A.; et al. Biological properties, bioactive constituents, and pharmacokinetics of some *Capsicum* spp. And capsaicinoids. *Int. J. Mol. Sci.* **2020**, *21*, 5179. [CrossRef] [PubMed]
43. Forero, M.D.; Quijano, C.E.; Pino, J.A. Volatile compounds of chile pepper (*Capsicum annuum* L. var. *Glabriusculum*) at two ripening stages. *Flavour. Frag. J.* **2009**, *24*, 25–30. [CrossRef]
44. Mares-Quiñones, M.D.; Valiente-Banuet, J.I. Horticultural aspects for the cultivated production of piquin peppers (*Capsicum annuum* L. var. *Glabriusculum*) a review. *HortScience* **2019**, *54*, 70–75. [CrossRef]
45. Reddy, K.K.; Ravinder, T.; Prasad, R.B.N.; Kanjilal, S. Evaluation of the antioxidant activity of capsiate analogues in polar, nonpolar, and micellar media. *J. Agric. Food Chem.* **2011**, *59*, 564–569. [CrossRef]
46. Sarwa, K.K.; Das, P.J.; Mazumder, B. A nanovesicle topical formulation of Bhut Jolokia (hottest *Capsicum*): A potential anti-arthritic medicine. *Expert Opin. Drug Del.* **2014**, *11*, 661–676. [CrossRef]
47. Almeida, M.; Nadal, J.; Klein, T.; De Paula, J.; Budel, J.; Novatski, A.; Campessato, E.A.; Farrago, P.V. Innovative phytoformulation containing capsaicinoids: Microparticles development, analytical method validation, and anti-ulcer effect. *Pharmacogn. Mag.* **2018**, *14*, 290–296. [CrossRef]
48. Kuzma, M.; Fodor, K.; Almási, A.; Mózsik, G.; Past, T.; Perjési, P. Toxicokinetic study of a gastroprotective dose of capsaicin by HPLC-FLD method. *Molecules* **2019**, *24*, 2848. [CrossRef]
49. Friedman, J.R.; Nolan, N.A.; Brown, K.C.; Miles, S.L.; Akers, A.T.; Colclough, K.W.; Seidler, J.M.; Rimoldi, J.M.; Valentovic, M.A.; Dasgupta, P. Anticancer activity of natural and synthetic capsaicin analogs. *J. Pharmacol. Exp. Ther.* **2018**, *364*, 462–473. [CrossRef]
50. Fattori, V.; Hohmann, M.S.N.; Rossaneis, A.C.; Pinho-Ribeiro, F.A.; Verri, W.A. Capsaicin: Current understanding of its mechanisms and therapy of pain and other pre-clinical and clinical uses. *Molecules* **2016**, *21*, 844. [CrossRef]
51. Cheok, C.Y.; Sobhi, B.; Adzahan, N.M.; Bakar, J.; Rahman, R.A.; Ab Karim, M.S.; Ghazali, Z. Physicochemical properties and volatile profile of chili shrimp paste as affected by irradiation and heat. *Food Chem.* **2017**, *216*, 10–18. [CrossRef] [PubMed]
52. Sricharoen, P.; Lamaiphan, N.; Patthawaro, P.; Limchoowong, N.; Techawongstien, S.; Chanthai, S. Phytochemicals in *Capsicum* oleoresin from different varieties of hot chilli peppers with their antidiabetic and antioxidant activities due to some phenolic compounds. *Ultrason. Sonochem.* **2017**, *38*, 629–639. [CrossRef]
53. Nuñez-Palenius, H.G.; Ochoa-Alejo, N. Effect of phenylalanine and phenylpropanoids on the accumulation of capsaicinoids and lignin in cell cultures of chili pepper (*Capsicum annuum* L.). *In Vitro Cell Dev. Biol. Plant.* **2005**, *41*, 801–805. [CrossRef]
54. Naves, E.R.; de Ávila Silva, L.; Sulpice, R.; Araújo, W.L.; Nunes-Nesi, A.; Peres, L.E.; Zsögön, A. Capsaicinoids: Pungency beyond *Capsicum*. *Trends Plant Sci.* **2019**, *24*, 109–120. [CrossRef] [PubMed]
55. Aza-González, C.; Núñez-Palenius, H.G.; Ochoa-Alejo, N. Molecular biology of capsaicinoid biosynthesis in chili pepper (*Capsicum* spp.). *Plant Cell Rep.* **2011**, *30*, 695–706. [CrossRef]
56. Stewart, C.; Mazourek, M.; Stellari, G.M.; O'Connell, M.; Jahn, M. Genetic control of pungency in *C. chinense* via the *Pun1* locus. *J. Exp. Bot.* **2007**, *58*, 979–991. [CrossRef] [PubMed]
57. Burgos-Valencia, E.; Echevarría-Machado, I.; Narváez-Zapata, J.A.; Martínez-Estévez, M. Gene expression related to the capsaicinoids biosynthesis in the *Capsicum* genus: Molecular and transcriptomic studies. *Braz. J. Bot.* **2020**, *43*, 201–212. [CrossRef]
58. Caicedo-Lopez, L.H.; Guevara-Gonzalez, R.G.; Ramirez-Jimenez, A.K.; Feregrino-Pérez, A.A.; Contreras-Medina, L.M. Eustress application trough-controlled elicitation strategies as an effective agrobiotechnology tool for capsaicinoids increase: A review. *Phytochem. Rev.* **2022**, *21*, 1941–1968. [CrossRef]
59. Díaz, J.; Pomar, F.; Bernal, Á.; Merino, F. Peroxidases and the metabolism of capsaicin in *Capsicum annuum* L. *Phytochem. Rev.* **2004**, *3*, 141–157. [CrossRef]
60. Arce-Rodríguez, M.L.; Ochoa-Alejo, N. An *R2R3-MYB* transcription factor regulates capsaicinoid biosynthesis. *Plant Physiol.* **2017**, *174*, 1359–1370. [CrossRef]
61. Kim, J.S.; Park, M.; Lee, D.J.; Kim, D. Characterization of putative capsaicin synthase promoter activity. *Mol. Cells* **2009**, *28*, 331–339. [CrossRef] [PubMed]
62. Díaz-Pérez, J.C. Bell pepper (*Capsicum annum* L.) crop as affected by shade level: Fruit yield, quality, and postharvest attributes, and incidence of phytophthora blight (caused by *Phytophthora capsici* Leon.). *HortScience* **2014**, *49*, 891–900. [CrossRef]
63. Jeeatid, N.; Techawongstien, S.; Suriharn, B.; Bosland, P.W.; Techawongstien, S. Light intensity affects capsaicinoid accumulation in hot pepper (*Capsicum chinense* Jacq.) cultivars. *Hortic. Environ. Biotech.* **2017**, *58*, 103–110. [CrossRef]
64. Jeeatid, N.; Suriharn, B.; Techawongstien, S.; Chanthai, S.; Bosland, P.W.; Techawongstien, S. Evaluation of the effect of genotype-by-environment interaction on capsaicinoid production in hot pepper hybrids (*Capsicum chinense* Jacq.) under controlled environment. *Sci. Hortic.* **2018**, *235*, 334–339. [CrossRef]
65. Nagy, Z.; Daood, H.; Neményi, A.; Ambrózy, Z.; Pék, Z.; Helyes, L. Impact of shading net color on phytochemical contents in two chili pepper hybrids cultivated under greenhouse conditions. *Korean J. Hortic. Sci.* **2017**, *35*, 418–430. [CrossRef]
66. Phimchan, P.; Chanthai, S.; Bosland, P.W.; Techawongstien, S. Enzymatic changes in phenylalanine ammonia-lyase, cinnamic-4-hydroxylase, capsaicin synthase, and peroxidase activities in *Capsicum* under drought stress. *J. Agric. Food Chem.* **2014**, *62*, 7057–7062. [CrossRef]
67. Murakami, K.; Ido, M.; Masuda, M. Fruit Pungency of "Shishito" pepper as affected by a dark interval in continuous fluorescent illumination with temperature alteration. *J. Soc. High Tech. Agric.* **2006**, *18*, 284–289. [CrossRef]

68. Yap, E.S.P.; Uthairatanakij, A.; Laohakunjit, N.; Jitareerat, P.; Vaswani, A.; Magana, A.A.; Morre, J.; Maier, C.S. Plant growth and metabolic changes in 'Super Hot' chili fruit (*Capsicum annuum*) exposed to supplemental LED lights. *Plant Sci.* **2021**, *305*, 110826. [CrossRef]
69. Quian-Ulloa, R.; Stange, C. Carotenoid biosynthesis and plastid development in plants: The role of light. *Int. J. Mol. Sci.* **2021**, *22*, 1184. [CrossRef]
70. Botella-Pavía, P.; Rodríguez-Concepción, M. Carotenoid biotechnology in plants for nutritionally improved foods. *Physiol. Plantarum.* **2006**, *126*, 369–381. [CrossRef]
71. Gómez-García, M.D.R.; Ochoa-Alejo, N. Biochemistry and molecular Biology of carotenoid biosynthesis in chili peppers (*Capsicum* spp.). *Int. J. Mol. Sci.* **2013**, *14*, 19025–19053. [CrossRef] [PubMed]
72. Nisar, N.; Li, L.; Lu, S.; Khin, N.C.; Pogson, B.J. Carotenoid metabolism in plants. *Mol. Plant.* **2015**, *8*, 68–82. [CrossRef] [PubMed]
73. Hernández-Pérez, T.; Gómez-García, M.D.R.; Valverde, M.E.; Paredes-López, O. *Capsicum annuum* (hot pepper): An ancient Latin-American crop with outstanding bioactive compounds and nutraceutical potential. A review. *Compr. Rev. Food Sci. F* **2020**, *19*, 2972–2993. [CrossRef] [PubMed]
74. Hassan, N.M.; Yusof, N.A.; Yahaya, A.F.; Rozali, N.N.M.; Othman, R. Carotenoids of *Capsicum* fruits: Pigment profile and health-promoting functional attributes. *Antioxidants* **2019**, *8*, 469. [CrossRef]
75. Mercy, E.R.; David, U. Potential health benefits of conventional nutrients and phytochemicals of *Capsicum* peppers. *Pharm. Pharmacol. Int. J.* **2018**, *6*, 62–69. [CrossRef]
76. Rodríguez-Rodríguez, E.; Sánchez-Prieto, M.; Olmedilla-Alonso, B. Assessment of carotenoid concentrations in red peppers (*Capsicum annuum*) under domestic refrigeration for three weeks as determined by HPLC-DAD. *Food Chem.* **2020**, *27*, 100092. [CrossRef]
77. Giuliano, G. Provitamin A biofortification of crop plants: A gold rush with many miners. *Curr. Opin. Biotech.* **2017**, *44*, 169–180. [CrossRef] [PubMed]
78. Pola, W.; Sugaya, S.; Photchanachai, S. Color development and phytochemical changes in mature green chili (*Capsicum annuum* L.) exposed to red and blue light-emitting diodes. *J. Agric. Food Chem.* **2020**, *68*, 59–66. [CrossRef]
79. Villa-Rivera, M.G.; Ochoa-Alejo, N. Chili pepper carotenoids: Nutraceutical properties and mechanisms of action. *Molecules* **2020**, *25*, 5573. [CrossRef]
80. Ghasemnezhad, M.; Sherafati, M.; Payvast, G.A. Variation in phenolic compounds, ascorbic acid and antioxidant activity of five coloured bell pepper (*Capsicum annum*) fruits at two different harvest times. *J. Funct. Foods* **2011**, *3*, 44–49. [CrossRef]
81. Fraser, P.D.; Schuch, W.; Bramley, P.M. Phytoene synthase from tomato (*Lycopersicon esculentum*) chloroplasts-partial purification and biochemical properties. *Planta* **2020**, *211*, 361–369. [CrossRef] [PubMed]
82. Yuan, H.; Zhang, J.; Nageswaran, D.; Li, L. Carotenoid metabolism and regulation in horticultural crops. *Hortic. Res.* **2015**, *2*, 15036. [CrossRef] [PubMed]
83. Yoo, H.J.; Kim, J.H.; Park, K.S.; Son, J.E.; Lee, J.M. Light-controlled fruit pigmentation and flavor volatiles in tomato and bellpepper. *Antioxidants* **2020**, *9*, 14. [CrossRef]
84. Llorente, B.; D'Andrea, L.; Ruiz-Sola, M.A.; Botterweg, E.; Pulido, P.; Andilla, J.; Loza-Alvarez, P.; Rodriguez-Concepcion, M. Tomato fruit carotenoid biosynthesis is adjusted to actual ripening progression by a light-dependent mechanism. *Plant J.* **2016**, *85*, 107–119. [CrossRef] [PubMed]
85. Liu, Z.; Lv, J.; Zhang, Z.; Li, H.; Yang, B.; Chen, W.; Dai, X.; Li, X.; Yang, S.; Liu, L.; et al. Integrative transcriptome and proteome analysis identifies major metabolic pathways involved in pepper fruit development. *J. Proteome Res.* **2019**, *18*, 982–994. [CrossRef] [PubMed]
86. Zhang, Z.; Cao, Y.; Yu, H.; Wang, L.; Zhang, B. Genetic control and metabolite composition of fruit quality in *Capsicum*. *Acta Hortic. Sin.* **2019**, *46*, 1825–1841. [CrossRef]
87. Díaz-Pérez, J.C.; St. John, K.; Kabir, M.Y.; Alvarado-Chávez, J.A.; Cutiño-Jiménez, A.M.; Bautista, J.; Gunawan, G.; Nambeesan, S.U. Bell Pepper (*Capsicum annum* L.) under Colored Shade Nets: Fruit Yield, Postharvest Transpiration, Color, and Chemical Composition. *HortScience* **2020**, *55*, 181–187. [CrossRef]
88. Wang, Y.; Gao, S.; He, X.; Li, Y.; Zhang, Y.; Chen, W. Response of total phenols, flavonoids, minerals, and amino acids of four edible fern species to four shading treatments. *PeerJ.* **2020**, *21*, e8354. [CrossRef]
89. Howard, L.R.; Wildman, R.E.C. Antioxidant vitamin and phytochemical content of fresh and processed pepper fruit (*Capsicum annuum*). In *Handbook of Nutraceuticals and Functional Foods*, 2nd ed.; CRS Press, Ed.; Taylor and Francis: New York, NY, USA, 2016; pp. 165–191.
90. Ambrózy, Z.; Daood, H.; Nagy, Z.; Darázsi Ledó, H.; Helyes, L. Effect of net shading technology and harvest times on yield and fruit quality of sweet pepper. *Appl. Ecol. Environ. Res.* **2016**, *14*, 99–109. [CrossRef]
91. Ombódi, A.O.; Pék, Z.; Szuvandzsiev, P.; Lugasi, A.; Ledóné Darázsi, H.; Helyes, L. Effect of coloured shade nets on some nutritional characteristics of a kapia type pepper grown in plastic tunnel. *Columella J. Agric. Environ. Sci.* **2016**, *25*, 33. [CrossRef]
92. Castillejo, N.; Martínez-Zamora, L.; Artés-Hernández, F. Postharvest UV radiation enhanced biosynthesis of flavonoids and carotenes in bell peppers. *Postharvest Biol. Technol.* **2022**, *184*, 111774. [CrossRef]
93. Kim, D.; Son, J.E. Adding far-red to red, blue supplemental light-emitting diode interlighting improved sweet pepper yield but attenuated carotenoid content. *Front. Plant Sci.* **2022**, *13*, 938199. [CrossRef] [PubMed]
94. *ISO 21348*; Definitions of Solar Irradiance Spectral Categories. ISO: Geneva, Switzerland, 2007.

95. Pérez-Ambrocio, A.; Guerrero-Beltrán, J.A.; Aparicio-Fernández, X.; Ávila-Sosa, R.; Hernández-Carranza, P.; Cid-Pérez, S.; Ochoa-Velasco, C. Effect of blue and ultraviolet-C light irradiation on bioactive compounds and antioxidant capacity of habanero pepper (*Capsicum chinense*) during refrigeration storage. *Postharvest Biol. Technol.* **2018**, *135*, 19–26. [CrossRef]
96. Dai, J.; Mumper, R.J. Plant phenolics: Extraction, analysis and their antioxidant and anticancer properties. *Molecules* **2010**, *15*, 7313–7352. [CrossRef] [PubMed]
97. Kokalj, D.; Hribar, J.; Cigić, B.; Zlatić, E.; Demšar, L.; Sinkovič, L.; Šircelj, H.; Bizjak, G.; Vidrih, R. Influence of yellow light-emitting diodes at 590 nm on storage of apple, tomato and bell pepper fruit. *Food Technol. Biotech.* **2016**, *54*, 228. [CrossRef]
98. Materska, M.; Perucka, I. Antioxidant activity of the main phenolic compounds isolated from hot pepper fruit (*Capsicum annuum* L.). *J. Agric. Food Chem.* **2005**, *53*, 1750–1756. [CrossRef]
99. Murthy, H.N.; Lee, E.J.; Paek, K.Y. Production of secondary metabolites from cell and organ cultures: Strategies and approaches for biomass improvement and metabolite accumulation. *Plant Cell Tiss. Org.* **2014**, *118*, 1–16. [CrossRef]
100. Meckelmann, S.W.; Riegel, D.W.; van Zonneveld, M.; Ríos, L.; Peña, K.; Mueller-Seitz, E.; Petz, M. Capsaicinoids, flavonoids, tocopherols, antioxidant capacity and color attributes in 23 native Peruvian chili peppers (*Capsicum* spp.) grown in three different locations. *Eur. Food Res. Technol.* **2014**, *240*, 273–283. [CrossRef]
101. Moreno-Ramírez, Y.D.R.; Martínez-Ávila, G.C.G.; González-Hernández, V.A.; Castro-López, C.; Torres-Castillo, J.A. Free radical-scavenging capacities, phenolics and capsaicinoids in wild piquin chili (*Capsicum annuum* var. *Glabriusculum*). *Molecules* **2018**, *23*, 2655. [CrossRef]
102. Lucci, P.; Saurina, J.; Núñez, O. Trends in LC-MS and LC-HRMS analysis and characterization of polyphenols in food. *TrA Trend Anal. Chem.* **2017**, *88*, 1–24. [CrossRef]
103. Lemos, V.C.; Reimer, J.J.; Wormit, A. Color for life: Biosynthesis and distribution of phenolic compounds in pepper (*Capsicum annuum*). *Agriculture* **2019**, *9*, 81. [CrossRef]
104. Bhattacharya, A.; Sood, P.; Citovsky, V. The roles of plant phenolics in defense and communication during Agrobacterium and Rhizobium infection. *Mol. Plant Pathol.* **2010**, *11*, 705–719. [CrossRef] [PubMed]
105. Barros, J.; Serrani-Yarce, J.C.; Chen, F.; Baxter, D.; Venables, B.J.; Dixon, R.A. Role of bifunctional ammonia-lyase in grass cell wall biosynthesis. *Nat. Plants* **2016**, *2*, 16050. [CrossRef] [PubMed]
106. Lahbib, K.; Dabbou, S.; Bok, S.E.L.; Pandino, G.; Lombardo, S.; Gazzah, M.E.L. Variation of biochemical and antioxidant activity with respect to the part of *Capsicum annuum* fruit from Tunisian autochthonous cultivars. *Ind. Crop Prod.* **2017**, *104*, 164–170. [CrossRef]
107. Angmo, P.; Dolma, T.; Phuntsog, N.; Chaurasia, O.P.; Stobdan, T. Effect of shading and high temperature amplitude on yield and phenolic contents of greenhouse *capsicum* (*Capsicum annuum* L.). *J. Biol. Pharm.* **2021**, *4*, 30–39. [CrossRef]
108. Andersen, O.M.; Markham, K.R. *Flavonoids. Chemistry, Biochemistry and Applications*; CRC Press: Boca Raton, FL, USA, 2006.
109. Pech, R.; Volná, A.; Hunt, L.; Bartas, M.; Červeň, J.; Pečinka, P.; Špunda, V.; Nezval, J. Regulation of Phenolic Compound Production by Light Varying in Spectral Quality and Total Irradiance. *Int. J. Mol. Sci.* **2022**, *23*, 6533. [CrossRef]
110. Kavga, A.; Strati, I.F.; Sinanoglou, V.J.; Fotakis, C.; Sotiroudis, G.; Christodoulou, P.; Zoumpoulakis, P. Evaluating the experimental cultivation of peppers in low-energy-demand greenhouses. An interdisciplinary study. *J. Sci. Food Agric.* **2019**, *99*, 781–789. [CrossRef]
111. Bae, J.H.; Park, Y.J.; Namiesnik, J.; Gülçin, I.; Kim, T.C.; Kim, H.-C.; Heo, B.-G.; Gorinstein, S.; Ku, Y.-G. Effects of artificial lighting on bioactivity of sweet red pepper (*Capsicum annuum* L.). *Int. J. Food Sci.* **2016**, *51*, 1378–1385. [CrossRef]
112. Vicente, A.R.; Pineda, C.; Lemoine, L.; Civello, P.M.; Martinez, G.A.; Chaves, A.R. UV-C treatments reduce decay, retain quality and alleviate chilling injury in pepper. *Postharvest Biol. Technol.* **2005**, *35*, 69–78. [CrossRef]
113. Ma, L.; Wang, Q.; Li, L.; Grierson, D.; Yuan, S.; Zheng, S.; Wang, Y.; Wang, B.; Bai, C.; Fu, A.; et al. UV-C irradiation delays the physiological changes of bell pepper fruit during storage. *Postharvest Biol. Technol.* **2021**, *180*, 111506. [CrossRef]
114. Lemessa, A.; Popardowski, E.; Hebda, T.; Jakubowski, T. The Effect of UV-C Irradiation on the mechanical and physiological properties of potato tuber and different products. *Appl. Sci.* **2022**, *12*, 5907. [CrossRef]

horticulturae

MDPI

Article

Supplemental LED Lighting Improves Fruit Growth and Yield of Tomato Grown under the Sub-Optimal Lighting Condition of a Building Integrated Rooftop Greenhouse (i-RTG)

Elisa Appolloni [1], Ivan Paucek [1], Giuseppina Pennisi [1], Gaia Stringari [2], Xavier Gabarrell Durany [2], Francesco Orsini [1,*] and Giorgio Gianquinto [1]

1 Alma Mater Studiorum, DISTAL—Department of Agricultural and Food Sciences, 48, Università di Bologna, 40127 Bologna, Italy

2 Institut de Ciència i Tecnologia Ambientals (ICTA-UAB), MdM Unit of Excellence (CEX2019-000940-M), Chemical, Biological and Environmental Engineering Department, Universitat Autònoma de Barcelona, 08193 Barcelona, Spain

* Correspondence: f.orsini@unibo.it

Abstract: The metabolism of a building can be connected to a rooftop greenhouse, exchanging energy, water and CO_2 flows, therefore reducing emissions and recycling cultivation inputs. However, integrating a rooftop greenhouse onto a building requires the application of stringent safety codes (e.g., fire, seismic codes), to strengthen and secure the structure with safety elements such as thick steel pillars or fireproof covering materials. These elements can shade the vegetation or reduce solar radiation entering the rooftop greenhouse. Nevertheless, application of additional LED light can help to overcome this constraint. The present study evaluated supplemental LED light application in an integrated rooftop greenhouse (i-RTG) at the ICTA-UAB research institute, located in Barcelona (Spain), for tomato cultivation (*Solanum lycopersicum* cv. Siranzo). The experiment explored the effects of three LED lighting treatments and a control cultivated under natural light only (CK). Applied treatments, added to natural sunlight, were: red and blue (RB), red and blue + far-red (FR) for the whole day, and red and blue + far-red at the end-of-day (EOD), each for 16 h d^{-1} (8 a.m.–12 a.m.) with an intensity of 170 $\mu mol\ m^{-2}\ s^{-1}$. The results indicate that LED light increased the overall yield by 17% compared with CK plants. In particular, CK tomatoes were 9.3% lighter and 7.2% fewer as compared with tomatoes grown under LED treatments. Fruit ripening was also affected, with an increase of 35% red proximal fruit in LED-treated plants. In conclusion, LED light seems to positively affect the development and growth of tomatoes in building integrated agriculture in the Mediterranean area.

Keywords: light emitting diode; rooftop greenhouse; building-integrated agriculture; *Solanum lycopersicum*; chilling injury

1. Introduction

Recent historical events, such as the COVID-19 pandemic and new geo-political arrangements, have demonstrated the need for more resilient food systems capable of ensuring food self-sufficiency, especially in response to sudden changes [1]. This need for resilience of the food supply chain particularly relates to the urban context, where half of the world's population is currently living [2]. Urban agriculture (UA) (e.g., home gardening, vertical farming, rooftop agriculture) has been identified as a viable solution to promote local production, ensure food security, reduce food waste and create more sustainable food systems [3].

In its most advanced form, UA is applied in or on urban structures as a building-integrated agriculture (BIA) system [4]. Rooftop farming is an example of BIA, applicable both in unprotected (open-air farms) and protected conditions (rooftop greenhouses) [5].

The latter case represents a sustainable method of building and cultivation integration, linking the metabolisms of the two systems and symbiotic exchange of CO_2, heat, and water, favoring the optimization of cultivation inputs, energy saving, and emission reduction [6]. In particular, an integrated rooftop greenhouse (i-RTG) was shown to be able to recycle about 342 kWh m^{-2} yr^{-1} from the main building, which, compared with a traditional fossil-fuel-heated greenhouse, can result in a CO_2 retention of about 114 kg CO_2 (eq) m^{-2} yr^{-1}, corresponding to about 20 EUR m^{-2} yr^{-1} in economic savings [7]. However, despite the potential environmental and economic benefits, production in an urban i-RTG may present some limitations due to the shading of surrounding buildings, as well as the bulky structural items, and the loss of transmissivity of fireproof covering materials (e.g., polycarbonate). In fact, given the integration in a city context, the structure must comply with the municipality's structural and fire safety codes, with consequent constraints affecting the greenhouse light environment [8,9].

Supplementary LED light is receiving wide application in various greenhouse production contexts characterized by reduced solar radiation, such as the cultivation of high-density crops, at high latitudes or during darker seasons [10–12]. In the Mediterranean region, supplementary lighting is still limited in use and applied mainly between October and March [13]. However, a more extensive application might be of interest also for the Southern European sector, especially considering the low transmissivity of the chosen plastic roofing materials (60% of PAR reduction) or the need to whitewash greenhouses to protect plants from excessive heat during the summer period [13]. These factors, together with increasing competition from Nordic countries, make it necessary for a technological upgrade in the Mediterranean greenhouse sector [14]. Although research has already begun to investigate the effects of supplementary LED light, even in Southern Europe [15–17], nothing has yet been demonstrated regarding its application in a low-solar-irradiance BIA context, whether in warm or cold climates. The present research aims to identify the potential benefits and limitations of supplementary LED lighting in the context of agriculture applied to buildings, with a specific focus on the Mediterranean.

2. Materials and Methods

2.1. Plants Growing Conditions

The experiment was performed in the i-RTG at the Institute of Environmental Science and Technology (ICTA-UAB) of the Universitat Autònoma de Barcelona (Catalunya, Spain) (41°49′78″ N, 2°10′89″ E) (Figure 1a). The i-RTG structure consisted of reinforced steel pillars and polycarbonate cladding to satisfy the Spanish Technical Code of Edification and fire safety laws [9].

Tomato plants (*Solanum lycopersicum* L. cv. 'Siranzo', Rijk Zwaan®) were cultivated in the south-east oriented corner of the building using a high-wire hydroponic system. Perlite bags (40 L) were used to support plants. Plant distance within rows was 30 cm, while the distance between rows was 80 cm, for a planting density of 3.1 plants m^{-2}. In total, 9 rows (5 m in length each) of plants were cultivated, out of which 5 rows were used as buffers to avoid light treatment pollution among blocks (Figure 1b). The plants were sown in peat in late January and transplanted into perlite bags in mid-March. The experiment was terminated at the end of July.

Plants were fertigated with a closed-loop drip irrigation system using the rainwater collected in an underground tank and transported to the top floor by a pump. Irrigation shifts and fertilizer amounts were adjusted during the experiment according to the phenological stage and climatic conditions, maintaining an average pH of 7 and an EC of 1.7 dS m^{-1}. Nutrient solution (Table 1), drainage, and rainwater were checked daily to maintain stable nutrition, pH and EC. Temperature (25 ± 4 °C), relative humidity (61 ± 14%), and outdoor (294 ± 344 W m^{-2}) and indoor (110 ± 136 W m^{-2}) radiation were constantly monitored using computer-monitored sensors. Passive ventilation was automatically adjusted according to environmental conditions by opening top and lateral windows. The residual heat coming from the lower floors of the building was used to warm-up the greenhouse during cool days.

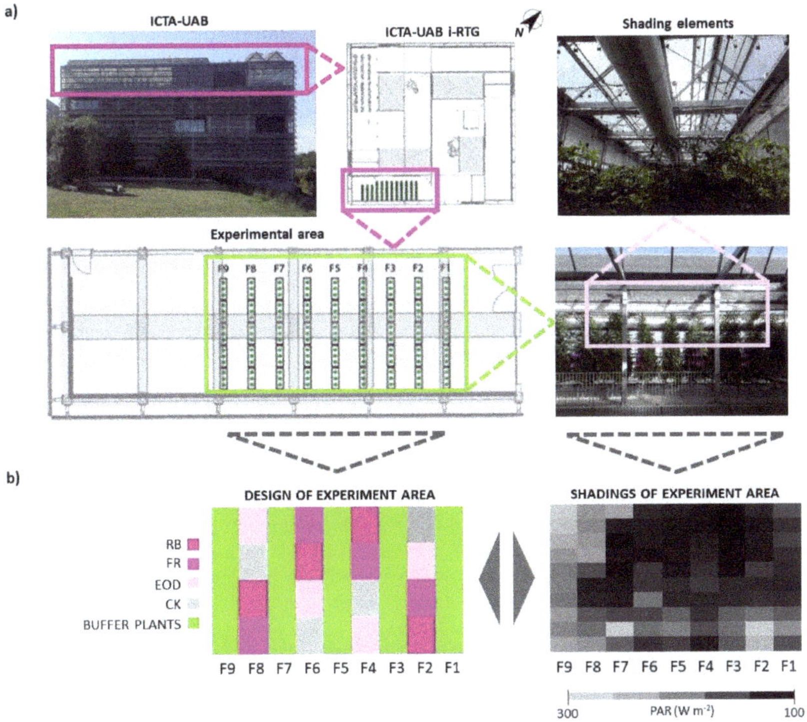

Figure 1. Location, experimental area, and shading elements of the i-RTG at ICTA-UAB (**a**). Experimental design and heatmap of internal shadings of experiment area (**b**).

Table 1. Composition of nutrient solutions during spring and summer period.

	Spring	*Summer*
Composition	**kg L^{-1}**	**kg L^{-1}**
KH_2PO_4	0.017	0.014
KNO_3	0.015	0.005
K_2SO_4	0.035	0.035
$Ca(NO_3)_2$	0.049	0.049
$CaCl_2$	0.014	0.014
$Mg(NO_3)_2$	0.011	0.015
Hortrilon	0.001	0.001
Sequestrene	0.001	0.001

2.2. Light Treatment and Experimental Design

Each light treatment was provided by a couple of LED inter-lighting lamps (Philips GreenPower LED, Philips®, Amsterdam, The Netherlands) located at 50 cm and 80 cm of height, and at 30 cm of distance from the stem. A control using natural light only (CK), plus three different supplemental LED lighting regimes, were evaluated. Lighting treatments consisted of:

(1) Red (660 nm) and blue (465 nm) light with R:B ratio of 3, a total photosynthetic photon flux density (PPFD) of 170 µmol m^{-2} s^{-1} (85 µmol m^{-2} s^{-1} per each lamp, measured at 30 cm from the plant) and a photoperiod of 16 h d^{-1} (8 a.m.–12 a.m.) (namely RB);

(2) RB treatment with an addition of 40 μmol m^{-2} s^{-1} of far-red light (730 nm) during the whole photoperiod (namely FR);
(3) RB treatment with an addition of 40 μmol m^{-2} s^{-1} of far-red, added only for 30 min right after the end of photoperiod (end-of-day) (namely EOD).

Lighting treatments started in mid-April until the end of the trial. A Latin square design was used to reduce systematic error determined by the shading of air conducts and structural elements on plants (Figure 1b). The experiment was divided into 4 replicates (n = 4) containing 3 plants per treatment (12 plants per treatment). Lines of buffer plants were used to screen the radiation coming from parallel rows, and 1 or 2 buffer plants were used between two adjacent treatments to avoid light interactions on the row (Figure 1b).

2.3. Plant Vegetative, Physiological and Biochemical Measurements

Stem and collar diameter were measured every two weeks from the beginning of lighting treatment until the end of May, at 1 cm under the fruit truss and 1 cm from the perlite bag, respectively. Internodes length was measured weekly as the distance between two consecutive fruit trusses. Apical growth and the number of clusters were measured every week until plant topping occurred in the last week of June. Final fresh weight of the entire plant biomass (e.g., leaves and stems) was measured with a digital scale at the end of the experiment.

Leaf area was evaluated on the last week of May using Easy Leaf Area software (Department of Plant Sciences, University of California, Davis, CA, USA) [18] on the first leaf above the third fruit truss of each plant. The same leaves were measured as fresh and after drying at 60 °C per 4 days. Weight measurements were used to evaluate leaf dry matter content (LDMC), as the ratio between leaf dry mass and leaf fresh mass (mg g^{-1}), and specific leaf area (SLA), as the ratio between leaf area and leaf dry mass (m^2 kg^{-1}) [19].

Chlorophyll content of leaves was evaluated in the first week of June, considering three points of the first leaf right under the third fruit truss of each plant. A Chlorophyll Content Meter CCM-200 (Opti-Sciences®, Hudson, NY, USA) was used to non-destructively estimate the content based on the ratio of light transmittance at 653 nm and at 931 nm [20].

Analysis of micro- and macro-elements (B, Mn, Fe, Cu, Zn, Na, Mg, K, P, S, and Ca) of plant biomass was carried out through a digestion process with HNO_3 and analyzed using ICP-OES optical spectrometry (Optima 4300DV, Perkin-Elmer, Waltham, MA, USA) as reported in Arcas-Pilz et al. [21]. The analysis was conducted on leaf samples collected weekly, and plant stems collected at the end of the experiment. Leaves and stems were analyzed separately.

2.4. Fruit Development and Yield

Fruit development was monitored on the proximal fruit of the first produced cluster, representative of spring development (Follow-up 1), and the fourth cluster, representative of summer development (Follow-up 2), by measuring the equatorial and polar diameter. Evaluations were taken two times per week during the first three weeks and once a week during the last three weeks, starting from 12 days after anthesis until stabilization of fruit growth (around the turning phase). A digital Vernier caliper was used to measure the polar and equatorial dimensions (cm), estimating the volume of the fruit as the volume of an ellipsoid of rotation $V = (4/3)\ \pi ab^2$, where a is one half of the polar radius and b is one half of the equatorial radius [22]. Ripening was evaluated on the same fruit two weeks before harvesting using a DA-Meter (SINTELEIA®, Bologna, Italy), which non-destructively evaluated the chlorophyll degradation and correlated it with a ripening index.

From the beginning of June until the end of the trial, fruits were harvested once per week (in total 6 clusters per plant). Fresh weight of the total clusters of each treatment was measured with a digital scale. The number of fruits per cluster was counted for each plant. At harvesting, fruits were divided, counted, and weighed as consumable and non-consumable, in which consumable was considered a mature red or dark orange tomato, and not consumable was considered a fruit backward in growth and ripening, mostly green and with small dimensions (<3 cm). This definition was not established on the basis of commercial parameters but rather on the assessment of the edibility of tomatoes to be used for local and direct consumption by the office workers. The average fresh weight of an individual fruit per treatment was estimated by dividing the total fresh weight by the total number of fruits. The total fresh mass of fruit produced by the plant was estimated by multiplying the number of fruits of each plant by the average fresh weight of an individual fruit for each treatment.

2.5. Fruit Quality and Biochemical Analysis

2.5.1. Fruit Qualitative Evaluation

Fruit qualitative evaluations were performed two times on mature tomatoes (spring development, Follow-up 1, and summer development, Follow-up 2), and one time on immature tomatoes at the turning stage (summer development, Follow-up 2). Mature tomatoes were selected considering a DA-Meter range index between 1.30 and 1.50, whereas immature tomatoes were selected considering a range index between 0.20 and 0.40.

Fruit hardness was evaluated using a fruit hardness tester (Turoni®, Forlì, Italy) on four opposite sides of the equatorial diameter of the fruit. The instrument non-destructively measured the elasticity of fruit exocarp, expressing it as an index ranging from 0 to 100. Soluble solid content and acidity were measured on fruit juice using a pocket Brix and acidity meter (PAL-BX | ACID3, Atago®, Tokyo, Japan). Fruit dry matter content (FDMC) was measured as the ratio between fruit fresh weight and dried fruit.

2.5.2. Lycopene and β-Carotene Content

Lycopene and β-carotene content were evaluated on tomato samples frozen at −20 °C, using the methodology described by Anthon and Barrett [23], with slight modifications. Briefly, an extraction solution was prepared by mixing hexane, acetone, and ethanol in a *v:v* proportion of 2:1:1 and 0.5 g L^{-1} of butylated hydroxytoluene (BHT). Thereafter, 0.5 g of frozen sample including exocarp and mesocarp, trying to avoid the green parts (e.g., petiole), were pestled and mixed with 10 mL of extraction solution. The material was left in darkness for 30 min and then centrifuged at 5000 rpm for 5 min. Finally, 1 mL of supernatant was read at 503 nm for lycopene and 444 nm for β-carotene with a spectrophotometer (8453 UV-Visible Spectrophotometer, HP®, Palo Alto, CA, USA).

The lycopene content was calculated using the following formula:

$$\text{lycopene (mg/kg)} = (X/Y) \times A503 \times 3.12, \quad (1)$$

where X is the amount of hexane (mL), Y the weight of the fruit tissue (g), A503 is the absorbance at 503 nm, and 3.12 is the extinction coefficient. β-Carotene was calculated with the following equation:

$$\beta\text{-carotene} = (9.38 \times A444 - 6.70 \times A503) \times 0.55 \times 537 \times V/W, \quad (2)$$

where A444 is the absorbance at 444 nm, A503 is the absorbance at 503 nm, 0.55 is the ratio of the final hexane layer volume to the volume of mixed solvents added for hexane:acetone:ethanol (2:1:1), V is the volume of mixed solvents added, W is the fresh weight of the sample, and 537 (g $mole^{-1}$) is the molecular weight of β-carotene.

2.5.3. Total Polyphenols and Total Antioxidant Capacity

Total polyphenols and total antioxidant capacity were evaluated on tomato samples frozen at −20 °C. The extraction was performed by placing 4 g of samples in tubes and adding 8 mL of extraction mixture (60% methanol, 30% H_2O, 10% acetone). The process was carried out by centrifugation at 5000 rpm for 10 min at 4 °C. Supernatant was collected and used for antioxidant and total phenol analysis.

Total antioxidant capacity was analyzed using the FRAP (Ferric Reducing Antioxidant Power) method, developed following the method of Benzie and Strain [24], with slight modifications. Briefly, a reaction mixture containing acetate buffer (pH 3.6), 300 mM of 2,4,6-Tris(2-pyridyl)-s-triazine (TPTZ) solution (in 40 mM HCl) and 20 mM $FeCl_3$ was prepared in a *v:v:v* proportion of 10:1:1 and incubated at 37 °C for 2 h in darkness. Then, 1.2 mL of reaction mixture was added to 20 μL of supernatant and incubated for 1 h at room temperature in darkness. Calibration standards were prepared by dissolving 28 mg of $FeSO_4$ in 10 mL of H_2O (10 mM) and subsequently diluting 1 mL of the solution with 1 mL of H_2O (5 mM). Six Eppendorf tubes containing 40 μL of H_2O were prepared and 20 μL of 5 mM solution were added to the first Eppendorf (2.5 mM). The operation was repeated moving 20 μL of solution, from one tube to another 6 times, in order to halve the concentration at each dilution. A blank with 40 μL of H_2O was also prepared. Samples and standards were read at 593 nm with a spectrophotometer (8453 UV-Visible Spectrophotometer, HP®, Palo Alto, CA, USA).

Total polyphenols were determined using the methodology described by Waterhouse [25], with slight modifications. Briefly, 50 μL of sample extract were added to 800 μL of extraction mixture (H_2O and Folin–Ciocalteu reagent in a *v:v* proportion of 15:1). Calibration standards were prepared by dissolving 80 mg of gallic acid in 10 mL of H_2O (8 mg mL^{-1}) and subsequently diluting 0.5 mL of the solution with 4.5 mL of H_2O (800 μg mL^{-1}). Six Eppendorf tubes containing 50 μL of H_2O were prepared and 50 μL of 800 μg mL^{-1} solution were added to the first (400 μg mL^{-1}). The operation was repeated moving 50 μL of solution from one tube to another for 6 times, in order to halve the concentration at each dilution. A blank with 50 μL of H_2O was also prepared. After an incubation of 5 min, samples and standards were added with 150 μL of 20% Na_2CO_3, incubated for 1 h at room temperature and then read at 765 nm with a spectrophotometer (8453 UV-Visible Spectrophotometer, HP®, Palo Alto, CA, USA).

2.6. Chilling Injury Analysis

In mid-June, both mature (DA-Meter Index 1.30–1.50) and immature tomatoes (DA-Meter Index 0.20–0.40) were harvested and stored for one week at 4 °C. After cold storage, fruits were moved into a dark room at 20 °C, 60% RH and 600–660 ppm of CO_2 for 2 weeks. Measurements took place right after 4 °C chilling (T1) and 10 days after 20 °C storage (T2).

Non-destructive analyses, including fresh weight loss, hardness, and ripening, were performed on the same tomatoes at T1 and T2 as described above. Fruit fresh weight was also evaluated before chilling. Destructive analyses, including soluble solids, lycopene content, β-carotene content, antioxidant activity, and total phenol content, were performed at the end (T2) of chilling injury evaluation as described above.

Chilling injury index was visually attributed at T2 according to Vega-Garcìa et al. [26] and Affandi et al. [27]. In particular, the criteria to evaluate the symptoms of chilling damage consisted of: uneven ripening and color development (U), pitting (P), and decay (D). A five-point scale was used to attribute the severity of the symptoms based on the percentage of affected fruit surface: 0 = no injury, $1 \leq 10\%$, 2 = 11% to 25%, 3 = 26% to 40%, and $4 \geq 40\%$.

2.7. Energy Cost Assessment

The energy cost assessment was calculated based on the actual consumption of a lamp with standard RB treatment (0.044 kW) applied for 16 h per day (0.704 kWh d^{-1}). Costs were estimated per plant per day, assuming that treatments were carried out with double lamps and that each pair of lamps covered about four plants. The price of electricity was acquired from EUROSTAT [28] dataset considering household electricity prices for Italy (0.176 € kWh^{-1}) and Spain (0.188 € kWh^{-1}) in 2021, the two main producers of tomato in the Mediterranean area.

2.8. Statistical Analysis

Statistical analysis was performed by applying one-way ANOVA and Tukey's test to compare means. Data were analyzed by using SPSS software. The Marascuilo procedure was used to compare multiple proportions in case of maturation degree and chilling injury index evaluation. Statistics considered a 5% significance level ($p \leq 0.05$).

3. Results

The influence of LED light on plants vegetative parameters was not significant. In particular, stem and collar diameter measurements showed no significant differences between light treatments and unlighted control at each time point (data not shown). Concurrently, an absence of differences was observed in the average internode length, although the third internode of CK plants showed a significant increase in length compared with RB treatment, featuring mean lengths of 31 ± 2 and 27 ± 4 cm, respectively. This difference disappeared in the following internodes. However, the average elongation of the plant apex showed a significantly greater length in the unilluminated control than in the RB treatment, with mean lengths of 29 ± 3 and 25 ± 3 cm, respectively. The final fresh weight of the entire plant, leaves and stems, did not show any difference (data not shown). In the same way, leaf area, LDMC, SLA, and chlorophyll content did not present statistically significant changes among treatments. Lighting regimes did not affect leaves' micro- and macro-elements content (Table 2). At the same time, stems presented significantly different accumulations of Fe, Cu, Mg, K, and P elements depending on light treatment. In particular, Cu and P were shown to have a greater accumulation in CK plants than in LED-treated plants, as reported in Table 2.

Table 2. Mineral element contents (mg L^{-1}) in leaves and stems of tomato plants grown under different lighting treatments (different letters indicate significant differences at $p \leq 0.05$, SE = Standard Error).

	B	Mn	Fe	Cu	Zn	Na	Mg	K	P	S	Ca
						Leaves					
CK	1.0 ^{a}	0.5 ^{a}	3.0 ^{a}	0.2 ^{a}	0.5 ^{a}	51.2 ^{a}	62.3 ^{a}	465.5 ^{a}	49.6 ^{a}	213.7 ^{a}	595.4 ^{a}
RB	0.9 ^{a}	0.4 ^{a}	3.7 ^{a}	0.3 ^{a}	0.5 ^{a}	42.2 ^{a}	64.3 ^{a}	391.1 ^{a}	38.5 ^{a}	195.5 ^{a}	573.8 ^{a}
FR	0.9 ^{a}	0.4 ^{a}	2.7 ^{a}	0.2 ^{a}	0.5 ^{a}	47.1 ^{a}	66.5 ^{a}	413.5 ^{a}	41.9 ^{a}	189.2 ^{a}	559.5 ^{a}
EOD	1.0 ^{a}	0.5 ^{a}	3.6 ^{a}	0.2 ^{a}	0.4 ^{a}	51.9 ^{a}	66.3 ^{a}	426.1 ^{a}	42.6 ^{a}	199.5 ^{a}	590.9 ^{a}
SE	0.1	0.0	0.6	0.1	0.0	6.1	4.6	32.1	4.9	15.3	40.1
						Stem					
CK	0.5 ^{a}	0.1 ^{a}	0.4 ^{b}	0.2 ^{a}	0.5 ^{a}	53.4 ^{a}	36.2 ^{b}	498.8 ^{ab}	62.3 ^{a}	30.2 ^{a}	139.7 ^{a}
RB	0.5 ^{a}	0.2 ^{a}	0.7 ^{a}	0.1 ^{b}	0.5 ^{a}	46.3 ^{a}	43.5 ^{a}	536.1 ^{a}	52.8 ^{ab}	34.4 ^{a}	154.2 ^{a}
FR	0.5 ^{a}	0.1 ^{a}	0.4 ^{b}	0.1 ^{b}	0.4 ^{a}	55.3 ^{a}	37.7 ^{ab}	421.8 ^{b}	45.7 ^{b}	28.6 ^{a}	117.6 ^{a}
EOD	0.5 ^{a}	0.1 ^{a}	0.4 ^{b}	0.1 ^{b}	0.3 ^{a}	48.3 ^{a}	32.0 ^{b}	442.6 ^{ab}	47.0 ^{b}	26.9 ^{a}	112.9 ^{a}
SE	0	0.1	0.1	0.0	0.1	9.0	2.5	40.3	5.3	3.5	19.9

The increase in volume of proximal fruits showed different trends depending on the season. In particular, fruits of Follow-up 1, representative of the spring season, were not affected by the lighting regimes (data not shown). In contrast, fruits of Follow-up 2, representative of the summer period, showed clear differences in growth, particularly comparing CK and RB treatments (Figure 2). Evaluation of ripening degree showed the same seasonal difference, demonstrating no difference for representative spring season fruit. In contrast, summer fruit seemed to present a faster ripening when exposed to LED light treatments, showing 35% more red fruit compared with CK plants (Figure 3).

Control tomatoes were approximately 9.3% lighter than tomatoes grown under light treatments (approximately 93.8 g for CK and 103.5 g in LED-treated plants). Fruit number was also lower (−7.2%) in control plants than in LED-treated plants (38.4 fruits plant^{-1} in CK and 41.4 fruits plant^{-1} in LED-treated plants). The statistical evaluation shows a significantly higher yield of LED-treated plants (RB, FR, and EOD) as compared with those grown with natural light only (CK). In particular, plants grown without supplemental LED light showed around 17% lower yield as compared with those grown with LED light (3.6 kg plant^{-1} in CK and 4.4 in LED-treated plants). Regarding yields for consumable and non-consumable tomatoes of each treatment, plants treated with LED lights produced a lower number of consumable red tomatoes (−9.3%) than control plants, although these fruits were larger in size (+10.5%). On the contrary, non-consumable green tomatoes of LED treatments were higher in number (+8.8%) than those in control, still presenting a larger average size (+13.7%).

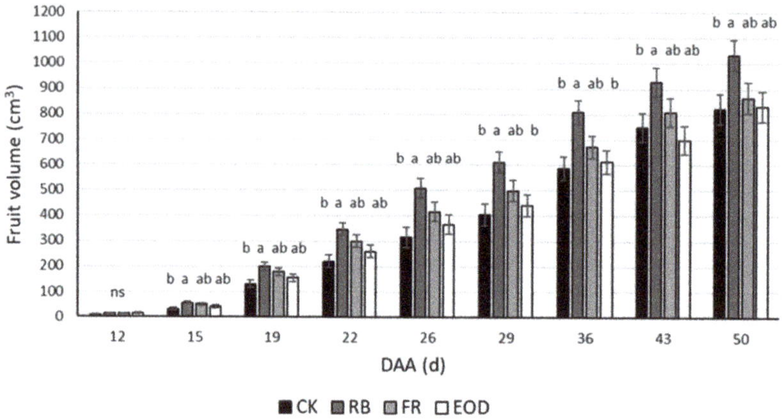

Figure 2. Volume development of representative fruit of summer period (Follow-up 2) depending on days after anthesis (DAA) (different letters indicate significant differences at $p \leq 0.05$).

a)

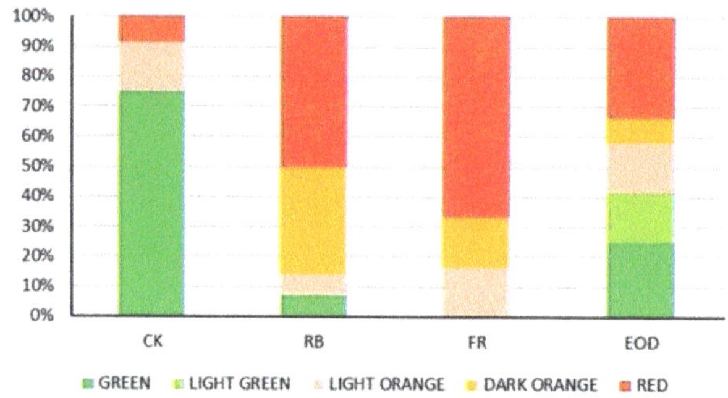

b)

	DA-METER INDEX	CK	RB	FR	EOD
1. GREEN	0	a	b	b	b
2. LIGHT GREEN	0 - 0.20	a	a	a	a
3. LIGHT ORANGE	0.20 - 0.80	a	a	a	a
4. DARK ORANGE	0.80 - 1.10	b	a	ab	ab
5. RED	1.10 - 1.90	b	ab	a	ab

c)

Figure 3. Comparison of multiple proportions (**a**) and respective significance ($p \leq 0.05$) (**b**) for five tomato maturation classes (**c**) depending on lighting treatment applied to fruits of Follow-up 2. (different letters indicate significant differences at $p \leq 0.05$).

Qualitative and biochemical parameters showed different trends depending on the period in which the fruits developed (e.g., spring or summer) and the ripening stage at harvest (e.g., mature or immature) (Figure 4). In particular, observing representative fruits from the spring period, no significant differences were observed concerning the evaluation of qualitative parameters. However, a significantly lower content of antioxidants and phenols was observed in control fruits' phenol content and antioxidant capacity compared with those grown with LED light treatments (Figure 4). In the case of summer fruits harvested at the mature stage, qualitative analyses showed lower acidity in fruits from the EOD treatment than the RB treatment (a reduction by 44.4% in EOD compared with RB), but no difference at the biochemical level. Finally, the evaluation of immature fruits, carried out only for the summer period, showed significantly higher hardness in fruits obtained from the RB treatment as compared with those produced by CK plants (11.5% less in CK compared with RB), and significantly higher soluble solid content in the RB than CK and EOD (13.7% less in CK and EOD compared with RB). For the biochemical analysis, the immature fruits from the summer period had a statistically significantly lower β-carotene content than RB and CK treatments (Figure 4). FDMC of selected fruits did not show any significant difference among treatments (data not shown).

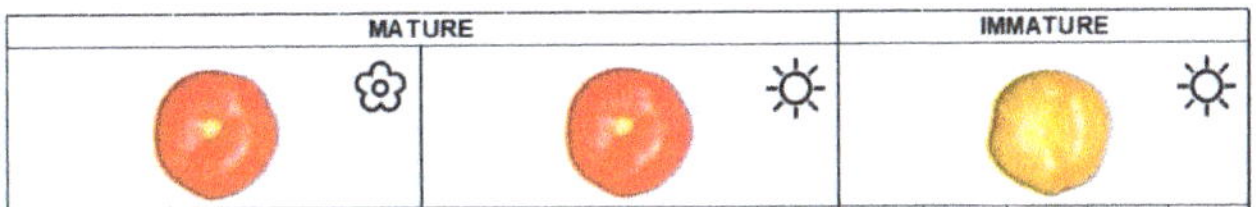

	MATURE (Spring)					MATURE (Summer)					IMMATURE (Summer)				
	CK	RB	FR	EOD	SE	CK	RB	FR	EOD	SE	CK	RB	FR	EOD	SE
QUALITATIVE ANALYSIS															
Fruit hardness (HI)	62.1^{a}	60.6^{a}	64.2^{a}	60.5^{a}	2.5	40.5^{a}	43.6^{a}	43.8^{a}	40.7^{a}	2.3	64.5^{b}	72.9^{a}	68.4^{ab}	70.7^{ab}	2.5
Soluble solids (Brix°)	5.0^{a}	5.3^{a}	5.2^{a}	5.0^{a}	0.2	4.5^{a}	4.7^{a}	4.7^{a}	4.5^{a}	0.2	4.4^{b}	5.1^{a}	4.8^{ab}	4.5^{b}	0.2
Acidity (%)	0.8^{a}	0.7^{a}	0.8^{a}	0.9^{a}	0.1	0.6^{ab}	0.9^{a}	0.8^{ab}	0.5^{b}	0.1	0.9^{a}	0.8^{a}	0.8^{a}	1.0^{a}	0.1
BIOCHEMICAL ANALYSIS															
Lycopene (mg kg^{-1})	7.5^{a}	9.3^{a}	7.6^{a}	8.4^{a}	0.6	9.7^{a}	8.4^{a}	8.6^{a}	9.3^{a}	0.5	5.8^{a}	6.7^{a}	4.7^{a}	3.7^{a}	1
β-carotene (mg kg^{-1})	3.7^{a}	4.4^{a}	3.7^{a}	4.1^{a}	0.4	5^{a}	3.8^{a}	3.9^{a}	4.7^{a}	0.4	3.3^{a}	3.8^{a}	2.4^{ab}	1.9^{b}	0.3
Antioxidant (mmol Fe^{2+} 100 g^{-1})	0.4^{b}	0.7^{a}	0.7^{a}	0.8^{a}	0.1	0.6^{a}	0.6^{a}	0.5^{a}	0.7^{a}	0.1	0.4^{a}	0.5^{a}	0.6^{a}	0.3^{a}	0.2
Phenols (mg GA 100 g^{-1})	15.4^{b}	32.2^{a}	30.2^{a}	32.6^{a}	3.7	20.9^{a}	9.9^{a}	5.4^{a}	20.6^{a}	7.1	7.3^{a}	9.4^{a}	25.5^{a}	10.0^{a}	6.2

= Spring = Summer

Figure 4. Qualitative and biochemical evaluation of mature and immature tomatoes representative of spring (Follow-up 1) and summer (Follow-up 2) periods (different letters indicate significant differences at $p \leq 0.05$, SE = Standard Error).

Cold storage showed different results on tomatoes depending on the ripeness degree. In particular, immature tomatoes showed no significant differences over time, except for polyphenol content. Indeed, polyphenols were found to be statistically lower in EOD and FR treatments compared with RB at T2, whereas no differences were observed among LED treatments and control (31.8 ± 8 mg GA 100 g^{-1} in CK, 39.2 ± 6 in RB, 26.1 ± 4 in FR and 19.9 ± 7 in EOD). Immature fruits also showed more irregular ripening with the application of additional LED treatments, whereas no uneven ripening was observed in CK fruit (Figure 5a). No significant statistical changes were observed in the case of pitting or decay evaluation for immature tomatoes (data not shown). On the other hand, ripe fruits showed significantly higher weight loss in CK compared with FR and EOD treatments immediately after 4 °C chilling (T1) (1.6 ± 0.1% in CK, 1.1 ± 0.4 in RB, 0.7 ± 0.3 in FR, and 0.6 ± 0.4 in EOD). However, differences among treatments disappeared after 10 days (T2). Differences in mature tomatoes were also observed in the case of β-carotene content, where RB treatment showed a significantly higher concentration than EOD treatment at T2 (4.9 ± 0.7 mg kg^{-1} in CK, 5.9 ± 0.9 in RB, 4.6 ± 0.7 in FR, and 4.1 ± 0.9 in EOD). Regarding chilling injury assessment, ripe tomatoes treated with RB light showed a significantly higher number of fruits in the intermediate class of decay compared with the other treatments 10 days after chilling (T2) (Figure 5b). However, although not significant, CK presented 80% more fruits in the highest decay class compared with fruits treated with RB light. No significant changes were observed in the case of pitting and uneven ripening evaluation for ripened tomatoes (data not shown).

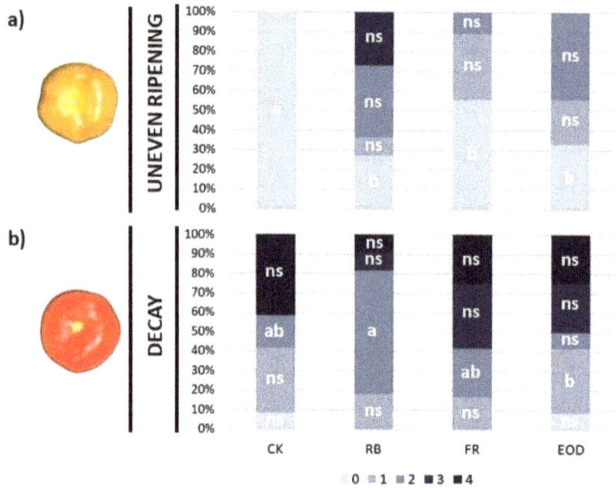

Figure 5. Comparison of multiple proportions and respective significance ($p \leq 0.05$) of five chilling injury classes (0–4) depending on lighting treatment. Reported chilling injury refers to uneven ripening of immature fruit (**a**) and decay of mature fruit (**b**). (different letters indicate significant differences at $p \leq 0.05$).

Evaluation of energy consumption per plant showed an average value of about 0.022 kW, which multiplied by 16 h of treatment results in a daily consumption of about 0.352 kWh per plant. Considering this consumption and the household electricity costs reported above [28], the daily cost to obtain a yield increase of 17% during the spring–summer period is about 0.06 EUR plant^{-1} d^{-1} for Italy and 0.07 EUR plant^{-1} d^{-1} for Spain, resulting in an increased cost by 1.27 EUR kg^{-1} in Italy and 1.35 EUR kg^{-1} in Spain.

4. Discussion

The amount of light radiation intercepted by plants depends on both the leaf area index (LAI, total leaf area per unit ground area) and a light extinction coefficient (k) influenced by morphological factors [29]. The capacity to modify plant architecture and consequently increase light interception is fundamental for improving photosynthetic rate. Specific wavelengths are known to influence and modify some vegetative and morphological traits of tomato plants. In particular, far-red radiation has been shown to have effects on the elongation of tomato internodes, leaf morphology, and inclination [12,30,31], being related to the so-called shade avoidance syndrome [32]. This phenomenon is determined by a low R:FR perception by plant phytochromes, triggering different responses that also involve leaf development, apical dominance, internode extension, chloroplast development, and assimilate partitioning, among others [33]. In natural conditions, without the artificial addition of far-red light, this phenomenon can be triggered by the far-red reflected by the leaves of the surrounding canopy as a signal of competition [34]. Furthermore, a natural low R:FR ratio can also occur at the end of the day [35], when the sun is low and longer wavelengths can travel further in the atmosphere.

In this report, the effects of different LED supplemental lighting spectral conditions, with or without far-red, have been evaluated on greenhouse tomato growth. Most vegetative and physiological traits were not affected by lighting regimes, contrary to what was reported by other researches, where some parameters, such as plant height or leaf area, were increased by far-red addition as compared with red and blue treatment alone [36]. The average apical elongation was the only parameter showing statistically significant differences, resulting in CK values higher than RB, but not than FR and EOD, as a possible consequence of the elongation effect given by far-red presence. On the other hand, the absence of a statistical difference among FR and EOD compared with RB may be attributed to the presence of the blue wavelength beside far-red light, known to have a possible dwarfing effect on plants, therefore lowering LED-treated plants' height [37]. Concerning mineral element evaluation, previous studies suggested that light treatments might impact their accumulation in some plant organs [38,39]. In particular, Samuolienė et al. [40] observed a higher accumulation of elements in tomato leaves grown with supplemental green LED light. In the present research, however, no differences were observed in leaves, although the different lighting treatments seem to have played different effects on the accumulation of specific elements in plant stems (Table 2).

Supplemental LED light can increase the photosynthetic rate of tomato plants and consequently influence fruit size and weight [41]. However, the effect of light on fruit growth can be affected by its position on the cluster, given the greater sink strength that a proximal fruit can have compared with a distal fruit. The sink strength of proximal fruit is related to higher cell division due to competitive assimilating processes triggered in the first phase of fruit development after pollination [42]. In the following stages, the fruit accumulates most of its dry matter and undergoes a process of cell enlargement until it reaches the final stage of ripening, where it stops accumulating carbohydrates [43]. In high-wire systems, where tomato plants are periodically lowered, fruits in the cell division stage are at the top of the plant, away from the LED lamps, whereas during the cell enlargement stage, fruits are at the bottom of the plant near the lamps. It might be possible to think that the lamps can only affect the second development stage, being in direct contact with the fruits. However, when comparing the volume growth over time of representative tomato fruits from spring (Follow-up 1) and summer (Follow-up 2) phases, different conclusions may be drawn. Whereas earlier in the season no significant differences were observed between treatments, later, RB treatment showed a significantly greater dimension of fruits than CK, already after the second measurement (T2) (Figure 2). Whereas the light treatment for spring fruits was started after anthesis, on already formed fruit, in summer fruits, the treatment was already in progress since the early flowering. From these observations, it can be deduced that, despite being distant from the lamps, summer fruits have been affected by LED light, accumulating a major quantity of assimilates since the cell division stage. A similar response was also reported by Paponov et al. [11], although differences were mainly observed in fruits at the intermediate position in the cluster.

The presence of far-red might influence the enhancement of tomatoes fresh weight [31,44]. However, as formerly observed in other studies performed in the Mediterranean area [17], the presence of additional far-red did not seem to have specifically affected plant yield. However, a general increase in average fruit weight and fruit number was observed in all LED treatments compared with the control. Total yield had a percentage increase (17%) very similar to those observed in other studies comparing the use of supplemental LED light with natural light alone [15,16,41,45,46]. By dividing the number of consumable fruits from non-consumable fruits, it was observed that plants under LED treatment produced fewer but larger red tomatoes, and more and bigger green tomatoes than the control. This observation seems to confirm the previous results by Paponov et al. [11] on the ability of LED light to increase the number of cells and thus the sink strength in the early developmental stage of fruit at a different height on the truss. During the ripening stage, although fewer in number than the control, proximal fruits of LED-treated plants were found to have an earlier ripening (Figure 3) as already observed by some authors [15,47]. The higher ripening rate in proximal fruits of LED-treated plants could be attributed to a greater competition and consequent accumulation of some biochemical compounds such as phytoene or melatonin, responsible for ripening processes [47,48].

Additional LED light is known to positively affect several quality attributes of tomatoes, although some inconsistencies among published research have been shown, probably determined by different environmental conditions and genotypes [39,49]. On ripe fruits, a significant increase in antioxidant capacity and phenol content in the LED-treated fruits was observed in spring, but no significant differences between lighting treatments was detected in the summer (Figure 4). The effect of LED light on qualitative attributes may also change according to the fruit ripening stage, as observed in immature tomatoes in relation to sugar content and fruit hardness. Accordingly, Fanwoua et al. [44] already observed that the effects of LED light on pericarp sugar concentrations could change based on fruit age, as a consequence of different fruit water content that would affect the sugar dilution.

To the best of our knowledge, limited research has investigated the effects of supplemental LED light applied at the cultivation stage on chilling injury and the post-harvest quality of tomatoes. In particular, Affandi et al. [27] observed that high levels of supplemental far-red light might improve cold tolerance in both mature green and red tomatoes, reducing pitting, decay, and weight loss. In our research, a significant effect on cold tolerance was not observed. However, far-red light, applied both throughout the day and at the end-of-the-day, would seem to confirm an ability to delay weight loss, especially during the chilling phase (T1). In contrast to a former report by Affandi et al. [27], mature green tomatoes treated with far-red light did not seem to have a faster turning to red compared with RB treatment, whereas the best performance was observed in the CK case.

Based on the results of our research, the application of additional LED light seems not economically feasible. In fact, considering the productive capacity of our experiment, the energy cost per kg is 1.27 EUR kg^{-1} in Italy and 1.35 EUR kg^{-1} in Spain, whereas selling prices reported by EUROSTAT [28] in 2021 are 1.15 EUR kg^{-1} and 0.74 EUR kg^{-1} for Italy and Spain, respectively. Other authors have already observed scarce economic feasibility, estimating energy consumption of about 28.8 kWh to increase tomato yield by 1 kg [50]. However, these observations need to be put into context. In particular, the country's socio-economic condition, geo-political situation, access to renewable energy, as well as the cultivation latitude, and greenhouse light accessibility and transmissivity, may all influence the feasibility of supplementary LED lighting application to increase tomato yield. Besides that, greenhouse design, management of cultivation period, and lighting strategy are other important aspects to adapt to maximize the economic benefit. For instance, it has been observed that in a Northern European context, a year-round production with high technology investment (e.g., supplemental lighting, heat pump) can provide the most elevated economic return [51]. This contextual approach could also be applied in the Mediterranean area, modifying some parameters according to the different climatic and socio-economic conditions. To give an example, our study looked only at the spring–summer time span, although the application during the cold-season could have resulted in more cost-effective results. Furthermore, our research referred to household on-grid electricity costs, applied to an urban BIA context intended for direct consumption by building users. Integration of alternative energy sources such as solar panels may drastically reduce the electricity cost, leading to a net profit increase both at domestic and commercial levels. Integrating the research with different lighting strategies (e.g., pulsed light, cold-season lighting, photoperiod reduction) should be considered better to estimate the economic benefits for the Mediterranean area.

5. Conclusions

The application of supplementary LED light for tomato cultivation in a BIA context, subjected to a higher reduction of the solar radiation entering the greenhouse, has been shown to increase yields by 17% regardless of the type of treatment used (with or without far-red, during the whole day or at the end-of-the-day). Additional light also showed an ability to increase the dimension and number of tomatoes, and speed up the ripening process, particularly in the case of proximal fruits. From an economic point of view, the application of supplementary light would seem to have low feasibility if applied during the spring–summer period. However, applying different and contextualized lighting strategies could lower energy costs. Therefore, future research should focus on more economically valuable methods of managing supplementary lighting, for example, the use during the winter period, different photoperiods and intensities, or techniques that can provide energy savings, such as pulsed light.

Author Contributions: E.A. designed and performed the experiment and data collection, and drafted the manuscript. I.P. and G.S., contributed to experiment setting and data collection, and critically revised the manuscript. G.P., F.O., X.G.D. and G.G. supervised the research and critically revised the manuscript. All authors contributed to the article and approved the submitted version. All authors have read and agreed to the published version of the manuscript.

Funding: The research leading to this publication has received funding from the European Union's Horizon 2020 research and innovation program under grant agreement No 862663, and from the Department of Research and Universities of the Generalitat of Catalonia in the framework of the project "PANDEMIA Project" (AGAUR 2020PANDE00021).

Institutional Review Board Statement: Not applicable.

Informed Consent Statement: Not applicable.

Acknowledgments: The authors gratefully acknowledge the TR Turoni Srl for providing the DA-Meter and Durofel instruments, UPC to provide the LED system from the Fertilecity project (CTM2016-75772-C3), and Ramiro Gonzelez Atero for his support in the experiment setting.

Conflicts of Interest: The authors declare that the research was conducted in the absence of any commercial or financial relationships that could be construed as a potential conflict of interest.

Disclaimer: The publication reflects the author's views. The Research Executive Agency (REA) is not liable for any use that may be made of the information contained therein.

References

1. Rivera-Ferre, M.G.; López-i-Gelats, F.; Ravera, F.; Oteros-Rozas, E.; di Masso, M.; Binimelis, R.; El Bilali, H. The two-way relationship between food systems and the COVID19 pandemic: Causes and consequences. *Agric. Syst.* **2021**, *191*, 103134. [CrossRef]
2. Knorr, D.; Khoo, C.S.H.; Augustin, M.A. Food for an urban planet: Challenges and research opportunities. *Front. Nutr.* **2018**, *4*, 73. [CrossRef]
3. Lal, R. Home gardening and urban agriculture for advancing food and nutritional security in response to the COVID-19 pandemic. *Food Secur.* **2020**, *12*, 871–876. [CrossRef]
4. Specht, K.; Siebert, R.; Hartmann, I.; Freisinger, U.B.; Sawicka, M.; Werner, A.; Dierich, A. Urban agriculture of the future: An overview of sustainability aspects of food production in and on buildings. *Agric. Hum. Values* **2014**, *31*, 33–51. [CrossRef]
5. Appolloni, E.; Orsini, F.; Specht, K.; Thomaier, S.; Sanye-Mengual, E.; Pennisi, G.; Gianquinto, G. The global rise of urban rooftop agriculture: A review of worldwide cases. *J. Clean. Prod.* **2021**, *296*, 126556. [CrossRef]
6. Montero, J.I.; Baeza, E.; Heuvelink, E.; Rieradevall, J.; Muñoz, P.; Ercilla, M.; Stanghellini, C. Productivity of a building-integrated roof top greenhouse in a Mediterranean climate. *Agric. Syst.* **2017**, *158*, 14–22. [CrossRef]
7. Nadal, A.; Llorach-Massana, P.; Cuerva, E.; López-Capel, E.; Montero, J.I.; Josa, A.; Royapoor, M. Building-integrated rooftop greenhouses: An energy and environmental assessment in the mediterranean context. *Appl. Energy* **2017**, *187*, 338–351. [CrossRef]
8. Muñoz-Liesa, J.; Royapoor, M.; López-Capel, E.; Cuerva, E.; Rufí-Salís, M.; Gassó-Domingo, S.; Josa, A. Quantifying energy symbiosis of building-integrated agriculture in a mediterranean rooftop greenhouse. *Renew. Energy* **2020**, *156*, 696–709. [CrossRef]
9. Muñoz-Liesa, J.; Toboso-Chavero, S.; Beltran, A.M.; Cuerva, E.; Gallo, E.; Gassó-Domingo, S.; Josa, A. Building-integrated agriculture: Are we shifting environmental impacts? An environmental assessment and structural improvement of urban greenhouses. *Resour. Conserv. Recycl.* **2021**, *169*, 105526. [CrossRef]
10. Lu, N.; Maruo, T.; Johkan, M.; Hohjo, M.; Tsukagoshi, S.; Ito, Y.; Shinohara, Y. Effects of supplemental lighting with light-emitting diodes (LEDs) on tomato yield and quality of single-truss tomato plants grown at high planting density. *Environ. Control Biol.* **2012**, *50*, 63–74. [CrossRef]
11. Paponov, M.; Kechasov, D.; Lacek, J.; Verheul, M.J.; Paponov, I.A. Supplemental light-emitting diode inter-lighting increases tomato fruit growth through enhanced photosynthetic light use efficiency and modulated root activity. *Front. Plant Sci.* **2020**, *10*, 1656. [CrossRef]
12. Zhang, Y.; Zhang, Y.; Yang, Q.; LI, T. Overhead supplemental Far-red light stimulates tomato growth under intra-canopy lighting with LEDs. *J. Integr. Agric.* **2018**, *17*, 62–69. [CrossRef]
13. Palmitessa, O.D.; Pantaleo, M.A.; Santamaria, P. Applications and development of LEDs as supplementary lighting for tomato at different latitudes. *Agronomy* **2021**, *11*, 835. [CrossRef]
14. Pardossi, A.; Tognoni, F.; Incrocci, L. Mediterranean greenhouse technology. *Chron. Horticult.* **2004**, *44*, 28–34.
15. Paucek, I.; Pennisi, G.; Pistillo, A.; Appolloni, E.; Crepaldi, A.; Calegari, B.; Gianquinto, G. Supplementary LED interlighting improves yield and precocity of greenhouse tomatoes in the Mediterranean. *Agronomy* **2020**, *10*, 1002. [CrossRef]
16. Palmitessa, O.D.; Paciello, P.; Santamaria, P. Supplemental LED increases tomato yield in Mediterranean semi-closed greenhouse. *Agronomy* **2020**, *10*, 1353. [CrossRef]

17. Palmitessa, O.D.; Leoni, B.; Montesano, F.F.; Serio, F.; Signore, A.; Santamaria, P. Supplementary Far-Red Light Did Not Affect Tomato Plant Growth or Yield under Mediterranean Greenhouse Conditions. *Agronomy* **2020**, *10*, 1849. [CrossRef]
18. Easlon, H.M.; Bloom, A.J. Easy Leaf Area: Automated digital image analysis for rapid and accurate measurement of leaf area. *Appl. Plant Sci.* **2014**, *2*, 1400033. [CrossRef]
19. Garnier, E.; Shipley, B.; Roumet, C.; Laurent, G. A standardized protocol for the determination of specific leaf area and leaf dry matter content. *Funct. Ecol.* **2001**, *15*, 688–695. [CrossRef]
20. Denis, A.; Desclee, B.; Migdall, S.; Hansen, H.; Bach, H.; Ott, P.; Tychon, B. Multispectral remote sensing as a tool to support organic crop certification: Assessment of the discrimination level between organic and conventional maize. *Remote Sens.* **2021**, *13*, 117. [CrossRef]
21. Arcas-Pilz, V.; Rufí-Salís, M.; Parada, F.; Petit-Boix, A.; Gabarrell, X.; Villalba, G. Recovered phosphorus for a more resilient urban agriculture: Assessment of the fertilizer potential of struvite in hydroponics. *Sci. Total Environ.* **2021**, *799*, 149424. [CrossRef]
22. Li, T.; Heuvelink, E.P.; Marcelis, L.F. Quantifying the source–sink balance and carbohydrate content in three tomato cultivars. *Front. Plant Sci.* **2015**, *6*, 416. [CrossRef]
23. Anthon, G.; Barrett, D.M. Standardization of a rapid spectrophotometric method for lycopene analysis. *X Int. Symp. Proces. Tomato* **2006**, *758*, 111–128. [CrossRef]
24. Benzie, I.F.; Strain, J.J. Ferric reducing/antioxidant power assay: Direct measure of total antioxidant activity of biological fluids and modified version for simultaneous measurement of total antioxidant power and ascorbic acid concentration. *Meth. Enzymol.* **1999**, *299*, 15–27. [CrossRef]
25. Waterhouse, A.L. Determination of total phenolics. *Curr. Protoc. Food Anal. Chem.* **2002**, *6*, I1-1. [CrossRef]
26. Vega-García, M.O.; López-Espinoza, G.; Ontiveros, J.C.; Caro-Corrales, J.J.; Vargas, F.D.; López-Valenzuela, J.A. Changes in protein expression associated with chilling injury in tomato fruit. *J. Am. Soc. Hortic. Sci.* **2010**, *135*, 83–89. [CrossRef]
27. Affandi, F.Y.; Verdonk, J.C.; Ouzounis, T.; Ji, Y.; Woltering, E.J.; Schouten, R.E. Far-red light during cultivation induces postharvest cold tolerance in tomato fruit. *Posthrvest Biol. Technol.* **2020**, *159*, 111019. [CrossRef]
28. EUROSTAT. 2021. Available online: https://ec.europa.eu/eurostat/en/ (accessed on 20 July 2022).
29. Hirose, T. Development of the Monsi–Saeki theory on canopy structure and function. *Ann. Bot.* **2005**, *95*, 483–494. [CrossRef]
30. Kurepin, L.V.; Yip, W.K.; Fan, R.; Yeung, E.C.; Reid, D.M. The roles and interactions of ethylene with gibberellins in the far-red enriched light-mediated growth of Solanum lycopersicum seedlings. *Plant Growth Regul.* **2010**, *61*, 215–222. [CrossRef]
31. Hao, X.; Little, C.; Zheng, J.M.; Cao, R. Far-red LEDs improve fruit production in greenhouse tomato grown under high-pressure sodium lighting. *VIII Int. Symp. Light Hortic.* **2016**, *1134*, 95–102. [CrossRef]
32. Franklin, K.A.; Whitelam, G.C. Red: Far-red ratio perception and shade avoidance. In *Light and Plant Development*, 1st ed.; Whitelam, G.C., Halliday, K.J., Eds.; Blackwell Publishing Ltd.: Oxford, UK, 2007; Volume 30, pp. 211–234. [CrossRef]
33. Smith, H.; Whitelam, G.C. The shade avoidance syndrome: Multiple responses mediated by multiple phytochromes. *Plant Cell Environ.* **1997**, *20*, 840–844. [CrossRef]
34. Ballaré, C.L.; Scopel, A.L.; Sánchez, R.A. Far-red radiation reflected from adjacent leaves: An early signal of competition in plant canopies. *Science* **1990**, *247*, 329–332. [CrossRef] [PubMed]
35. Holmes, M.G.; Smith, H. The function of phytochrome in the natural environment—I. Characterization of daylight for studies in photomorphogenesis and photoperiodism. *Photochem. Photobiol.* **1977**, *25*, 533–538. [CrossRef]
36. Kalaitzoglou, P.; Van Ieperen, W.; Harbinson, J.; Van der Meer, M.; Martinakos, S.; Weerheim, K.; Marcelis, L.F. Effects of continuous or end-of-day far-red light on tomato plant growth, morphology, light absorption, and fruit production. *Front. Plant Sci.* **2019**, *10*, 322. [CrossRef]
37. Ahmad, M.; Grancher, N.; Heil, M.; Black, R.C.; Giovani, B.; Galland, P.; Lardemer, D. Action spectrum for cryptochrome-dependent hypocotyl growth inhibition in Arabidopsis. *Plant Physiol.* **2002**, *129*, 774–785. [CrossRef]
38. Amoozgar, A.; Mohammadi, A.; Sabzalian, M.R. Impact of light-emitting diode irradiation on photosynthesis, phytochemical composition and mineral element content of lettuce cv. Grizzly. *Photosynthetica* **2017**, *55*, 85–95. [CrossRef]
39. Palmitessa, O.D.; Durante, M.; Caretto, S.; Milano, F.; D'Imperio, M.; Serio, F.; Santamaria, P. Supplementary light differently influences physico-chemical parameters and antioxidant compounds of tomato fruits hybrids. *Antioxidants* **2021**, *10*, 687. [CrossRef]
40. Samuolienė, G.; Miliauskienė, J.; Kazlauskas, A.; Viršilė, A. Growth Stage Specific Lighting Spectra Affect Photosynthetic Performance, Growth and Mineral Element Contents in Tomato. *Agronomy* **2021**, *11*, 901. [CrossRef]
41. Jiang, C.; Johkan, M.; Hohjo, M.; Tsukagoshi, S.; Ebihara, M.; Nakaminami, A.; Maruo, T. Photosynthesis, plant growth, and fruit production of single-truss tomato improves with supplemental lighting provided from underneath or within the inner canopy. *Sci. Hortic.* **2017**, *222*, 221–229. [CrossRef]
42. Bertin, N.; Gautier, H.; Roche, C. Number of cells in tomato fruit depending on fruit position and source-sink balance during plant development. *Plant Growth Regul.* **2002**, *36*, 105–112. [CrossRef]
43. Ho, L.C.; Hewitt, J.D. Fruit development. In *The Tomato Crop*, 1st ed.; Atherton, J.G., Rudich, J., Eds.; Springer: Dordrecht, The Netherlands, 1986; pp. 201–239. [CrossRef]
44. Fanwoua, J.; Vercambre, G.; Buck-Sorlin, G.; Dieleman, J.A.; de Visser, P.; Génard, M. Supplemental LED lighting affects the dynamics of tomato fruit growth and composition. *Sci. Hortic.* **2019**, *256*, 108571. [CrossRef]

45. Pepin, S.; Fortier, E.; Béchard-Dubé, S.A.; Dorais, M.; Ménard, C.; Bacon, R. Beneficial effects of using a 3-D LED interlighting system for organic greenhouse tomato grown in Canada under low natural light conditions. *II Int. Symp. Org. Greenh. Hortic.* **2013**, *1041*, 239–246. [CrossRef]
46. Hao, X.; Zhang, Y.; Guo, X.; Little, C.; Zheng, J. Dynamic temperature control strategy with a temperature drop improves responses of greenhouse tomatoes and sweet peppers to long photoperiods of supplemental lighting and saves energy. *Int. Symp. New Technol. Environ. Control. Energy Sav. Crop Prod. Greenh. Plant* **2017**, *1227*, 291–298. [CrossRef]
47. Zhang, J.; Zhang, Y.; Song, S.; Su, W.; Hao, Y.; Liu, H. Supplementary red light results in the earlier ripening of tomato fruit depending on ethylene production. *Environ. Exp. Bot.* **2020**, *175*, 104044. [CrossRef]
48. Li, Y.; Liu, C.; Shi, Q.; Yang, F.; Wei, M. Mixed red and blue light promotes ripening and improves quality of tomato fruit by influencing melatonin content. *Environ. Exp. Bot.* **2021**, *185*, 104407. [CrossRef]
49. Appolloni, E.; Orsini, F.; Pennisi, G.; Durany, X.G.; Paucek, I.; Gianquinto, G. Supplemental LED Lighting Effectively Enhances the Yield and Quality of Greenhouse Truss Tomato Production: Results of a Meta-Analysis. *Front. Plant Sci.* **2021**, *12*, 596927. [CrossRef] [PubMed]
50. Yan, W.; Zhang, Y.; Zhang, Y.; Cheng, R.; Zhang, Y.; Yang, Q.; Li, T. Effects of supplementary artificial light on growth of the tomato (Solanum lycopersicum) in a chinese solar greenhouse. *Hortic. J.* **2018**, *87*, 516–523. [CrossRef]
51. Naseer, M.; Persson, T.; Righini, I.; Stanghellini, C.; Maessen, H.; Ruoff, P.; Verheul, M.J. Bioeconomic evaluation of extended season and year-round tomato production in Norway using supplemental light. *Agric. Syst.* **2022**, *198*, 103391. [CrossRef]

MDPI AG
Grosspeteranlage 5
4052 Basel
Switzerland
Tel.: +41 61 683 77 34

Horticulturae Editorial Office
E-mail: horticulturae@mdpi.com
www.mdpi.com/journal/horticulturae

www.ingramcontent.com/pod-product-compliance
Lightning Source LLC
Chambersburg PA
CBHW061609210926
51850CB00009B/2271

* 9 7 8 3 7 2 5 8 4 6 5 9 7 *